EXS 50:
**Experientia Supplementum
Vol. 50**

Birkhäuser Verlag
Basel · Boston · Stuttgart

Cadmium in the Environment

Edited by

Hans Mislin
Oscar Ravera

1986

Birkhäuser Verlag
Basel · Boston · Stuttgart

Parts of this review were published previously in 2 issues of the journal
EXPERIENTIA, Vol. 40, No. 1, pp. 1–52, 1984 and
EXPERIENTIA, Vol. 40, No. 2, pp. 117–164, 1984.

Library of Congress Cataloging in Publication Data

Cadmium in the environment.

   (Experientia. Supplementum ; vol. 50)
   Parts of this review were published previously in:
Experientia ; vol. 40, no. 1–2, 1984.
   Includes bibliographies and index.
   1. Cadmium–Toxicology. 2. Cadmium–Environmental
aspects. I. Mislin, Hans, 1907–   . II. Ravera, O.
III. Series: Experientia. Supplementum ; v. 50.
[DNLM: 1. Cadmium–adverse effects. 2. Cadmium–
analysis. 3. Environmental Pollution-analysis.
W1 EX23 v. 50 / QV 290 C1241]
RA1231.C3C34 1986   363.7'384   86-8299
ISBN 3-7643-1760-4

CIP-Kurztitelaufnahme der Deutschen Bibliothek

Cadmium in the environment / ed. by Hans Mislin ;
Oscar Ravera. – Basel ; Boston ; Stuttgart :
Birkhäuser, 1986.
   (Experientia : Supplementum ; Vol. 50)
   ISBN 3-7643-1760-4

NE: Mislin, Hans [Hrsg.]; Experientia /
Supplementum

©1986 Birkhäuser Verlag Basel
Printed in Switzerland by Birkhäuser AG, Graphisches Unternehmen, Basel
ISBN 3-7643-1760-4

Contents

Part I: Cadmium in the Environment

Geochemistry of cadmium

by I. Thornton

Applied Geochemistry Research Group, Department of Geology, Imperial College, London SW7 2BP (England)

Cadmium, a transition metal, is a member of Group IIB of the periodic table, which also includes zinc and mercury. Pure cadmium is a bluish-white metal but does not occur as such in nature. It was first discovered in 1817 by a German chemist, F. Stromeyer, as a constituent of the zinc ore smithsonite ($ZnCO_3$). Cadmium is mainly found in zinc, lead-zinc and lead-copper-zinc ores, and its concentration is usually related to their zinc content. It is also found in varying amounts as a natural component of the surface environment in rocks, overburden and soils, water, air, plant and animal tissues. Its geochemical behavior is similar to that of zinc because of the similar electron structures and ionization potentials of the two elements. In nature cadmium is nearly always present in the Cd^{2+} oxidation state and occurs as eight stable isotopes as shown in table 1. ^{112}Cd and ^{114}Cd are the most common. Radioactive isotopes with mass numbers 104, 105, 107, 109, 111, 113, 115, 117, 118 and 119 have been made artificially, of which ^{113}Cd has the longest half-life of 5.1 years[27].

Cadmium in rock forming minerals

Cadmium has an ionic radius of 0.97 Å, similar to that of Ca^{++} (0.99 Å) and Na^+ (0.98 Å) and could be expected to be found in their minerals. Goldschmidt[10] has shown detectable substitution of cadmium for calcium in bytownite feldspar of anorthosite and in monoclinic augite phenocysts of basalts and has also shown crystal structures of cadmium oxide and carbonate closely resembling those of similar calcium compounds. Vinogradov[36] comments on the relatively large ionic radius of Cd^{++} and that it probably entered into minerals of later crystallization. Cadmium, like zinc, is found in ferro-magnesian minerals. It seldom becomes enriched in igneous rocks[24] though is present in acid igneous rocks mostly in biotite and in traces in apatite[25].
Cadmium has a markedly chalcophile nature and forms two common sulphides – greenockite and hawleyite (hexagonal and isometric forms respectively of CdS). The former is frequently found under natural conditions as a

Table 2. Cadmium content of minerals (in μg/g, except where noted)

Mineral	Composition	Range
Sphalerite	$(Zn,Cd)S$	0.0001–2%
Greenockite	CdS	77.8%
Hawleyite	CdS	77.8%
Chalcopyrite	$CuFeS_2$	< 0.4–110
Marcasite	FeS_2	< 0.3– < 50
Arsenopyrite	$FeAsS$	< 5
Galena	PbS	< 10.–3000
Pyrite	FeS_2	< 0.06–42
Pyrrhotite	$Fe_{(1-x)}S$	Trace
Tetrahedrite	$(Cu,Fe,Zn,Ag)_{12}SbAs_4S_{13}$	80–2000
Magnetite	Fe_3O_4	0–0.31
Cadmium oxide	CdO	87.5%
Limonite	Hydrous iron oxides	< 5–1000
Wad and manganese oxides	Hydrous manganese oxides	< 10–1000
Anglesite	$PbSO_4$	120– > 1000
Barite	$BaSO_4$	< 0.2
Anhydrite and gypsum	$CaSO_4$; $CaSO_4 \cdot 2H_2O$	< 0.2
Calcite	$CaCO_3$	< 1–23
Smithsonite	$ZnCO_3$	0.1–2.35%
Otavite	$CdCO_3$	65.18%
Pyromorphite	$Pb_5Cl(PO_4)_3$	< 1–8
Scorodite	$FeAsO_4 \cdot 2H_2O$	< 1–5.8
Beudantite	$PbFe_3(AsO_4)(SO_4)(OH)_6$	100–1000
Apatite	$Ca_5(F,Cl)(PO_4)_3$	0.14–0.15
Bindheimite	$Pb_2Sb_2O_6(O,OH)$	100–1000
Silicates		0.03–5.8

(from Boyle and Jonasson[7]).

yellow coating on weathering sphalerite (ZnS). Cadmium is a constituent of several zinc minerals, particularly the sulphide, sphalerite which is the principal cadmium bearing mineral in primary deposits. Cadmium forms an oxide and carbonate under natural conditions of which the former is the more common. Both cadmium and zinc are strongly enriched in hydrothermal rocks and minerals found at relatively low temperatures. Cadmium is also found in some secondary minerals, particularly smithsonite ($ZnCO_3$) and in hydrous manganese and iron oxides formed in the oxidized zones of zinc deposits. A more comprehensive review of cadmium containing minerals is given by Holmes[12], who provides a comprehensive listing of the ranges of cadmium concentrations found in sulphide, sulphate, oxide, carbonate, silicate and non-specific minerals. The cadmium content of the more common minerals found in sulphide and other deposits has been tabulated by Boyle and Jonasson[7] and this listing is reproduced in table 2. The principal cadmium minerals formed from the oxidation of primary sphalerite and other cadmium bearing minerals are greenockite and hawleyite. Cadmium oxide and octavite are rare.
Waketa and Schmitt[27] have also compiled published data for the cadmium content of rock-forming minerals, including those for sphalerites from the United States, the Soviet Union, Sweden and Vietnam ranging from 500 to 18,500 μg/g Cd.

Table 1. Stable isotopes of cadmium with natural abundances

Isotope	%
^{106}Cd	1.21
^{108}Cd	0.88
^{110}Cd	12.39
^{111}Cd	12.75
^{112}Cd	24.07
^{113}Cd	12.26
^{114}Cd	28.86
^{116}Cd	7.58

Cadmium in rocks

The average concentration of cadmium in the earths crust has been reported as 0.15 µg/g[38] and 0.11 µg/g[5] and in the upper lithosphere as 0.5 µg/g with a zinc:cadmium ratio of around 250:1[10]. A similar zinc:cadmium ratio has been reported for American magmatic rocks[25] and for 'terrestrial' rocks[7]. The average cadmium content of igneous rocks has been reported as 0.18 µg/g with a zinc:cadmium ratio exceeding 400:1[25].

Page and Bingham[21] have condensed and tabulated data for the abundance of cadmium in igneous, sedimentary and metamorphic rocks reported by Waketa and Schmitt[37] as shown in table 3. Further tabulated data are given by Boyle and Jonasson[7]. There is little difference in the cadmium content of igneous rocks which rarely contain more than 1 µg/g and usually very much less. For example, the mean contents of granites have been variously reported as ranging from 0.09 to 0.24 µg/g and basalts from 0.13 to 0.22 µg/g. Of the sedimentary rocks, bituminous and carbonaceous shales (sometimes referred to as black shales) may contain abnormally large concentrations of cadmium, which may in turn be reflected in the weathering cycle and provide significant pathways to plants and animals. A detailed study of cadmium in marine black shales in Britain has been undertaken by Holmes[12], who reports concentrations ranging up to 219 µg/g. Data for cadmium in these black shales are summarized in table 4. Concentrations of cadmium ranging from 100 to 1000 µg/g has been found in the Mansfield copper-shale ('kupferschiefer') in Germany[8], although the accuracy of analysis undertaken at this time may be open to question. The mean content of cadmium in bituminous shale has been previously reported as 0.80 µg/g[37] and, in shale 0.22 µg/g, sandstone 0.05 µg/g and limestone 0.028 µg/g[40]. Metamorphic rocks rarely exceed 1 µg/g Cd, though hornfels and skarn may range up to 5 µg/g and schists up to 3 µg/g[7].

The cadmium content of coals ranges widely and is reported as 0.01 to 180 µg/g for the U.S.[35], < 0.01 to 22 µg/g worldwide[5], 0.3 to 2.0 µg/g[7], up to 0.2 µg/g in bituminous coals from Australia and 0.1 to 65 µg/g in Illinois basin coals[28]. Concentrations of cadmium in peats have been reported as < 1 to 3 µg/g (mean 0.25 µg/g) and up to 50 µg/g in the ash of enriched bogs[24]. Cadmium in crude oil ranges widely and in Russia has been reported as below detection limit to over 1000 µg/g[23].

Cadmium has been determined in marine sediments, ranging from 0.1 to 1.0 µg/g in the Atlantic and Pacific oceans[20]. Oceanic sediments from and flanking the mid-Atlantic ridge have been found to contain almost twice the cadmium as those away from ridge systems, with average carbonate-free cadmium contents of 0.650 and 0.335 µg/g respectively[3]. Manganese nodules are enriched in cadmium, with a mean content of 8 µg/g reported by Bowen[5] (from other published data). Phosphorites may also be enriched, ranging from 0.01 to 25 µg/g Cd[5].

Cadmium in soils

The main sources of trace metals in unpolluted soils are the parent materials from which they are derived. These materials usually comprise weathered bedrock or overburden that has been transported by wind, water or glacial activity. 95 percent of the earth's crust is made up of igneous rocks, and 5% sedimentary rocks, of which about 80% are shales, 15% sandstones and 5% limestones[19]. Sedimentary materials tend to overlie the igneous rocks from which they were derived and are thus more common in the surface weathering environment.

Soils tend to reflect the chemical composition of the parent materials from which they were derived. Where residual soils are formed in situ from the underlying bedrock, the trace metal content of the soil may be directly

Table 3. Abundance of cadmium in common rocks*

Rock type	Cadmium (µg/g)		No. of samples
	Range	Mean	
Igneous			
Granite	0.001–0.60	0.12	44
Granodiorite	0.016–0.10	0.07	5
Biotite-granite	< 0.05–0.50	–	9
Quartz monzonite	1.4 –1.8	–	–
Pitchstone	0.05–0.34	0.17	24
Rhyolite	0.05–0.48	–	8
Obsidian	0.22–0.29	0.25	2
Andesite	–	0.017	2
Syenite	0.04–0.32	0.16	6
Basalt	0.006–0.6	0.22	39
Gabbro	0.08–0.20	0.11	8
Sedimentary			
Bituminous shale	< 0.3–11	0.80	84
Bentonite	< 0.3–11	1.4	10
Marlstone	0.4–10	2.6	8
Shale and claystone	< 0.3–8.4	1.0	66
Limestone	–	0.035	–
Metamorphic			
Eclogite	0.04–0.26	0.11	6
Garnet schist	–	1.0	–
Grey gneiss	0.12–0.16	0.14	2

* Condensed from data cited by Waketa and Schmitt[37].

Table 4. The cadmium content of some black shales in England and Wales (taken from Holmes[12])

Formation	Locality	Age	Range	Mean*
Lower Worston shale group	Bowland Forest Lancashire	B1–2	< 1–32	4.4 (46)
Lower Bowland shale group	Bowland Forest Lancashire	P1	1–105	16.2 (35)
Lower Bowland shale group	Bowland Forest Lancashire	P2	1–158	16.5 (20)
Upper Bowland shale group	Bowland Forest Lancashire	E1	1–219	16.6 (59)
Edale shales	North Derbyshire	E2	1–39	5.2 (48)
Edale shales	North Derbyshire	H1	1–50	6.0 (14)
Edale shales	North Derbyshire	H2	< 1–91	14.8 (25)
Edale shales	North Derbyshire	R1	1–32	6.0 (11)
Dove shales	Southwest Derbyshire	E2	< 1–25	6.5 (45)
Mixon limestone and shales	North Staffordshire	P1–2	< 1–65	12.8 (31)
Onecote sandstone and shales	North Staffordshire	P1–2	1–39	9.3 (8)
Onecote sandstone and shales	North Staffordshire	E2	1–2	1.4 (12)
Crackington formation	Devon/Cornwall	H1	< 1–5	1.3 (39)
Crackington formation	Devon/Cornwall	R1	< 1–4	1.7 (55)
Crackington formation	Devon/Cornwall	R2	< 1–3	1.5 (29)
Coal measures	Glamorgan	d5	< 1–5	1.0 (9)
Coal measures	Chesterfield	d5	< 1–3	1.5 (15)

* Number of samples in parenthesis.

Table 5. Amounts of cadmium, zinc and lead in some British soils (µg/g, 0–15 cm)

	No. of samples	Cadmium	Zinc	Lead
Derbyshire				
Mining area	13	1.1–34	94–8000	230–48000
Control areas	5	0.9–3.8	82–241	69–290
Shipham				
Mining area	12	29–800	2520–62400	720–9600
Control areas	6	2.0–10	208–740	128–344
Carboniferous black shales				
Valley bottoms	7	3.4–24	170–740	120–1480
Hill tops and slopes	17	1.5–5.0	55–460	70–500

(from Marples and Thornton[15]).

related to bedrock geochemistry. However, this relationship may be modified to varying degrees by pedogenetic processes leading to the mobilization and redistribution of elements both within the soil profile and between neighboring soils. Although during weathering, cadmium goes readily into solution, the main factor determining the cadmium content of soil is the chemical composition of the parent rock[13]. Kabata-Pendias and Pendias[13], listing the ranges of cadmium contents of surface soils from many parts of the world, report that the average contents lie between 0.07 and 1.1 µg/g and consider that all values over 0.5 µg/g reflect anthropogenic inputs. A national survey in Japan in 1972 showed a mean concentration of cadmium in unpolluted paddy soils of 0.4 µg/g (n = 2746) and in polluted soils 0.9 µg/g (n = 300); in areas surrounding zinc mines and smelters the mean value found in another survey in 1970 was 4.49 µg/g Cd (n = 797) with a maximum value in paddy soil of 68.7 µg/g[41].
The cadmium content of British soils was found to range from 0.08 to 10 µg/g (median < 1.0 µg/g, n = 659)[2]. However, where cadmium is found in association both with lead-zinc mineralization and with marine black shales, soils sometimes contain very much larger concentrations of the metal, as shown in table 5[14,15]. Cadmium in surface soils derived from carboniferous black shales in Derbyshire ranged from 1.5 to 24 µg/g[15,16]. It has been postulated that the smaller amounts of cadmium found in the soils compared to rocks in some areas underlain by cadmium-rich black shales in Britain is because of mobilization and leaching of cadmium in the course of soil formation[12,15]. Ratios of cadmium in rock:soil of 7:1 to 2:1 have been shown in several such areas in Britain[12]. Because of

the mobility of cadmium in acidic soils, it has been proposed that migration down the soil profile is more likely than accumulation at the surface under the influence of humid climates[13]. Redistribution of cadmium and zinc has been clearly shown both in the soil profile and between neighboring soils in a catenary square developed over Carboniferous black shale in Derbyshire, England (table 6). In this instance, cadmium has been leached from imperfectly drained slightly acid soils on the hill top and neighboring slopes and accumulated in waterlogged organic soils at the base of the slope. Metals are markedly enriched in soils of several mineralized areas in Britain with a past history of mining. As shown in table 5, cadmium, together with lead and zinc, are present in elevated concentrations in soils of the historical mining area of Derbyshire where galena (PbS) and sphalerite (ZnS) were worked for several hundred years. Mining of zinc in the form of smithsonite ($ZnCO_3$) at Shipham in southwest England has resulted in extremely high concentrations of both cadmium and zinc in soils (table 5) and even soils developed from the host rock, dolomitic conglomerate of Carboniferous age, contain 10 µg/g Cd or more and extend over some 8 km^2. Metals in two soils profiles illustrate both a) surface contamination on reclaimed land over the old zinc workings and b) the influence of underlying mineralized parent materials (table 7). Further detailed studies on the distribution of cadmium and other metals in the soil profiles of this area have been reported[16,29].

Geochemical cycling of cadmium

A detailed description of the sources and cycling of cadmium in the environment falls outside the scope of this chapter and the subject has been discussed by several other authors in this volume. The influence of geochemical parameters commences during the weathering of rocks and mineral deposits containing cadmium. During weathering cadmium mainly enters the surface environment as soluble compounds, mainly as Cd^{2+}, and the most important factors controlling its mobility are pH and oxidation potential[13]. It may also form several complex ions ($CdCl^+$, $CdOH^+$, $CdHCO_3^+$, $CdCl_3^-$, $CdCl_4^{2-}$, $Cd(OH)_3^-$) and $Cd(OH)_4^{2-}$) and a variety of chelated and organometallic complexes resulting from the decay of plants and animal matter[7,13].
Computer-based models using chemical equilibrium programs have been developed to predict the chemical speciation of potentially hazardous trace metals in soils and

Table 6. Cd, Zn and Pb in typical soil profiles overlying carboniferous black shale in the vicinity of Tissington, Derbyshire (Marples)[14]

		Cd	Zn	Pb
		(µg/g, dry soil)		
a) Pelostagnogley, top of slope				
0–10 cm	Clay loam	4	248	316
10–20 cm	Clay	4	228	252
20–50 cm	Clay	3	240	160
50–75 cm	Clay	3	356	44
75–100 cm	Clay/shale	9	536	100
b) Earthy eutro-amorphous peat, waterlogged valley bottom				
0–10 cm	Fibrous peat	10	312	204
40–50 cm	Amorphous peat	14	660	414
70–80 cm	Peaty clay loam	36	1000	328
80–90 cm	Peaty clay loam	52	2280	400

Table 7. Cd, Zn and Pb in typical soil profiles in the vicinity of old zinc workings at Shipham, Somerset (Marples and Thornton[15])

		Cd	Zn	Pb
		(µg/g, dry soil)		
a) Brown earth, reclaimed land over zinc workings				
0– 2 cm	Root mat	503	48000	4440
2–20 cm	Silt loam	559	54000	4600
20–45 cm	Silt loam	216	22000	1960
b) Brown earth, 300 m from nearest workings				
0– 2 cm	Root mat	48	3360	960
2–20 cm	Silt loam	58	4760	920
20–45 cm	Silt loam	84	8040	1640
45–60 cm	Silt loam	188	9720	1560

waters[18,26]. One such model, GEOCHEM, applied to two Californian soils predicted the principal aqueous species of cadmium as Cd^{2+}, $CdSO_4^\circ$ and $CdCl^{+}$[26]. This approach has yet to be applied widely and may prove difficult in environments where organic compounds are frequent. Under strongly oxidizing conditions, cadmium may form oxide and carbonate minerals (CdO, $CdCO_3$), and under reducing conditions can also utilize H_2S produced by bacteria during the decay of organic residues and precipitate as a sulphide. Cadmium precipitated in soils and sediments may be mobilized again by bacteria[7], probably in both inorganic and organic forms, and released to soil, air and water.

Most unpolluted waters contain very small amounts of cadmium and values of less than 1 ng/g have been reported; however concentrations may exceed 10 ng/g[9]. Data compiled by Boyle and Jonasson[7] are listed in table 8; waters in the vicinity of cadmium-bearing mineral deposits may range up to 1000 µg/l or more. Inshore waters in the U.K. have been shown to range from 0.14 to 4.20 µg Cd/l[1], reflecting metal-rich inputs from industrial and old-mining sources. Rivers containing present-day mining effluent and adit drainage from old mines in Cornwall, southwest England, have been shown to range up to 6 µg Cd/l[4,40].

Factors influencing the availability and uptake of cadmium into biological systems are detailed elsewhere in this volume. The principles of environmental geochemistry which govern the chemical characteristics in the surface environment and the complex interactions in the system rock-water-air-life are outlined by Plant and Raiswell[22]. A generalized geochemical cycle for cadmium has been compiled by Boyle and Jonasson[7] and is reproduced in figure 1.

Table 8. Cadmium content of natural waters (µg/l (parts per billion))

Description	No. of samples	Range	Mean
Rainwater, snow (virgin areas, high contents are near mineralised zones)	46	< 0.5 –2.0	< 0.5
Rainwater, snow (urban area, Ottawa)	12	< 0.2 –1.0	0.3
Normal stream, river and lake waters	74	< 0.01–5.0	0.3
Stream and river waters near cadmium deposits	10	< 0.01–1000	
Oceans and seas	52	0.24–0.48	0.14
Normal groundwaters	22	< 0.01–1.0	0.05
Groundwaters and mine waters near polymetallic sulfide deposits	29	Up to 1140	

As listed by Boyle and Jonasson[7].

Geochemical maps

The first map to show the distribution of cadmium on a regional/national scale was initially published in 1973[33] and then in 1978 as part of the Wolfson Geochemical Atlas of England and Wales[39]. Based on the analysis of 50,000 stream sediment samples, maps for 21 elements were presented as smoothed data plotted by computer. For cadmium (fig. 2) 83% of the area is covered by the lowest class interval (less than 1 µg/g in the < 204 µm fraction), corresponding to normal base-line concentrations in rocks and soils. Anomalous patterns are shown by the darker areas and indicate three main sources of the metal:

1) Cadmium associated with zinc ores in mineralized areas where past mining and smelting of lead and zinc has led to contamination of surface drainage and soils. These areas include the mining districts in Derbyshire and around Shipham, Somerset, mentioned earlier in the chapter.

2) Cadmium (usually associated with zinc and/or lead

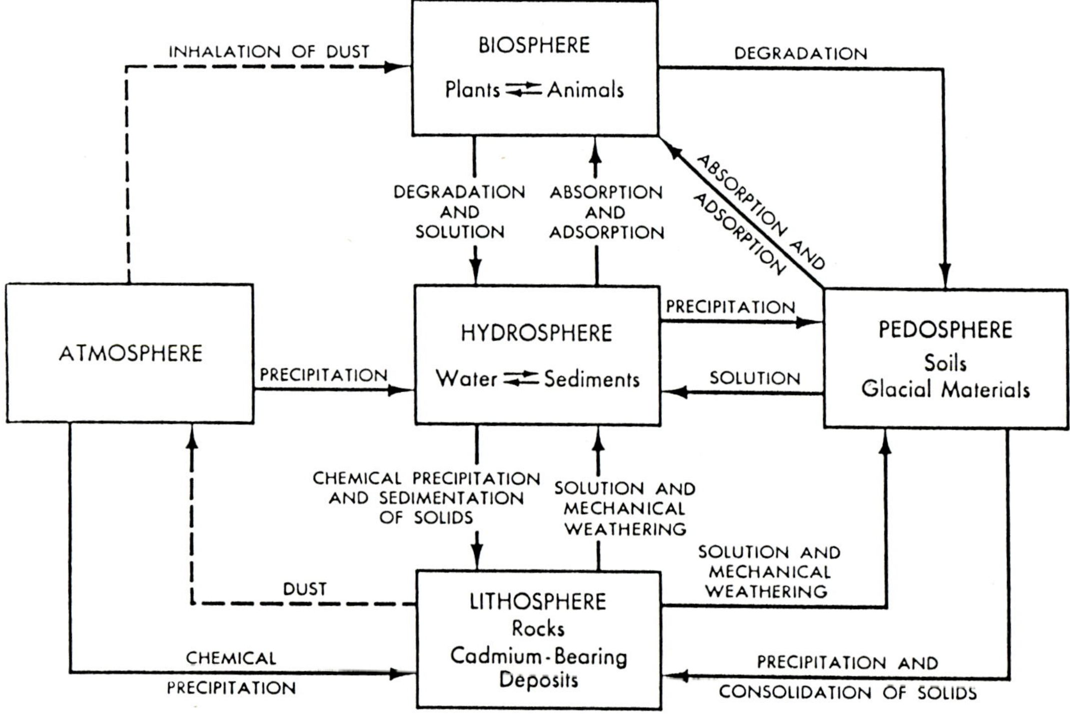

Figure 1. Generalized geochemical cycle of cadmium (from Boyle and Jonasson[7]).

and other metals) dispersed by wind and water from industrial activities, including the world's largest zinc-lead smelter at Avonmouth. Haloes of raised metal values are also found around some large industrial conurbations, particularly Birmingham, England's second largest city.

3) Cadmium derived from the weathering of naturally metal-rich marine black shales of Carboniferous age which outcrop mainly in parts of central and northwest England and are referred to in table 5.

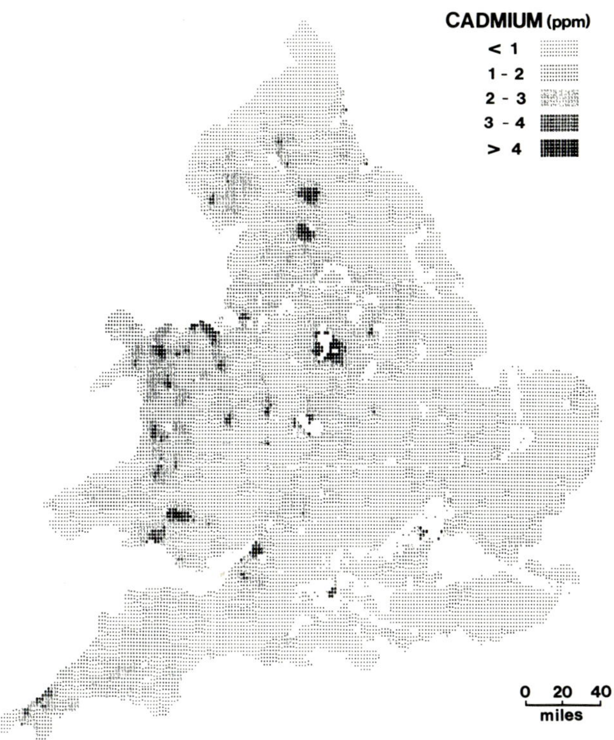

Figure 2. Map showing the distribution of cadmium in stream sediments over England and Wales (µg/g) (compiled by the Applied Geochemistry Research Group as part of the Wolfson Geochemical Atlas of England and Wales; Webb et al.[39]).

The total area within England and Wales in which anomalous patterns of cadmium of 4 µg/g or more occur (compared to the background of less than 1 µg/g) extends to some 1200 km^2 or 0.33% of the area covered[15]. The results of several inter-related studies into the sources, dispersion, distribution and pathways of the metal in the rock-soil-plant-animal (including human) system and rock-stream sediment-water system have been published in a number of post-graduate theses and scientific articles[4, 11, 12, 14–17, 29, 31–34].

Regional geochemical maps have now been published for many other parts of the world. Their application to the understanding of cadmium distribution is of considerable significance both in highlighting those areas in which levels of the metal are elevated as a result of pollution and in providing valuable information on background concentrations of the element resulting from the chemical composition of bedrock and overburden, a knowledge of which is essential in order to assess the degree and extent of man-made contamination.

1 Abdullah, M. I., Royle, L. G., and Morris, A. W., Heavy metal concentration in coastal waters. Nature 235 (1972) 158–160.
2 Archer, F. C., Trace elements in soils in England and Wales, in: Inorganic Pollution and Agriculture Reference Book 326, pp. 184–190. Ministry of Agriculture Fisheries and Food, H.M.S.O., London 1980.
3 Aston, S. R., Chester, R., Griffiths, A., and Riley, J. P., Distribution of cadmium in North Atlantic deep sea sediments. Nature 239 (1972) 393.
4 Aston, S. R., Thornton, I., Webb, J. S., Purves, J. B., and Milford, B. L., Stream sediment composition: an aid to water quality assessment. Water, Air Soil Poll. 3 (1974) 321–325.
5 Bowen, H. J. M., Environmental Chemistry of the Elements. Academic Press, London 1979.
6 Boyle, R. W., and Jonasson, I. R., Geochemistry of cadmium. Department of Energy, Mines and Resources, Ottawa 1972.
7 Boyle, R. W., and Jonasson, I. R., Geochemistry of cadmium, in: Effects of Cadmium in the Canadian Environment, pp. 15–32. Publication No. NRCC 16743, National Research Council of Canada, 1979.
8 Cissarz, A., Quantitative spectroanalytical investigation of a Mansfield Kupferschiefer profile. Chemie der Erde 5 (1930) 48–75.
9 Friberg, L., Piscator, M., Nordberg, G. F., and Kjellstrom, T., Cadmium in the Environment, 2nd edn. CRC Press, Boca Raton, Florida 1974.
10 Goldschmidt, V. M., Geochemistry. Oxford University Press, Oxford 1958.
11 Goodman, J. G., The dispersion of cadmium, lead and zinc in agricultural soils in the vicinity of old zinc mines at Shipham, Somerset. M.Sc. Thesis, University of London, 1979.
12 Holmes, R., The regional distribution of cadmium in England and Wales. Ph.D. Thesis, University of London, 1975.
13 Kabata-Pendias, A., and Pendias, H., Trace Elements in Soils and Plants. CRC Press, Boca Raton, Florida 1984.
14 Marples, A. E., The occurrence and behaviour of cadmium in soils and its uptake by pasture grasses in industrially contaminated and naturally metal-rich environments. Ph.D. Thesis, University of London, 1979.
15 Marples, A. E., and Thornton, I., The distribution of cadmium derived from geochemical and industrial sources in agricultural soils and pasture herbage in parts of Britain, in: CADMIUM 79, pp. 74–79. Proceedings 2nd Int. Cadmium Conference, Cannes 1979. Metal Bulletin Ltd., London 1980.
16 Matthews, H., The distribution of cadmium and associated elements in the soil-plant system at sites in Britain contaminated by mining, smelting and metal-rich bedrock. Ph.D. Thesis, University of London, 1982.
17 Matthews, H., and Thornton, I., Agricultural implications of zinc and cadmium contaminated land at Shipham, Somerset, in: Trace Substances in Environmental Health, XIV, pp. 478–488. Ed. D. D. Hemphill. University of Missouri, Columbia 1980.
18 Mattigod, S. V., and Page, A. L., Assessment of metal pollution in soils, in: Applied Environmental Geochemistry, pp. 355–394. Ed. I. Thornton. Academic Press, London 1983.
19 Mitchell, R. L., Trace elements in soils, in: Chemistry of the Soil, 2nd edn. Ed. F. E. Bear. Reinhold Publ. Co., New York 1964.
20 Mullin, J. B., and Riley, J. P., The occurrence of cadmium in seawater and in marine organisms and sediments. J. mar. Res. 15 (1956) 103–122.
21 Page, A. L., and Bingham, F. T., Cadmium residues in the environment. Residue Reviews, vol. 48, pp. 1–43. Ed. F. A. Gunther. Springer-Verlag, New York 1973.
22 Plant, J. A., and Raiswell, R., Principles of environmental geochemistry, in: Applied Environmental Geochemistry, pp. 1–39. Ed. I. Thornton. Academic Press, London 1983.
23 Portfir'ev, V. B., Krayushkin, V. A., and Kazakov, S. B., Ash composition of crude oils from the Frasnian stage of the Pripyat Basin. Dopov. Akad. Nawk. UKr. RSR, Ser. B. 32 (1970) 121–125.
24 Rankama, K., and Sahama, Th. G., Geochemsitry. University of Chicago Press, Chicago 1950.
25 Sandell, E. B., and Goldrich, G. S., The rarer metallic constituents of some American igneous rocks. J. Geol. 51 (1943) 167.
26 Sposito, G., The chemical forms of trace metals in soils, in: Applied Environmental Geochemistry, pp. 123–170. Ed. I. Thornton. Academic Press, London 1983.
27 Strominger, D., et al., Table of isotopes. Rev. mod. Phys. 30 (1958) 585.
28 Swaine, D. J., Trace elements in coal, in: Trace Substances in Environmental Health, XI, pp. 107–116. Ed. D. D. Hemphill. University

of Missouri, Columbia 1977.

29 Swarbrick, Z. A., The distribution of cadmium and associated elements within selected soil profiles at Shipham, Somerset. M. Sc. Thesis, University of London, 1980.

30 Thorne, L. T., Metcalfe, A., and Thornton, I., A geochemical investigation of Cd, Pb and Zn in waters and sediments of the River Gannel, Cornwall, in: Trace Substances in Environmental Health, XVI, pp. 500–506. Ed. D. D. Hemphill, University of Missouri, Columbia 1980.

31 Thornton, I., Geochemistry applied to agriculture, in: Applied Environmental Geochemistry, pp. 231–266. Ed. I. Thornton. Academic Press, London 1983.

32 Thornton, I., and Plant, J., Regional geochemical mapping and health in the United Kingdom. J. geol. Soc. London 137 (1979) 575–586.

33 Thornton, I., and Webb, J. S., Environmental geochemistry: some recent studies in the United Kingdom, in: Trace Substances in Environmental Health, VII, pp. 89–98. Ed. D. D. Hemphill. University of Missouri, Columbia 1973.

34 Thornton, I., John, S., Moorcroft, S., Watt, J., Strehlow, C. D., Barltrop, D., and Wells, J., Cadmium at Shipham – a unique example of environmental geochemistry and health, in: Trace Substances in Environmental Health, XIV, pp. 27–37. Ed. D. D. Hemphill. University of Missouri, Columbia 1980.

35 Valkovic, V., Trace Elements in Coal, vol. 1. CRC Press, Boca Raton, Florida 1983.

36 Vinogradov, A., Geochemistry of Rare and Dispersed Metals in Soils. Chapman and Hall Ltd., London 1959 (translation from Russian).

37 Waketa, H., and Schmitt, R. A., Cadmium, in: Handbook of Geochemistry, vol. 11–2. Ed. K. H. Wedepohl 1970.

38 Weast, C., Handbook of Chemistry and Physics, 50th edn. Chemical Rubber Co., Cleveland, Ohio 1969.

39 Webb, J. S., Thornton, I., Howarth, R. J., Thompson, M., and Lowenstein, P. L., The Wolfson Geochemical Atlas of England and Wales. Oxford University Press, 1978.

40 Wedepohl, K. H., Handbook of Geochemistry. Springer-Verlag, New York 1969–74.

41 Yamagata, N., Cadmium in the environment and in humans, in: Cadmium Studies in Japan: a Review, pp. 19–43. Ed. K. Tsuchiya. Kodansha Ltd., Tokyo and Elsevier/North Holland Biomedical Press, Amsterdam 1978.

Evaluation of methods for the speciation of cadmium

by M. Astruc

Analytical Chemistry Laboratory, Faculty of Sciences, University of Pau, Avenue de l'Université, F–64000 Pau (France)

Introduction

The development of analytical techniques during the last decades has been such that it became clear ten years ago that neither the biogeochemical cycle nor the ecotoxicity of trace or ultratrace elements could be understood on the basis of total concentration data, introducing thus the necessity of speciation methods. Florence[24] has defined the speciation of one element as the determination of the individual physicochemical species which together represent its total concentration in a sample. Considerable efforts have since been devoted to the development of methods allowing speciation of toxic trace elements in the aquatic environment. Large advances have been made in the study of water[41, 69], yet important problems remain unsolved[39]. Studies on sediment and suspended solids developed very slowly but have now reached a very active stage[28, 41]. In contrast, very little effort has been made in the field of atmospheric chemistry of the majority of trace elements including cadmium[68]. Research on heavy metal speciation in water is in fact relevant to two differing philosophies:

– the short term one uses operationally defined procedures to try to evaluate the composition – toxicity relationship to provide information for environmental authorities;

– the long term one aims at improving the general knowledge of the physicochemical behavior of toxic elements in water and of their interaction with living organisms. Both approaches are necessary, but are sometimes conflicting.

Aquatic chemistry of cadmium

The aquatic chemistry of an element is governed by its ability to participate in the classical elementary processes of wet chemistry (proton, electron or ligand exchange, precipitation), to take part in surface reactions involving colloids, suspended matter and sediment or to form organometallic compounds. Interactions with the biota are certainly not negligible but must involve at least one of the preceding chemical processes. In some instances these interactions are even suspected to control the overall biogeochemical cycle of the element. The first step of speciation is a separation of dissolved species from the whole element on a size basis[4].

The aquatic chemistry of cadmium is rather simple as compared to that of some other elements due to the absence of oxidoreduction processes and the lack of any pronounced ability to form covalent bonds with carbon in the aquatic environment[46]. Cadmium is characterized by oxyphilic and sulfophilic properties. In coordination reactions it behaves as a soft acid acceptor thus preferring sulfur, selenium or nitrogen donors[46].

Inorganic dissolved species

A theoretical speciation of cadmium by equilibrium calculations using one of the numerous programs available[49] is thus simple, provided that reliable thermodynamic data are available. This assumption is true in some instances but definitely wrong when unidentified particulate matter or dissolved organic matter is present (i.e.

quite often!). If only inorganic components are considered, equilibrium calculations demonstrate that free aqueous ion, hydroxylated and chlorinated dissolved species predominate, depending on pH and salinity[46]. In oxic sea-water cadmium is almost entirely bound in chloride complexes ($CdCl_2^0$, $CdCl^+$, $CdCl_3^-$) and the free aqueous cation concentration is only a few percent of the total[47]. But even with these simplified assumptions the precision of the equilibrium constants is poor enough to introduce a sizeable uncertainty in the calculation (± 0.06 in pCd at 35‰ salinity[70]). In reducing aquatic environments such as pore waters, sulfide species ($Cd(HS)_2^0$, $Cd(HS)_3^-$, $Cd(HS)_4^{2-}$, $CdHS^+$) are predominant, preventing precipitation of cadmium sulfide[19], and may explain the unexpectedly high level of dissolved labile cadmium in pore waters[7]. In fresh waters only a few percent of the dissolved cadmium is complexed[47] ($CdOH^+$ essentially, $CdSO_4^0$, $CdCl^+$ ...) and the better part remains as the free cationic species.

Influence of the dissolved organic matter (DOM)

DOM is usually a very complex mixture of unknown chemicals. Several categories of potentially important organic ligands have been identified in natural waters[17].
Amino acids, occurring at submicromolar concentrations, cannot compete with inorganic ligands; this statement, true for all the trace metals, is especially clear for cadmium as the stabilities of its complexes with amino acids are relatively low[65]. Exceptions to this rule could occur for sulfur containing amino acids such as cysteine. Humic substances (HS) usually constitute the greater part of the DOM. It must be remembered that under this general appellation are gathered very different chemicals; their molecular weights vary widely, as do the density and nature of complexing sites. Fresh water and coastal sea-water contain HS derived mainly from degraded lignins[14] with a high degree of aromaticity. In open ocean waters, HS are autochtonous compounds condensed from smaller molecules with little aromatic character[36]. It is thus not surprising that very diverse conclusions have been drawn from experimental studies of Cd-HS interactions even without considering the diversity of the techniques used. Stability constants in the range 6.0–7.7 at high pH values have been reported[13], so that although it is rather probable that complexing of cadmium by HS in DOM rich waters (rivers, lakes, estuaries, coastal sea waters) is an important phenomenon it may be quite negligible in open ocean waters[54]. It has been reported[31] that cadmium-HS complexes are sometimes more stable than the copper equivalents, which is contrary to what has normally been stated[42]. The polyelectrolyte nature of humic substances[13] may explain many strange variations in the data on heavy metal complexation.
The production and release of strong chelating agents by aquatic organisms has been widely demonstrated in the laboratory in recent years. The organisms concerned so far include phytoplankton, zooplankton, bacteria, fungi and even large species such as algae[22, 48], and snails[18], but this list is certainly not complete. A field demonstration of the release of chelating agents by diatoms has been performed[38].
These experiments have been mainly linked with studies of copper, zinc or iron behavior and very few data exist for cadmium complexation by these strong chelators. Most cell exudates complex copper strongly, but cadmium seems to remain essentially in labile forms. These exudates do certainly modify very seriously the speciation of trace metals in the immediate vicinity of cellular membranes but virtually no information is presently available in this fascinating domain.
Complexation/dissociation kinetics are essentially determined by the water exchange rate constant of the metal ion and the concentration of the reactants. In environmental conditions the complexation rates are thus usually low, as well as those of the dissociation of stable chelates. Little precise direct information on the behavior of cadmium complexes is available[58]. Comparison of the water exchange rate constants of Cu^{2+} ($9 \cdot 10^8$ $M^{-1} \cdot s^{-1}$) and Cd^{2+} ($9 \cdot 10^7$ $M^{-1} \cdot s^{-1}$) indicates that the dissociation of cadmium complexes should be faster than those of copper equivalents, in agreement with experimental evidence.
However, the metal-DOM interactions are dynamic processes and it is difficult to avoid changes in the composition of samples during transport and storage[51], owing to rapid ligand evolution; this illustrates the difficulties of comparison between field and laboratory studies.

Water-solid interface

The interface between water and suspended solids has a tremendous importance in the speciation of trace elements, especially in freshwater, estuaries and coastal sea-water, but seems to be unimportant in the open ocean[11, 45, 49]. A multitude of laboratory studies has elucidated the interaction of heavy metals with carefully defined suspended solids. Extrapolation to natural situations remains difficult as particulates are usually heterogeneous mixtures of organic and inorganic matter[15]. Reactions involved include[12] metal adsorption, ligand adsorption and formation of surface complexes[16], ion exchange, adsorption of metal complexes[60].
The adsorption of cadmium seems weaker than those of copper or zinc ions[12]; nevertheless, a decisive role has been attributed to organic colloids in the speciation of cadmium in lake waters[1] and humic substances were proven to be responsible for cadmium adsorption in river mud[30]. Cadmium adsorption and organic matter in Ottawa river sediment are very well correlated[53], but selective extractions of other river sediments[63] demonstrated that cadmium bound to organic matter was undetectable. A recent general review is available[29] and specific applications to cadmium are included in this volume[28, 56].

Main physicochemical species in natural waters

Form	Examples	Diameter (nm)
Particulate	Clays; detritus; biota; fecal pellets	> 450
Colloids	Clays; metal oxides; humic substances	10–500
Dissolved complexes	With organic ligands such as fulvic acids, fatty acids, amino acids or inorganic ligands such as CO_3^{--}, Cl^-, OH^-, HS^-...	1–10
Aqueous ion		1

Experimental speciation methods for cadmium in water

Many different approaches have been developed throughout the world[3] providing what may be called 'operationally defined' classifications[4]. Much effort is still needed to refine, compare and standardize these speciation procedures and their utility by biological testing. As cadmium in water seems not to be engaged in organometallic compounds, no account of speciation methods specially designed for organometallic species will be taken here.

Very few analytical techniques are able to detect selectively metal species in natural waters, due to the very low concentrations involved. Thus separation and pretreatment procedures are used to isolate some of them prior to a determination of the total concentration in these fractions. It must be strongly emphasized at this point that every separation procedure, whatever its physicochemical principle, disturbs solution equilibrium[2] and may lead to simultaneous or subsequent uncontrolled changes in the metal speciation.

Size separation methods

An excellent review appeared recently on this subject[61]. *Filtration* with 0.45 μm membrane filters is universally agreed to separate 'particulate' matter from 'dissolved' (including colloids). Care must be taken not to overload the membrane with a substantial amount of deposit so that pore diameter is seriously reduced.

Ultrafiltration with selective membranes allows species separations down to the 500-dalton size, contamination problems being now at least partially resolved. In phytoplankton cultures[32], for example, cadmium is mainly present in the < 500-da fraction, whereas copper is essentially associated with the 500–1000-da fraction. A similar behavior of Cd and Cu in coastal sea-water has been reported[37], Cd being essentially labile, and associated with the < 1000-da fraction.

Dialysis is a simple and inexpensive size separation method and its use is thus especially interesting for field experiments[9]. The membranes used have a molecular weight cut-off in the range 10^3–10^4 dalton. The major shortcoming is that owing to the time scale of the experiment the displacement of chemical equilibria at the interface may be important as well as contamination or adsorption by the membrane[61]. In situ sampling by dialysis[8] where dialyzed species may be continuously removed by some process[35, 50] presents interesting features that have not yet been fully investigated; the limitations discussed above are much less stringent due to the large time scale of the experiment and volume of equilibrating water.

Gel permeation chromatography has been used mainly to study interactions between humic substances and metals[43]. Adsorption and contamination phenomena seem to preclude application to natural waters.

Other separation methods

Ion exchange resins (cation and or anion exchangers) allow a separation of species depending on their charge[21]. But inorganic complexes are usually labile enough for considerable equilibrium displacement to occur. In DOM free water, where about 10^{-6}% of cadmium was calculated to be present as dissolved anionic species, 33% was retained on an anion exchanger[40].

Chelating resins provide more significant information as they fix the metal ion from free and labile metal species, excluding macromocules and colloids[25]. Comparison of batch and column experiments[20] allows an insight into the dissociation kinetics (lability) of metal complexes.

Polymeric adsorbents may retain metal-organic complexes[62] but none of those tested so far has demonstrated a general efficiency for retention of DOM and a lack of adsorption of free metal ions[15].

Solvent extraction with a solvent such as 9:1 hexane-butanol mixture[26], simulating the lipid bilayer of cell membranes, indicates the amount of species able to penetrate the living cell without the help of carrier and may produce very interesting insights in toxicity studies[27].

The MnO_2 *equilibration technique*[15] has been designed so that only inorganic species are adsorbed while ligands or organic complexes are not, due to the high negative charge surface of suspended MnO_2 at common pH values. After completion of equilibrium and phase separation the measurement of the dissolved metal by GFAA or DPASV is a direct evaluation of the amount of metal-DOM complexes in the sample. This new procedure has not yet been applied to cadmium speciation to my knowledge.

Electrochemical measurements

Electrochemical techniques have unique features as they are able to discriminate between electroactive and electroinactive metal species for a wide range of elements including cadmium. Many environmental studies rely thus on the most sensitive of these methods, anodic stripping voltammetry (A.S.V.); ion selective electrodes (I.S.E.) are useless except in laboratory simulations due to their insufficient sensitivity.

Anodic stripping voltammetry is a two-step procedure. At first the potential of the mercury electrode (hanging drop or thin film) is fixed at a suitable negative value so that sufficiently labile metal species are reduced, thus concentrating metals in the electrode as amalgams. Then the electrode is slowly anodized (continuous or pulsed potential variation) while the amalgam reoxidation current is registered. Extremely sensitive, this method allows a direct study of toxic metals even in unpolluted waters. The presence of ligands generating labile metal complexes[64] (Cl^-, CO_3^-, amino acids …) produces a shift of the peak potential that can be used to evaluate the conditional stability constant[50]. When inert (non-labile) metal complexes are present they remain untouched throughout the experiment. It is therefore possible by comparison with a determination of the total metal concentration to evaluate the amount of metal engaged in non-labile complexes, corresponding to strong organic chelators such as humic substances, cell exudates, etc. and also, possibly, particulates and microorganisms if suspended matter has not been eliminated from the sample. The presence of quasi-labile metal complexes may lead to difficulties as their partial dissociation in the diffusion layer will occur and a mixed, time-scale dependant, signal will be registered[67]. Systematic A.S.V. investigations on the speciation of cadmium in oceans and freshwater have been conducted[44, 50]. The main conclusions are that in open ocean water, with typically DOM < 1 mg·l^{-1} and humic content < 0.2 mg·l^{-1}, cadmium is not noticeably engaged in organic complexes[50], its speciation remaining purely inorganic. Different conclusions could be reached for waters containing higher levels of DOM, specially in freshwaters, estuaries, coastal seawaters[54].

Speciation schemes

Several attempts have been made to design an experimental set of separation and determination methods so that a rational insight into metal speciation could be obtained.

Batley and Florence[5], using A.S.V. distinction of labile and non labile species, combined with passage through a chelating resin and U.V. irradiation, were able to distinguish seven operational classes of metal species. Application of this scheme to seawater and freshwater samples indicated that respectively 33% and 5% of total cadmium were linked to colloids[6, 23].

Hart and Davies[34] defined classes by combining filtration, batch equilibration with a chelating resin and dialysis in a much simpler scheme.

Figura and McDuffie[20] combined A.S.V. to column and batch exchange with a chelating resin, defining four classes of metal species on a kinetic basis.

Lecomte[40] evaluated a combination of A.S.V., weak acidification and anionic and cationic ion exchangers in a water treatment oriented speciation scheme. Dissolved cadmium in river water samples (60–100% of total cadmium) was demonstrated as free or engaged on labile species (40%), the remainder being easily liberated by weak acidification and the totality retained by a cationic resin.

Harrison and Laxen[33] developed a scheme based on multiple ultrafiltration steps combined with A.S.V.

Results on the physicochemical speciation of cadmium in waters

Reports on river water usually present a 60–100% proportion of total cadmium in filtrable species[35, 40]. Retentions were about 75% on chelating resins[35] and 80–95% on cationic resins[40]. Free and labile species represented about 40%[40] or 75%[35] of total dissolved cadmium[40]. The remaining species were liberated by weak acidification[40]. In lakes[57], free and labile species were widely dominant, except in situation where high autochtonous DOM conditions prevailed with a resulting 53% complexing of cadmium by the organic ligands.

In one estuary[35] the speciation of cadmium differed from that in the upper riverine situation in that total and filterable cadmium were 50% lower and non-dialyzable species were present in high proportions.

In open ocean waters the total concentration of cadmium is so low that speciation studies are often beyond the capacities of the most sensitive analytical techniques. Particulates and DOM being very low, except in areas of high biological productivity, simple calculations indicate a 100% speciation of total cadmium as soluble free or labile inorganic species[50].

Rational links between the speciation of cadmium and its toxicity towards living organisms are very scarce[55, 59]. Complexation by natural or artificial strong chelators has been demonstrated as reducing toxicity towards one unicellular alga[52].

Conclusion

Trace element speciation in water has recently been the subject of a considerable research effort leading to an enormous improvement in our knowledge of the aquatic chemistry of these toxic substances. The amount of information concerning dissolved cadmium speciation is rather low compared to that for other metals such as copper or zinc, which are recognized as much less toxic elements. Two main explanations may be suggested:
– The total concentration of cadmium is often very low in water and the dissolved fraction even lower, and the sensitivities of available speciation techniques are sometimes insufficient to deal with the problem.
– The chemical properties of cadmium are such that it forms usually relatively weaker and more labile complexes with DOM, as compared to other toxic metals. These complexes are thus difficult to study and evaluate. Much effort is still required, especially with regard to accumulating experimental data on the speciation of cadmium in various aquatic environments and to investigating the toxicological aspects of cadmium speciation.

1 Allen, H. E., Noll, K. E., Jamjun, O., and Boonlayangoor, C., Reaction of cadmium in Lake Michigan: kinetics and equilibria. Proc. Am. chem. Soc. 176th Nat. Meeting, Florida 1978, Abstract 14.

2 Astruc, M., Metal forms and speciation. Proc. Int. Conf. 'Heavy metals in the environment'. London 1979, pp. 439–445.

3 Astruc, M., and Pinel, R., La spéciation des éléments en traces dans les systèmes aquatiques. L'Actualité chim. May 1985, pp. 29–35.

4 Batley, G. E., The current status of trace element speciation studies in natural waters, in: Trace Elements Speciation in Surface Waters and its Ecological Implication, pp. 17–35. Ed. G. G. Leppard. NATO Conf. Series I, vol. 6. Plenum, New York 1983.

5 Batley, G. E., and Florence, T. M., Determination of the chemical forms of dissolved cadmium, lead and copper in seawater. Mar. Chem. 4 (1976) 347.

6 Batley, G. E., and Gardner, D., A study of copper, lead and cadmium speciation in some estuarine and coastal marine waters. Estuarine coast. mar. Sci. 7 (1978) 59.

7 Batley, G. E., and Giles, M. S., A solvent displacement technique for the separation of sediment interstitial waters, in: Contaminants and Sediments, vol. 2. Ed. R. A. Baker. Ann Arbor Science Publ., Ann Arbor, Michigan 1980.

8 Beneš, P., Semi-continuous monitoring of truly dissolved forms of trace elements in streams using dialysis in-situ. I – Principle and conditions. Water Res. 14 (1980) 511.

9 Beneš, P., and Steinnes, E., In situ dialysis for the determination of the state of trace elements in natural waters. Water Res. 8 (1974) 947.

10 Boulegue, J., Trace metals (Fe, Cu, Zn, Cd) in anoxic environments, in: Trace Metals in Sea Water, pp. 563–578. Eds C.S. Wong, E. Boyle, K. W. Bruland, J. D. Burton and E. D. Goldberg. NATO Conf. Series IV, vol. 9. Plenum, New York 1980.

11 Bourg, A. C. M., Effect of ligands at the solid-solution interface upon the speciation of heavy metals in aquatic systems. Proc. Int. Conf. 'Heavy metals in the environment'. London 1979, pp. 446–449.

12 Bourg, A. C. M., Role of freshwater/sea water mixing on trace metal adsorption phenomena, in: Trace Metals in Sea Water, pp. 195–208. Eds C.S. Wong, E. Boyle, K. W. Bruland, J. D. Burton and E. D. Goldberg. NATO Conf. Series IV, vol. 9. Plenum, New York 1983.

13 Cleven, R. F. M. J., Heavy metal/polyacid interaction. Thesis Agricultural Univ., Wageningen, The Netherlands 1984.

14 Crawford, R. L., Lignin biodegradation and transformation. Wiley, New York 1981.

15 Davis, J. A., and Leckie, J. O., Effect of adsorbed complexing ligands

on trace metal uptake by hydrous oxydes. Envir. Sci. Technol. *12* (1978) 1309.

16 Davis, J. A., and Leckie, J. O., Surface ionization and complexation at the oxyde/water interface: surface properties of amorphous iron oxyhydroxyde and adsorption of metal ions. J. Colloïd. Interface Sci. *67* (1978) 90–107.

17 Dawson, R., Symposium on concepts in marine organic chemistry, Edinburgh 1976, cited in: Principles of Aquatic Chemistry, p. 267. Ed. F. M. M. Morel. Wiley, New York 1983.

18 El Mednaoui, H., Castetbon, A., and Astruc, M., unpublished results.

19 Emerson, S., Jacobs, L., and Tebo, B., The behaviour of trace metals in marine anoxic waters: solubilities at the oxygen-hydrogen sulfide interface, in: Trace Metals in Sea Water, pp. 579–608. Eds C. S. Wong, E. Boyle, K. W. Bruland, J. D. Burton and E. D. Goldberg. NATO Conf. Series IV, vol. 9. Plenum, New York 1983.

20 Figura, P., and McDuffie, B., Determination of the labilities of soluble trace metal species in aqueous environmental samples by anodic stripping voltammetry and Chelex column and batch methods. Analyt. Chem. *52* (1980) 1433–1440.

21 Filby, R. H., Shah, K. R. and Funk, W. H., Role of neutron activation analysis in the study of heavy metal pollution of a lake-river system. Proc. 2nd Int. Conf. 'Nuclear Methods in Environmental Research'. Eds J. R. Vogt and W. Meyer. NTIS, Springfield, Va. 1974.

22 Fisher, N. S., and Fabris, J. G., Complexation of Cu, Zn and Cd by metabolites excreted by marine diatoms. Mar. Chem. *11* (1982) 245–255.

23 Florence, T. M., Trace metal species in freshwaters. Water Res. *11* (1977) 681.

24 Florence, T. M., Trace element speciation and aquatic toxicology. Trends analyt. Chem. *2* (1983) 162–166.

25 Florence, T. M., and Bathley, G. E., Chemical speciation in natural waters. CRC Crit. Rev. analyt. Chem. *9* (1980) 219.

26 Florence, T. M., and Batley, G. E., A new scheme for chemical speciation of copper, lead, cadmium and zin in seawater. Proc. Int. Conf. 'Heavy metals in the environment'. Amsterdam 1981, pp. 599–602.

27 Florence, T. M., Lumsden, B. G., and Fardy, J. J., Algae as indicators of copper speciation, in: Complexation of Trace Metals in Natural Waters, pp. 411–418. Eds C. J. M. Kramer and J. C. Duinker. Junk Pub., The Hague 1984.

28 Förstner, U., Cadmium in sediments. This volume.

29 Förstner, U., and Salomons, W., Trace element speciation in surface waters: interactions with particulate matter, in: Trace Elements Speciation in Surface Waters and its Ecological Implication, pp. 245–270. Ed. G. G. Leppard. NATO Conf. Series I, vol. 6. Plenum, New York 1983.

30 Gardiner, J., The chemistry of cadmium in natural water. I – A study of cadmium complex formation using the cadmium specific ion electrode. Water Res. *8* (1974) 23–30.

31 Giesy, J. P., Control of trace metal equilibria, in: Trace Elements Speciation in Surface Waters and its Ecological Implication, pp. 195–210. Ed. G. G. Leppard. NATO Conf. Series I, vol. 6. Plenum, New York 1983.

32 Gnassia-Barelli, M., Harstedt-Romeo, M., and Nicolas, E., Copper and cadmium speciation in different phytoplankton culture media, in: Complexation of Trace Metals in Natural Waters, pp. 425–428. Eds C. J. M. Kramer and J. C. Duinker. Junk Publ., The Hague 1984.

33 Harrison, R. M., and Laxen, D. P. H., Physicochemical speciation of lead in drinking water. Nature *286* (1980) 791.

34 Hart, B. T., and Davies, S. H. R., A new dialysis-ion exchange technique for determining the forms of trace metals in water. Aust. J. mar. Freshwater Res. *28* (1977) 105.

35 Hart, B. T., and Davies, S. H. R., A study of the physico-chemical forms of trace metals in natural waters and waste waters. Australian Water Resources Council, technical paper No. 35, 1978.

36 Harvey, G. R., Boran, D. A., Chesal, L. S., and Tokar, J. M., cited in: Principles of Aquatic Chemistry, p. 282. Wiley, New York 1983.

37 Hasle, J. R. and Abdullah, M. I., Analytical fractionation of dissolved copper, lead and cadmium in coastal seawater. Mar. Chem. *10* (1981) 487–503.

38 Imber, B., Robinson, M. G., and Pollehne, F., Complexation by diatom exudates in culture and in the field, in: Complexation of Trace Metals in Natural Waters, pp. 429–440. Eds C. J. M. Kramer and J. C. Duinker. Junk Publ., The Hague 1984.

39 Kramer, C. J. M., and Duinker, J. C., Ed., Complexation of Trace Metals in Natural Waters. Junk. Publ., The Hague 1984.

40 Lecomte, Y., Etude d'un schéma de spéciation du cadmium et du plomb. Application à l'étude de la préozonation d'eaux de rivière. Thesis, University of Pau, France 1981.

41 Leppard, G. G., Trace elements speciation in surface waters and its ecological implication. NATO Conf., Séries I, vol. 6, Plenum, New York 1983.

42 Mantoura, R. F. C., Organo-metallic interactions in natural waters, in: Marine Organic Chemistry. Eds E. K. Duursma and R. Dawson. Elsevier, Amsterdam 1981.

43 Mantoura, R. F. C., and Riley, J. P., The use of gel filtration in the study of metal binding by humic acids and related compounds. Analyt. chim. Acta *78* (1975) 193–200.

44 Mart, L., and Nürnberg, H. W., The distribution of cadmium in the sea. This volume.

45 Martin, J. M., and Whitfield, M., The significance of the river input of chemical elements to the ocean, in: Trace Metals in Sea Water, pp. 265–296. Eds C. S. Wong, E. Boyle, K. W. Bruland, J. D. Burton and E. D. Goldberg. NATO Conf. Series IV, vol. 9. Plenum, New York 1983.

46 Moore, J. W., and Ramamoorthy, S., Heavy Metals in Natural Waters, p. 38. Springer Verlag, New York 1984.

47 Morel, F. M. M., Principles of Aquatic Chemistry, p. 263. Wiley, New York 1983.

48 Morel, F. M. M., Principles of Aquatic Chemistry, pp. 273–274. Wiley, New York 1983.

49 Nordstrom, D. K., and Ball, J. W., Chemical models, computer programs and metal complexation in natural waters, in: Complexation of Trace Metals in Natural Waters, pp. 149–164. Eds C. J. M. Kramer and J. C. Duinker. Junk Publ., The Hague 1984.

50 Nürnberg, H. W., Voltammetric studies on trace metal speciation in natural waters. Part. II: Application and conclusions for chemical oceanography and chemical limnology, in: Trace Elements Speciation in Surface Waters and its Ecological Implication, pp. 211–229. NATO Conf. Series I, vol. 6, Plenum, New York 1983.

51 Piotrowicz, S. R., Harvey, G. R., Springer-Young, M., Courant, R. A., and Boran, D. A., Studies of cadmium, copper and zinc interactions with marine fulvic and humic materials in seawater using anodic stripping voltammetry, in: Trace Metals in Sea Water, pp. 699–718. Eds C. S. Wong, E. Boyle, K. W. Bruland, J. D. Burton and E. D. Goldberg. NATO Conf. Series IV, vol. 9. Plenum, New York 1983.

52 Premazzi, G., Bertone, R., Freddi, A., and Ravera, O., Combined effects of heavy metals and chelating substances on *Selenastrum* cultures. Proc. Seminar on Ecological Tests Relevant to the Implementation of Proposed Regulations Concerning Environmental Chemicals: Evaluation and Research Needs, Berlin 1977, pp. 169–187.

53 Ramamoorthy, S., and Rust, B. R., Heavy metal exchange processes in sediment-water systems. Envir. Geol. *2* (1978) 165–172.

54 Raspor, B., Nürnberg, H. W., Valenta, P., and Branica, M., Significance of heavy metal speciation in natural waters, in: Complexation of Trace Metals in Natural Waters, pp. 317–327. Junk Publ., The Hague 1984.

55 Ravera, O., Cadmium in freshwater ecosystems. This volume.

56 Salomons, W., and Kerdijk, H. N., Cadmium in fresh and estuarine waters. This volume.

57 Shephard, B. K., McIntosh, A. W., Atchison, G. J., and Nelson, D. W., Aspects of the aquatic chemistry of cadmium and zinc in a heavy metal contaminated lake. Water Res. *14* (1980) 1061–1066.

58 Simoes Goncalves, M. L. S., and Correia dos Santos, M. M., Kinetics of the dissociation of cadmium-glutamic acid complex, in: Complexation of Trace Metals in Natural Waters, pp. 367–370. Eds C. J. M. Kramer and J. C. Duinker. Junk Publ., The Hague 1984.

59 Smies, M., Biological aspects of trace element speciation in the aquatic environment, in: Trace Elements Speciation in Surface Waters and its Ecological Implication, pp. 177–191. Ed. G. G. Leppard. NATO Conf. Series I, vol. 6, Plenum, New York 1983.

60 Spivack, A. J., Husted, S. S., and Boyle, E. A., Copper, nickel and cadmium in the surface waters of the Mediterranean, in: Trace Metals in Sea Water, pp. 505–512. Eds C. S. Wong, E. Boyle, K. W. Bruland, J. D. Burton and E. D. Goldberg. NATO Conf. Series IV, vol. 9. Plenum, New York 1983.

61 Steinnes, E., Physical separation techniques in trace element speciation studies, in: Trace Elements Speciation in Surface Waters and its Ecological Implication, pp. 37–46. Ed. G. G. Leppard. NATO Conf. Series I, vol. 6, Plenum, New York 1983.

62 Sugimura, Y., Suzuki, Y., and Miyake, Y., Chemical forms of minor metallic elements in the ocean. J. Oceanogr. Soc. Japan *34* (1978) 93.

63 Tessier, A., Campbell, P. G. C., and Bisson, M., Trace metal speciation in the Yamaska and St. François rivers (Quebec). Can. J. Earth Sci. *17* (1980) 90–105.
64 Valenta, P., Voltammetric studies on trace metal speciation in natural waters. Part. I: Methods, in: Trace Elements Speciation in Surface Waters and its Ecological Implication, pp. 49–67. Ed. G. G. Leppard. NATO Conf. Series I, vol. 6. Plenum, New York 1983.
65 Valenta, P., Simoes Goncalves, M. L. S., and Sugawara, M., Voltammetric studies on the speciation of cadmium and zinc by amino acids in sea water, in: Complexation of Trace Metals in Natural Waters, pp. 357–366. Eds C. J. M. Kramer and J. C. Duinker. Junk Publ., The Hague 1984.
66 Van den Berg, C. M. G., Determination of complexing capacities and conditional stability constants using ion exchange and ligand competition techniques, in: Complexation of Trace Metals in Natural Waters, pp. 17–32. Eds C. J. M. Kramer and J. C. Duinker. Junk Publ., The Hague 1984.
67 Varney, M. S., Turner, D. R., Whitfield, M., and Mantoura, R. F. C., The use of electrochemical techniques to monitor complexation capacity titrations in natural waters, in: Complexation Trace Metal in Natural Waters, pp. 33–46. Eds C. J. M. Kramer and J. C. Duinker. Junk Publ., The Hague 1984.
68 Williams, C. R., and Harrison, R. M., Cadmium in the atmosphere. This volume.
69 Wong, C. S., Boyle, E., Bruland, K. W., Burton, J. D., and Goldberg, E. D., Eds, Trace Metals in Sea Water. NATO Conf. Series IV, vol. 9. Plenum, New York 1983.
70 Zuehlke, R. W., and Byrne, R. H., Thermodynamic and analytical uncertainties in trace metal speciation calculations, in: Complexation of Trace Metals in Natural Waters, pp. 181–186. Eds C. J. M. Kramer and J. C. Duinker. Junk Publ., The Hague 1984.

Cadmium in the atmosphere

by C. R. Williams[1] and R. M. Harrison[2]

Department of Environmental Sciences, University of Lancaster, Lancaster LA1 4YQ (England)

Summary. Cadmium is present naturally in the air mainly as a result of volcanic emissions and release by vegetation. Anthropogenic sources, which overall give rise to emissions one order of magnitude greater than natural sources, are largely primary non-ferrous metals production and waste incineration. Measured concentrations of airborne cadmium are typically < 1 ng m^{-3} at remote sites, 0.1–10 ng m^{-3} at rural sites and 1–100 ng m^{-3} at urban and industrial sites, dependent upon the nature and proximity of local sources. Particle sizes are generally < 2 µm, and often considerably smaller, consistent with an anthropogenic source and a long atmospheric lifetime. Cadmium deposition to the land occurs with fluxes varying from 0.05 ng cm^{-2} month^{-1} in Greenland to circa 1000 ng cm^{-2} month^{-1} in the vicinity of major industrial sources. The possible significance of a motor vehicular source of airborne cadmium is also reviewed.

Introduction

The atmosphere plays an important role in the dispersal of cadmium within the environment. Despite the fact that the largest natural source of cadmium in air (volcanic emissions) as well as the major anthropogenic sources are very localized, their influence may be seen at remote sites throughout the Northern Hemisphere. Thus within this article, the major sources will be reviewed, as well as the reported levels of cadmium measured in ambient air.

Sink processes for atmospheric cadmium are wet and dry deposition, which lead to cadmium enrichment of soil, vegetation and surface waters. The atmospheric source may provide an important cadmium input into these media, and consequently contributes to human exposure through food and drink, as well as more directly by the breathing of polluted air. Data on cadmium deposition fluxes are also reviewed to enable quantitative evaluation of this pathway in relation to cadmium enrichment of these media from other sources.

I. Sources of atmospheric cadmium emissions

On a global scale, anthropogenic discharges of cadmium to atmosphere exceed natural sources by

Table 1. Worldwide annual atmospheric emissions of cadmium (source: Nrіagu[56])

Source	Global production of particulate ($\times 10^9$ kg year^{-1})	Worldwide Cd emission rate ($\times 10^6$ kg year^{-1})
Natural sources[a]		
Wind-blown dusts	500	0.1
Forest fires	36	0.012
Volcanogenic particles	10	0.52
Vegetation	75	0.2
Seasalt sprays	1000	~ 0.001
Total		0.83
Anthropogenic sources[b]		
Mining, non ferrous metals	16	0.002
Primary non-ferrous metal production:		
Cd	0.0017	0.11
Cu	7.9	1.6
Pb	4.0	0.20
Zn	5.6	2.8
Secondary non-ferrous metal production	4.0	0.60
Iron and steel production	1300	0.07
Industrial applications	–	0.05
Coal combustion	3100	0.06
Oil (incl. gasoline) combustion	2800	0.003
Wood combustion	640	0.2
Waste incineration	1500	1.4
Manufacture, phosphate fertilizers	118	0.21
Total		7.3

[a] Based on 'most acceptable' production figures; [b] for the year 1975.

18

an order of magnitude; this is apparent in the global emission inventory for Cd which is given in table 1. Although the estimates used in this inventory are necessarily subject to considerable uncertainties, the values are of a similar magnitude to those reported by other workers; Galloway et al.[25] state that the total emission rate from natural sources is 0.29×10^6 kg year^{-1}, whilst that from anthropogenic emissions is 5.5×10^6 kg year^{-1}.

From table 1, volcanic emissions represent about 60% of the Cd emitted from natural sources, reflecting the high enrichment of the metal in volcanogenic aerosols[56]. Vegetative exudates and windblown dust of crustal origin also form a significant proportion of the natural emission; for the former, it has been concluded[3] that the earth's vegetation-covered land mass contributes generally to the trace metal composition of the atmosphere.

Turning to the anthropogenic sources listed in table 1, non-ferrous metals production and use accounts for 67% of the pollutant Cd released. For comparison, Faoro et al.[20] suggest that, in the U.S.A., 43% of the Cd emitted by stationary sources arises from smelters and metallurgical processing. The importance of these sources reflects the occurrence of Cd in polymetallic ores; of especial note in table 1 is the high emission from primary Zn production, where Cd metal is manufactured as a by-product[37].

It appears that the incineration of wastes is a substantial source of airborne Cd on a global scale. Indeed, for the United States, this may be the principal emission source[20]. This reflects the dispersive end-uses to which Cd is put in plastic stabilizers and pigments. The average level of Cd in all plastics is about 37 µg g^{-1}, and the total Cd concentration in municipal waste is often taken as 12 µg g^{-1}, of which plastics contribute about 2 µg g^{-1} [14,73]. Incineration can produce Cd concentrations of 319 µg g^{-1} in the particulate emitted by a residential incinerator[66].

Of the remaining anthropogenic sources in table 1, it now appears that iron and steel production is a relatively minor emitter of Cd on a global scale. It had been suggested that iron blast furnaces account for major atmospheric Cd emissions[70]. However, Prater[60] has carried out a comprehensive analysis of the behavior of cadmium in the processes used by the UK steel industry, and has concluded that the atmospheric emission of Cd is slight (1.7 t annual total for the UK) when modern steel-making processes (basic oxygen and electric arc) are considered. Similarly, the cadmium emissions from combustion of coal appear to be relatively small. In the UK, where 120×10^6 t of coal are burnt per annum, of which 81×10^6 t are used for electricity generation, it has been estimated that an annual atmospheric discharge of 2.5 t Cd occurs[14]. For the German Federal Republic, it has been suggested that the 140×10^6 t of brown coal burnt per annum for power production results in a Cd emission of 0.39 t[35]. However, the Cd emission figures from coal combustion may be subject to considerable errors, due to the wide range of Cd concentrations in coal (up to 10 µg g^{-1}, but very variable) and to the fact that domestic coal burning emissions have not been studied separately[14].

In table 1, no figures are given for Cd emissions from mobile sources, although some authors have considered that emissions from motor vehicles may be significant. Cadmium is a component of tyre rubber, and tyre debris was ranked seventh in a list of 16 major sources of atmospheric Cd emissions in the United States[23]. Vehicle emissions will be considered in detail in section III (c).

II. *Particle size of emissions in relation to atmospheric transport and deposition*

Cadmium exists in particulate form within the atmosphere. The dynamics of airborne particles depend upon the nature and locations of emission sources, as well as the mechanisms of removal of material following its entry to the atmosphere. When particles are very small, they may remain suspended in the air for prolonged periods of time. The resulting two-phase system, or aerosol, has a stability which depends upon the particle size[22].

Three particle size modes are observed in atmospheric aerosols. Firstly, the 'transient nuclei mode', of the order of 0.01 µm, is observed only when relatively fresh combustion aerosols are present[71]. Particles tend to remain in this size range for less than 1 h; they grow rapidly by coagulation or condensation to the 2nd, 'accumulation', mode having a size range from 0.1 to 1.0 µm. These fine particles may remain airborne for days; their gravitational settling velocity is very low[5]. They may be raised by updraughts and turbulence, and thereby be transported in the zonal circulation of the troposphere[10]. In contrast, particles of diameter greater than about 10 µm in the 3rd, 'mechanical aerosol', mode settle rapidly and have a short average lifetime in the atmosphere.

Sedimentation, impaction and Brownian diffusion comprise the mechanisms of dry deposition by which aerosols are transferred to the earth's surface; removal from the atmosphere occurs also by wet deposition via precipitation scavenging. For any element, a washout factor, W, and a dry deposition velocity, Vg, may be defined[11] as:

$$W = \frac{\text{concentration in rain (µg kg}^{-1}\text{)}}{\text{concentration in air (µg kg}^{-1}\text{)}}$$

$$Vg \text{ (cm sec}^{-1}\text{)} = \frac{\text{rate of dry deposition (µg cm}^{-2}\text{ sec}^{-1}\text{)}}{\text{concentration in air (µg cm}^{-3}\text{)}}$$

A high value of W is associated with greater dispersion in altitude and therefore a distant source, low W with a more local source. Vg is related to such factors as particle size, surface roughness and wind velocity; in the case of particle size, there is a flat minimum for Vg in the size range from 0.1 to 1.0 µm[11].

From the above discussion, it will be clear that the effect of Cd emissions depends critically upon the particle size; coarse particle emissions will tend to be deposited close to source and have a local influence only, whereas fine particle emissions may have a more widespread effect. In this regard, referring to table 1, much of the globally-emitted Cd consists predominantly of fine particles. Volcanogenic particles are mainly < 40 μm in diameter; their residence time in the atmosphere is lengthened by the injection height for the buoyant plume which has been recorded as 3.8 km for an Alaskan volcano[38]. Vegetation emits particles in the sub-micrometer size range[3]. The major anthropogenic emissions are associated mainly with high-temperature combustion processes. Flue dusts from the stacks of nonferrous smelters may consist of particles circa 0.1–0.2 μm in diameter, forming 'fused assemblages'[19]. Lee and Duffield[45] have estimated that, in the U.S.A., 46.6% of the total particulate emitted by metallurgical processing and 67.3% of the particulate emitted by municipal waste incineration, is in the size range 0.1–2 μm. Heinrichs[35] has found that the Cd emitted by a coal-fired power station is not retained well by the electrostatic precipitators used for emission control. An explanation is that, during the combustion process, Cd is associated with particulate matter only to a limited extent: due to the volatility of elemental cadmium and its compounds, condensation of the metal may occur on fly ash, with preferential adsorption of Cd on the inefficiently collected sub-micrometer sized fly ash due to the large surface area presented by this particle size group.

Hence, most of the Cd which is emitted to atmosphere on a global scale is of such a particle size as to favor its transport over long distances in the atmosphere.

III. *The interpretation of environmental measurements*

a) *Levels of cadmium in ambient air*

Some measured concentrations of Cd in ambient air, and rates of Cd deposition from the atmosphere, are listed in tables 2 and 3 respectively. In general, these data show that levels of Cd in air are higher in urban areas than in rural areas. Heavily industrialized cities, and sites close to non-ferrous metal smelting and refining activities, show the highest levels.

When comparing the Cd concentrations listed in table 2, it should be noted that the range of values found at a particular site will depend, inter alia, upon the duration of sampling; for short-duration samples, the variability of results will tend to increase. Also, in table 3, most of the quoted deposition rates were obtained using deposit gauges or other 'artificial' collection surfaces. Generally, such devices will not match

Table 2. Some measurements of airborne cadmium concentrations

Site	[Cd] (ng m^{-3})	Sampling period/notes	Reference
Remote/rural			
North Atlantic	0.003–0.6		Duce et al.[17]
South Norway:			
Birkenes	0.11–0.34	Monthly means (6 samples)	Thrane[65]
Vasser	0.21–0.96		
Leiston, U.K.	1.7	June 1972–April 1973	Cawse[9]
Wraymires, U.K.	< 4	Jan. 1971–Dec. 1971	Cawse and Peirson[10]
Hazelrigg, U.K.	1.93	Mean and range for 30 × 24-h samples	
	0.1–8.2		Harrison and Williams[34]
	1.28	Mean and range for 56 × 7-day samples	
	0.1–4.7		
Shap Summit, M6 Motorway, U.K.	0.1–0.5	Rural roadside site; 2 × 7-day samples	Harrison and Williams[32]
Missouri, U.S.A.	3.7	16 days	Dorn et al.[16]
Arizona, U.S.A.	2.0	Annual mean for 3 sites	Moyers et al.[54]
Urban/industrial			
Bombay, India	4–46	9 sites	Khandekar et al.[41]
Rio de Janeiro, Brazil	1.8–3.1	2 sites	Trindade et al[66]
Tel Aviv, Israel	40		Donagi et al.[15]
Tucson, Arizona, U.S.A.	2.4	Annual mean for 11 sites	Moyers et al[54]
Lancaster, U.K.	3.33	Urban roadside; mean and range for 30 × 24-h samples	
	0.3–14.8		Harrison and Williams[34]
	1.17	Mean and range for 56 × 7-day samples	
	0.1–3.3		
Glasgow, U.K.	13.6	13 months average at 14 sites	McDonald and Duncan[50]
20 sites in U.K.	4.4	April 1976–March 1977	McInnes[51]
	3.4	April 1977–March 1978	McInnes[52]
Birmingham, U.K.	46	Aug.–Oct. 1973; 8 sites within 2 km of non-ferrous metal factory	Turner[67]
Birmingham, U.K.	10–15	Feb.–May 1974; range of site means at 6 sites near battery factory	Turner and Killick[68]
Industrial cities in Belgium	95–310	During pollution episodes when [SO$_2$] ≥ 30 ppb	Ronneau and Desaedeleer[62]
Missouri, U.S.A.	24.8	24 days; near Pb smelter	Dorn et al.[16]
Cockle Creek, Australia	60	90 days; at point of maximum ground-level Cd concentrations near Zn-Pb smelter	Smith et al.[64]
Avonmouth, U.K.	108	5 days; 0.7 km from Zn-Pb smelter, in prevailing wind direction	Williams[72]

Table 3. Deposition rates for airborne cadmium

Site	Cd deposition rate ($ng\ cm^{-2}\ month^{-1}$)	Sampling period/notes	Reference
Remote/rural			
Greenland Ice Sheet	0.05	1 year	Davidson et al.[13]
Wraymires, U.K.	< 80		
Leiston, U.K	< 25	1973 annual mean	Cawse[9]
Hazelrigg, U.K	< 0.6–3.4	Range for 12 × 1-month samples, 1978/79	Harrison and Williams[34]
Chester, U.S.A	2–3	3 × 1-month samples, 1978	Feely and Larsen[21]
Ardennes, Belgium	11.4	2-monthly samples. Mean of 8 sites	Hallet et al.[28]
Urban/industrial			
New York City, 1.	3–37		
U.S.A. 2.	10–66	Ranges for 22 × 1-month samples, 1972–1974	Kleinmann et al.[42]
(3 locations) 3.	6–46		
Pasadena, U.S.A	3.9		Huntzicker et al.[40]
Liege, Belgium	780	Near zinc works	Duhameau and Noël[18]
Avonmouth, U.K.	3000	1 month. Near Zn-Pb smelter. 'Moss bag' deposit collectors	Parry et al.[59]
Avonmouth, U.K.	113	1 month. 0.7 km from Zn-Pb smelter, in prevailing wind direction	Williams[72]
Charleroi, Belgium	16.2	2-monthly samples, sub-urban site	Hallet et al.[28]

the aerodynamics and collection efficiency of a natural surface, and hence will provide only an approximate measure of the deposition rate of Cd on the land.

b) Relating environmental measurements to stationary emission sources

In agreement with the discussion of particle sizes for emitted Cd (section II, above) it has been found that the metal exists at highest concentrations in the smallest particles collected from ambient air[55]. A technique which has been widely used to assess particle size distributions for airborne metals in that of fractionating airborne particles according to their aerodynamic diameters by means of a cascade impactor[16,32,34,63]. Using this method, a mass median aerodynamic diameter (MMAD) of 0.4 μm for airborne Cd in a rural environment, distant from industrial sources of the metal, has been reported[34]. Similar measurements at urban sites have shown MMAD values of 0.1–0.2 μm[63] and 0.9 μm[34]. At a site near a lead smelter, Dorn et al.[16] found that airborne Cd was associated mainly with particles of aerodynamic diameter smaller than 2 μm.

There is now a considerable amount of data which evidences the transport of airborne Cd over long distances in the atmosphere. The transport of atmospheric particulate pollutants from the European continent to the United Kingdom was noted by Lee et al.[44]. Thrane[65] has suggested that airborne Cd may be transferred from the U.K. to Scandinavia. Galloway et al.[25], from analyses of lake sediment cores in the eastern United States, note that an increase in Cd deposition rates began about 30 years ago: this indicates both an increase in anthropogenic emissions, and the importance of long-range atmospheric transport. Recently, Harrison and Williams[34] investigated airborne Cd at rural and urban sites in North-West England, in an area lacking any local industrial emis-

sions of the metal. They found that, at all their environmental monitoring sites, airborne Cd was derived primarily from distant sources in South-East England and continental Europe. This view was supported as follows: 1. For a given sampling period, the airborne Cd levels at the urban and the rural sites were very similar. 2. Airborne sulphate and nitrate, which are considered to be indicative of distant, rather than local, pollutant emissions[30] showed strong correlations with airborne Cd. 3. Meteorological analyses indicated that the highest Cd levels at all sites were associated with winds having a southerly component and with air masses having a continental origin.

At very long distances from emission sources, the impact of pollutant Cd appears not to be measurable: although analyses of Greenlandic ice indicate an increased deposition of Pb and Zn in snowfall between 1971–73 compared with pre-1900 deposition, and hence an enhanced pollutant input, no such increase is apparent for Cd[36]. At this remote location, deposition of cadmium occurs entirely by wet processes[13], as might be anticipated from the discussion in section II, above. From the analysis of snow layers deposited at the South Pole between 1929 and 1977, Boutron[6] found no change in the concentration of Cd (and all metals apart from Pb for which a local source was possible) over the past 50 years. He concluded that the influence of global atmospheric pollution is probably still negligible in the remote areas of the Southern Hemisphere.

By investigating the associations between Cd and other components of airborne particulate, an indication may be gained of likely sources of pollution, although the results of such analyses alone tend not to be specific. For example, Moyers et al.[54] have used correlation analysis to investigate the relations between the chemical components of the atmosphere at Tucson, Arizona, U.S.A. In urban air, Cd was not apparently related with any other chemical component; in rural air, Cd, Cu, Pb and Zn showed a significant correlation both with each other and with SO_4^{2-} and NO_3^-. It was suggested that this could indicate similarities in particle size distribution (transport and residence time phenomena), identical sources, or a chemical association between cadmium and sulphate ions. From a statistical analysis of these same data using pattern recognition and factor analysis, Gaarenstroom et al.[24] suggested that airborne Cd might be associated with distant and/or diffuse sources, small particles, and combustion sources. In the U.K., McInnes[53] has found that strong inter-site correlations exist for airborne pollutant metals in urban atmospheres, indicating diffuse rather than localized emission sources.

Hallet et al.[28] measured deposition fluxes for a wide range of airborne metals in and around the industrial town of Charleroi in Belgium and in the rural Ardennes region. Cadmium, together with inter alia Pb and Zn displayed a rather diffuse behavior relative to many other metals, which were deposited primarily close to the urban or industrial sources. This was attributed to the diffuseness of their emitters (e.g. incineration plants for Cd).

A quantity sometimes used in assessing the source of a metal in the atmosphere is the enrichment factor, E, relative to a crustally-derived species such as aluminium[54], scandium[10] or iron[50]. Thus, for Cd,

$$E = \frac{\dfrac{\text{airborne concentration of Cd}}{\text{airborne concentration of Al}}}{\dfrac{\text{crustal concentration of Cd}}{\text{crustal concentration of Al}}}$$

Whilst airborne metals whose source is predominantly crustal have enrichment factors close to unity, the value of E for Cd is typically $>> 100$ at urban and rural sites, suggesting the anthropogenic origin of the airborne metal[34,50,54]. However, it should be noted that volcanic emissions and sea spray may provide material which is enriched in certain elements to the same degree as industrial emissions[13].

One aspect of airborne cadmium, upon which it appears that there is little information, is that of the atmospheric chemistry of the metal. This contrasts with the chemistry of airborne lead emitted from motor vehicles, which has been investigated by Biggins and Harrison[4]. By means of X-ray powder diffraction spectrometry, CdO has been positively identified in the emissions from a zinc refinery[72]. Also, from the thermal stability criteria for sulphates, it has been postulated that the predominant cadmium compound emitted from the sintering of zinc-lead ores may be $CdSO_4$[33,72]. Where ambient air samples are concerned, it is likely that the levels of cadmium will invariably be too low for this type of compound-specific analysis to be considered. However, in a study of metal solubility in a sample of airborne particulate from an urban site, Lum et al.[48] state that all the cadmium present in the sample was extractable by 1 M $MgCl_2$ at pH 7, indicating that the cadmium was present in air as a soluble compound.

Where specific known emission sources are concerned, ambient air data generally support the relative importance of sources which was expounded in section I, above. Thus, at an otherwise rural site, Lindberg[46] states that the concentration of Cd in rainfall is not elevated by local coal-fired power plant emissions, indicating that these emissions are not a significant source of airborne Cd. Conversely, the incineration of wastes does appear to have a significant effect upon levels of Cd in ambient air: Trindade et al.[66] state that the concentrations of airborne Cd in Rio de Janeiro declined after the incineration of garbage had been prohibited. In the case of primary nonferrous metallurgical operations, a well-documented example of the nature of Cd emissions is provided by the primary zinc-lead smelting works at Avonmouth, U.K. Here, emissions are made to atmosphere via stacks which are 50–60 m high[61]; emission control is mainly by wet gas-cleaning methods. Studies of airborne Cd deposition to land in the vicinity of this smelter indicate that dispersion and deposition are dominated by the prevailing wind direction[8,47,59]; at a distance of 15 km downwind from the smelter, deposition rates[8] for Cd exceed background levels by a factor of a least 15[8]. This suggests that a substantial portion of emitted Cd is associated with fine particles. Parry et al.[59] have found that blowage of coarse, metal-rich dusts may occur from, for example, smelting waste tips. However, it appears that in the case of an operating smelter, fine particle emissions from the stacks of the smelter comprise the major percentage of the total emissions, with blowage of coarse dusts being of very limited significance[72].

c) Cadmium from motor vehicles

It is well established that motor vehicles contribute to the Cd deposited in the environment close to highways. Enhanced levels of Cd have been found in roadside soils and vegetation[12,27,43,49]. Cd concentrations in street dusts are higher than the values typical of local soils[29]. Ward et al.[69] note a significant correlation between the traffic flow on a highway and the Cd concentrations in surface soil from a 1-m-wide central strip between carriageways. The heaviest deposition of Cd occurs close to the road; at a site with 48,000 vehicles day^{-1}, Lagerwerff and Specht[43] have found that the level of Cd in surface soil falls from 0.94 µg g^{-1} at a distance of 8 m from the road to 0.24 µg g^{-1} at 32 m.

The most obvious source of this cadmium is as an impurity in technical-grade zinc compounds; zinc dithiophosphate is present as an antioxidant in lubricating oil, and zinc oxide or zinc-diethyl or -dimethyl carbamate is used in the vulcanisation of tyre rubber[43]. Hence there are potential sources of airborne Cd both from exhaust emissions and from tyre wear; both of these sources are evidenced from physico-chemical analyses of deposited dust from an urban roadway[39]. However, the major contribution to deposited Cd appears to be from tyre wear[12]. Cadmium exists in American tyres[43] at concentrations between 20 and 90 µg g^{-1}, but at much lower levels, circa 0.14 µg g^{-1} in

Australian tyre rubber[12]. In the U.K., the level may be 5–6 µg g^{-1} [14]. In the case of vehicle exhausts, the concentration of Cd in emitted particulate[57] is between 1.3 and 3.8 µg g^{-1}.

It has been suggested that these vehicular Cd emissions might contribute substantially to levels of the metal in urban air. For example, Harrison et al.[31] note variations in the deposition of Cd from the atmosphere over London, in the absence of any local industrial emissions. In Bombay, the highest levels of airborne Cd are observed at sites having a high density of motor traffic[41]. Ondov et al.[58] have found that the Cd level in the air inside a road tunnel is 77 ng m^{-3}, which is seven times the concentration outside the tunnel. In this lastmentioned example, however, it is likely that the re-suspension of deposited dust due to turbulence would be greater than for a non-enclosed highway.

Contrary findings, namely that motor vehicles do not contribute substantially to Cd in ambient air, have been reported recently by Harrison and Williams[34]. From a lengthy programme of air sampling conducted simultaneously at rural sites and at a site immediately alongside a heavily-trafficked urban roadway, it was concluded that the influence of local vehicular sources of Cd was very slight: the advection of polluted air masses was the dominant source of airborne Cd at all sites. Overall, then, it seems that vehicular sources of airborne Cd are of very limited significance, even near to roads. Large particles up to 30 µm in diameter dominate the mass of tyre wear[7]. These Cd-containing particles would be deposited close to source, and their contribution to air pollution would be slight. It should, of course, be borne in mind that the magnitude of the vehicular source will depend critically upon the Cd content of tyres and motor oil.

IV. *Further research needs*

Cadmium has received less attention than some other airborne metals, most notably lead. There is, therefore, a case for considerable further measurement of ambient airborne concentrations, their spatial and temporal variation, and of deposition fluxes.

Separate measurements of wet and dry deposition of cadmium are lacking, and the use of more realistic dry deposition collectors is desirable though experimentally difficult, so as to obtain a better measure of deposition to natural surfaces. This point is highly relevant to the important question of the relative contribution of direct deposition and of soil uptake to the ultimate cadmium content of growing crop plants. Fuller studies of cadmium deposition are also necessary to aid budget studies for cadmium in lakes (where atmospheric deposition can be a substantial input) and for studies of cadmium cycling in ecosystems.

Virtually no information is available on the atmospheric chemistry of cadmium. Since the chemical form of cadmium affects the respiratory absorption efficiency, and will also influence the environmental mobility in terrestrial and aquatic systems, data in this area are crucial to a fuller comprehension of the environmental pathways of the metal.

1   Present address: British Nuclear Fuels Ltd, Springfields Works, Salwick, Preston PR4 OXJ, England.
2   To whom correspondence should be addressed.
3   Beauford, W., Barber, J., and Barringer, A.J., Release of particles containing metals from vegetation into the atmosphere. Science *195* (1972) 571–573.
4   Biggins, P.D.E., and Harrison, R.M., Atmospheric chemistry of automotive lead. Envir. Sci. Technol. *13* (1979) 558–565.
5   Bloom. H., and Noller, B.N., Application of trace analysis techniques to the study of atmospheric metal particulates, in: Proceedings of a Symposium on Analytical Techniques for the Determination of Air Pollutants, pp. 98–111. Clean Air Society of Australia and New Zealand, 1977.
6   Boutron, C., Atmospheric trace metals in the snow layers deposited at the South Pole from 1928–1977. Atmos. Envir. *16* (1982) 2451–2459.
7   Cadle, S.H., and Williams, R.L., Gas and particle emission from automobile tyres in laboratory and field studies. J. Air Pollut. Control Assoc. *28* (1978) 502–507.
8   Cameron, A.J., and Nickless, G., Use of mosses as collectors of airborne heavy metals near a smelting complex. Water, Air, Soil Pollut. *7* (1977) 117–125.
9   Cawse, P.A., A survey of atmospheric trace elements in the U.K. (1972–73). United Kingdom Atomic Energy Authority Report AERE-R7669. H.M.S.O., London 1974.
10 Cawse, P.A., and Peirson, D.H., An analytical study of trace elements in the atmospheric environment. United Kingdom Atomic Energy Authority Report AERE-R7134. H.M.S.O., London 1972.
11 Chamberlain, A.C., Aspects of the deposition of radioactive and other gases and particles. Int. J. Air Pollut. *3* (1960) 63–68.
12 David, D.J., and Williams, C.H., Heavy metal contents of soils and plants adjacent to the Hume Highway near Marulan, New South Wales. Aus. J. exp. Agric. Anim. Husb. *15* (1975) 414–418.
13 Davidson, C.I., Chu, L., Grimm, T.C., Nasta, A., and Qamoos, M.P., Wet and dry deposition of trace elements onto the Greenland ice sheet. Atmos. Envir. *15* (1981) 1429–1437.
14 Department of the Environment, Cadmium in the environment and its significance to man. Pollution Paper No. 17. H.M.S.O., London 1980.
15 Donagi, A., Ganor, E., Shenhar, A., and Cember, H., Some metallic trace elements in the atmospheric aerosols of the Tel-Aviv urban area. J. Air Pollut. Control Assoc. *29* (1979) 53–54.
16 Dorn, C.R., Pierce, J.O., Phillips, P.E., and Chase, G.R., Airborne Pb, Cd, Zn and Cu concentration by particle size near a Pb smelter. Atmos. Envir. *10* (1976) 443–446.
17 Duce, R.A., Hoffman, G.L., and Zoller, W.H., Atmospheric trace metals at remote Northern and Southern hemisphere sites. Science *187* (1975) 59–61.
18 Duhameau, W., and Noël, R., Five years of recording the atmospheric fall-out in the industrial region of Liege, in: Atmospheric Pollution 1978, Proceedings of the 13th International Colloquium, Paris, France, pp. 29–32. Elsevier, Amsterdam 1978.
19 Eatough, D.J., Eatough, N.L., Hill, M.W., Mangelson, N.F., Ryder, J., and Hansen, L.D., The chemical composition of smelter flue dusts. Atmos. Envir. *13* (1979) 489–506.
20 Faoro, R.B., and McMullen, T.B., National trends in trace metals in ambient air. Report No. 450/1-77-003. U.S. Environmental Protection Agency, 1977.
21 Feely, H.W., and Larsen, R., Measurements of the chemical composition of wet, dry and total atmospheric deposition. Environmental Measurements Laboratory Report EML 347, pp. 7–20. U.S. Department of Energy, 1978.
22 First, M.W., Aerosols in nature. Archs intern. Med. *131* (1972) 24–32.
23 Fleischer, M., Sarofim, A.F., Fassett, D.W., Hammond, P., Shacklette, H.T., Nisbet, I.C.T., and Epstein, S., Environmental impact of cadmium, a review by the Panel on Hazardous Trace Substances. Envir. Hlth Perspect. *7* (1974) 252–323.
24 Gaarenstroom, P.D., Perone, S.P., and Moyers, J.L., Application of pattern recognition and factor analysis for characterisation of atmospheric particulate composition in southwest desert atmosphere. Envir. Sci. Technol. *11* (1977) 795–800.
25 Galloway, J.N., Eisenreich, S.I., and Scott, B.C., U.S. National Atmospheric Deposition Report No. NC-141. Dept. of Environmental Sciences, University of Virginia 1980.

26 Galloway, J.N., Thornton, J.D., Norton, S.A., Volchok, H.L., and McLean, R.A.N., Trace metals in atmospheric deposition: A review and assessment. Atmos. Envir. *16* (1982) 1677–1700.

27 Gish, C.D., and Christensen, R.E., Cadmium, lead and zinc in earthworms from roadside soil. Envir. Sci. Technol. *7* (1973) 1060–1062.

28 Hallet, J.P., Lardinois, P., Ronneau, C., and Cara, J., Elemental deposition as a function of distance from an industrial zone. Sci. total Envir. *25* (1982) 99–109.

29 Harrison, R.M., Toxic metals in street and household dusts. Sci. total Envir. *11* (1979) 89–97.

30 Harrison, R.M., and McCartney, H.A., Some measurements of ambient air quality at a coastal site in rural north-west England. Atmos. Envir. *14* (1980) 233–244.

31 Harrison, R.M., Perry, R., and Wellings, R.A., Lead and cadmium in precipitation: their contribution to pollution. J. Air Pollut. Control Assoc. *25* (1975) 627–630.

32 Harrison, R.M., and Williams, C.R., Atmospheric cadmium pollution at rural and urban sites in north-west England, in: Proceedings of the International Conference, Management and Control of Heavy Metals in the Environment, pp.262–266. CEP Consultants Ltd., London 1979.

33 Harrison, R.M., and Williams, C.R., Characterisation of airborne heavy metals within a primary zinc-lead smelting works. Envir. Sci. Technol. *15* (1981) 1197–1204.

34 Harrison, R.M., and Williams, C.R., Airborne cadmium, lead and zinc at rural and urban sites in North-west England. Atmos. Envir. *16* (1982) 2669–2681.

35 Heinrichs, H., Trace element discharge from a brown coal fired power plant. Envir. Technol. Lett. *3* (1982) 127–136.

36 Herron, M.M., Langway, C.C., Weiss, H.V., and Cragin, J.M., Atmospheric trace metals and sulphate in the Greenland ice sheet. Geochim. cosmochim. Acta *4* (1977) 915–920.

37 Hiatt, V., and Huff, J.E., The environmental impact of cadmium: an overview. Int. J. envir. Stud. *7* (1975) 277–285.

38 Hobbs, P.V., Radke, L.F., and Stith, J.L., Eruptions of the St.Augustine Volcano; airborne measurements and observations. Science *195* (1977) 871–873.

39 Hopke, P.K., Lamb, R.E., and Natusch, D.F.S., Multielemental characterisation of urban roadway dust. Envir. Sci. Technol. *14* (1980) 164–172.

40 Huntzicker, J.J., Friedlander, S.K., and Davidson, C.I., Air-Water-Land Relationships for Selected Pollutants, pp.82–115. Final Report to the Rockefeller Foundation, USA, 1975.

41 Khandekar, R.N., Kelkar, D.N., and Vohra, K.G., Lead, cadmium, copper, zinc and iron in atmosphere of Greater Bombay, Atmos. Envir. *14* (1980) 457–461.

42 Kleinman, M.T., Kneip, T.J., Bernstein, D.M., and Eisenbud, M., Fallout of toxic trace metals in New York City, in: Biological Implications of Metals in the Environment, CONF-740929, pp.144–152. Energy Research and Development Administration, Washington D.C. 1977.

43 Lagerwerff, J.V., and Specht, A.W., Contamination of roadside soil with cadmium, nickel, lead and zinc. Envir. Sci. Technol. *4* (1970) 583–586.

44 Lee, R.E., Caldwell, J., Akland, G.E., and Fankhauser, R., The distribution and transport of airborne particulate matter and inorganic components in Great Britain. Atmos. Envir. *8* (1974) 1095–1109.

45 Lee, R.E., and Duffield, F.V., Sources of environmentally important metals in the atmosphere, in: Advances in Chemistry Series No.172, Ultratrace Metal Analysis in Biological Sciences and Environment, ch.11, pp.146–171. American Chemical Society, 1979.

46 Lindberg, S.E., Factors influencing trace metal, sulphate and hydrogen ion concentration in rain. Atmos. Envir. *16* (1982) 1701–1709.

47 Little, P., and Martin, M.H., A survey of zinc, lead and cadmium in soil and natural vegetation around a smelting complex. Envir. Pollut. *3* (1972) 241–254.

48 Lum, K.R., Betteridge, J.S., and Macdonald, R.R., The potential availability of P, Al, Cd, Co, Cr, Cu, Fe, Mn, Ni, Pb and Zn in urban particulate matter. Envir. Technol. Lett. *3* (1982) 57–62.

49 Mankovska, B., The content of Pb, Cd and Cl in forest trees caused by the traffic of motor vehicles. Biologia, Bratisl. *32* (1977) 477–489.

50 McDonald, C., and Duncan, H.J., Variability of atmospheric levels of metals in an industrial environment. J. envir. Sci. Hlth *A13* (1978) 687–695.

51 McInnes, G., Multi-element survey: report on first year's results, Report No.LR277(AP). Warren Spring Laboratory, U.K. Department of Industry, 1978.

52 McInnes, G., Multi-element survey: report on second year's results, Report No.LR289(AP). Warren Spring Laboratory, U.K. Department of Industry, 1978.

53 McInnes, G., Multi-element survey: analysis of the first two years' results. Report No.LR305(AP). Warren Spring Laboratory, U.K. Department of Industry, 1979.

54 Moyers, J.L., Ranweiler, L.E., Hopf, S.B., and Korte, N.E., Evaluation of particulate trace species in southwest desert atmosphere. Envir. Sci. Technol. *11* (1977) 789–794.

55 Natusch, D.F.S., Wallace, J.R., and Evans, C.A., Toxic trace elements. Preferential concentration in respirable particles. Science *183* (1974) 202–204.

56 Nriagu, J.O., Global inventory of natural and anthropogenic emissions of trace metals to the atmosphere. Nature *279* (1979) 409–411.

57 Oakes, T.W., Furr, A.K., Adair, D.J., and Parkinson, T.F., Neutron activation analysis of automotive exhaust pollutants. J. radioanalyt. Chem. *37* (1977) 881–888.

58 Ondov, J.M., Zoller, W.H., and Gordon, G.E., Trace metal emissions on aerosols from motor vehicles. Envir. Sci. Technol. *16* (1982) 318–328.

59 Parry, G.D.R., Goodman, G.T., and Smith, S., A simple technique for the monitoring of airborne heavy metals prior to revegetation, in: Environmental Management of Mineral Wastes, pp.273–295. Eds G.T. Goodman and M.J. Chadwick. Sijthoff and Noordhoff, Amsterdam 1978.

60 Prater, B.E., The environmental significance of cadmium in U.K. steelmaking operations. Envir. Technol. Lett. *3* (1982) 179–192.

61 Roach, S.A., and Wakeley, R.W., Environmental control and hygiene at the Avonmouth zinc-lead smelter, in: Proceedings of a Symposium, Minerals and the Environment, pp.505–513. Institution of Mining and Metallurgy, London 1975.

62 Ronneau, C., and Desaedeleer, G., Fall-out of pollutants from distant industrial areas in a rural environment. Paper IAEA-SM-206-11, in: Measurement Detection and Control of Environmental Pollutants, pp.165–71. International Atomic Energy Agency, Vienna 1976.

63 Skogerboe, R.K., Dick, D.L., and Lamothe, P.J., Evaluation of filter inefficiencies under low loading conditions. Atmos. Envir. *11* (1977) 243–259.

64 Smith, B.C., Coppock, B.W., Scott, T.C., and Firkin, G.R., Pollution control on an ISP complex, in: Conference Proceedings, South Australia, pp.277–287. Australasian Institute of Mining and Metallurgy, 1975.

65 Thrane, K.E., Background levels in air of lead, cadmium, mercury and some chlorinated hydrocarbons measured in south Norway. Atmos. Envir. *12* (1978) 1155–1161.

66 Trindade, H.A., Pfeiffer, W.C., Londres, H., and Costa-Ribeiro, C.L., Atmospheric concentration of trace metals and total suspended particulates in Rio de Janeiro. Envir. Sci. Technol. *15* (1981) 84–89.

67 Turner, A.C., The Midlands Metal Survey: Concentrations of 15 elements in the air around factories in Birmingham and the Black Country – Part 1, Report No.LR303(AP). Warren Spring Laboratory, U.K. Department of Industry, 1979.

68 Turner, A.C., and Killick, C.M., The Midlands Metal Survey: Concentrations of 15 elements in the air around factories in Birmingham and the Black Country – Part 2, Report No.LR304(AP). Warren Spring Laboratory, U.K. Department of Industry, 1979.

69 Ward, N.I., Brooks, R.R., Roberts, E., and Boswell, E., Heavy-metal pollution from automotive emissions and its effect on roadside soils and pasture species in New Zealand. Envir. Sci. Technol. *11* (1977) 917–920.

70 Wilcox, S.L., Keitz, E.L., and Duncan, L.J., Establishing safe ambient air quality levels for eighteen hazardous pollutants, in: Recent Advances in the Assessment of the Health Effects of Environmental Pollution, vol.III, pp.1241–1251. Commission of the European Communities, 1975.

71 Willeke, K., and Whitby, K.T., Atmospheric aerosols: size distribution interpretation. J. Air Pollut. Control Assoc. *25* (1975) 529–534.

72 Williams, C.R., Investigations into airborne cadmium, lead and zinc at rural, urban and industrial sites. Ph.D. Thesis, University of Lancaster, Lancaster, U.K. 1981.

73 Wilson, D.C., Young, P.J., Hudson, B.C., and Baldwin, E., Leaching of cadmium from pigmented plastics at a landfill site. Envir. Sci. Technol. *16* (1982) 560–566.

Cadmium in fresh and estuarine waters

by W. Salomons and H. N. Kerdijk

Delft Hydraulics Laboratory, c/o Institute for Soil Fertility, P.O. Box 30003, 9750 RA Haren (The Netherlands)

Introduction

Although the volume of fresh and estuarine waters is small with respect to the water-mass in the oceans, these waters have their particular importance as our freshwater supply, and estuaries are important biological areas in themselves. In addition, the estuaries are the interface between the continents and the oceans. Processes in estuaries affect the total amount of cadmium entering the oceans.

With respect to the behavior of cadmium, fresh and estuarine waters differ from oceans in the concentration of suspended matter and in depth. Particulate concentrations in fresh water and in estuaries are typically orders of magnitude higher compared with the oceans[6]. Concentrations in rivers and estuaries range from 10 to 1000 mg/l or even higher, depending on stream velocity and tidal action. The values in oceans are below 1 mg/l. The particulate matter is of great importance in controlling the concentration of dissolved cadmium. The shallowness of most estuaries and freshwater systems makes the interaction between sediments and overlying water more important than for oceans.

The various interactions taking place during transport of cadmium in the hydrological cycle are presented in figure 1.

In this overview the movement of cadmium in fresh and estuarine waters will be discussed. The processes affecting cadmium will be emphasized. Additional information on the movement of cadmium, and of other trace metals, in the hydrological cycle can be found in Salomons and Foerstner[18].

Cadmium in rivers

A summary of global fluxes of cadmium and other trace metals has been presented by Yeats and Bewers[22]. Few data were available on dissolved cadmium concentrations for rivers. Yeats and Bewers[22] quote values of 0.11, 0.09 and 0.39 μg/l for the St. Lawrence, the Mississippi and the Rhine respectively. A mean value of 0.02 μg/l for rivers was used by Martin and Whitfield[11]. The data from Yeats and Bewers[22] show that cadmium and zinc exhibit a relative high percentage of the total flux in the dissolved state. Co, Mn, Ni and Cu have intermediate mobilities whereas Al and Fe are the least mobile and are transported mainly in particulate form.

For a number of rivers it has been found that the concentration of particulate cadmium (and in fact that of most trace metals) depends on the discharge. An example is shown in figure 2 for the river Rhine. During periods of high discharge the particulate cadmium concentrations are low, whereas during low discharge the cadmium concentrations may increase by a factor of 5. This phenomenon is probably caused by the following effects:

– Proportional dilution as the discharge increases, assuming a constant load of cadmium entering the river system;

– Dilution as the increased erosion during high discharge (surface runoff) causes a mixing of contaminated fluvial particulates with soil particles;

– Difference in grain size composition. At low discharge suspended matter is relatively finer. The coarser particles are settled on the river floor. Cadmium (and most contaminants) is preferentially associated with the finer particles;

– Difference in residence time. The residence time of the particles in the river during high discharge will be low compared with that in periods of low discharge. As the

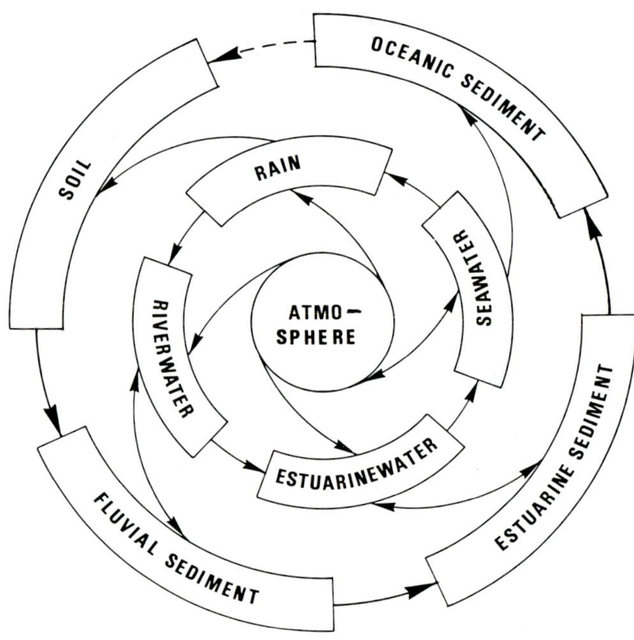

Figure 1. The movement of contaminants in the hydrological cycle (Salomons and Foerstner[18]).

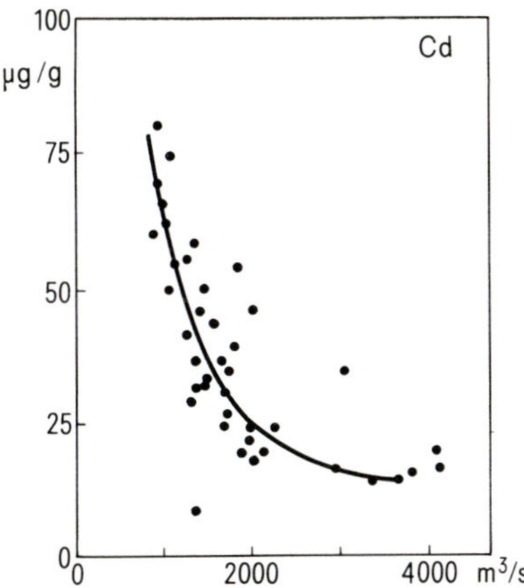

Figure 2. Influence of discharge on cadmium concentrations in suspended matter from the river Rhine.

amount of cadmium which can adsorb to the particles depends on their residence time, relatively small amounts will be available for adsorption during periods of high discharge.

In rivers containing locks the relationship between particulate cadmium concentration and discharge looks somewhat different. During low discharge the suspended matter with relatively high Cd-concentrations settles in front of these locks. This sediment is eroded at higher discharge, with the result that the Cd-concentration does not differ much compared with that at low discharge. Only after longer periods of high discharge when all the sediment is removed does the same relation exist as in non lock-regulated rivers. An example is the river Meuse. The distribution of cadmium between the particulate and dissolved phases depends on a number of parameters. Important in this respect (apart from the nature of the suspended matter) is the presence of complexing agents. The speciation of cadmium and other trace metals in river systems has been studied in detail for the Minnesota and Mississippi rivers using ultra-filtration. Most trace metals were found to be associated with molecular weigth fractions between 1000 and 10,000[5]. In a study of the metal-binding capacity of ultrafiltrable fractions of several surface waters of the southeastern United States it was found that both in hard and soft waters the binding of cadmium is controlled by inorganic ligands[1].

The speciation of dissolved cadmium shows changes on time scales of days to weeks as a consequence of changes in the nature of organic matter inputs in the river Rhine. Seasonal variation in cadmium toxicity towards algae in lakes has been attributed to seasonal changes in the qualitative composition of dissolved organic matter[7]; the dissolved organic matter changes the speciation and hence the toxicity of cadmium.

Much research has been carried out on the influence of nitrilo-tri-acetic acid (NTA) on a potential release of cadmium from sediments/suspended matter. The extent of this effect depends on the concentration of NTA, the pH value, the mode of occurrence of cadmium in the sediments and on competition by other cations[20]. The influence is twofold:

1) NTA may actively desorb cadmium from the suspended matter and the deposited sediments.

2) When both NTA (or any other complexing agent) and cadmium are discharged into a river system, NTA may negatively influence the naturally occurring adsorption processes, causing enhanced dissolved cadmium concentrations.

Studies on expected NTA-concentrations in the river Rhine showed that concentrations may rise to 580 µg/l. This maximum value is expected for low discharge conditions and low temperature. At higher discharge (more dilution) and higher temperatures (more decay of NTA) lower values are expected. Laboratory experiments on adsorption and desorption showed that concentrations greater than 200 µg/l may affect adsorption processes and may cause a release of cadmium from the sediments[20]. The influence is small compared with that on zinc. Zinc forms stronger complexes with NTA, and dissolved zinc levels are already affected at very low NTA-concentrations. Not only NTA but also other organic and inorganic complexing agents may cause similar effects. One known example is chloride, which forms strong complexes with cadmium. The chloride concentrations in the Rhine (due to salt discharge) are sufficiently high to affect the distribution of cadmium over the dissolved and particulate phases. Other heavy metals are much less affected, because of their weaker complexes with chloride.

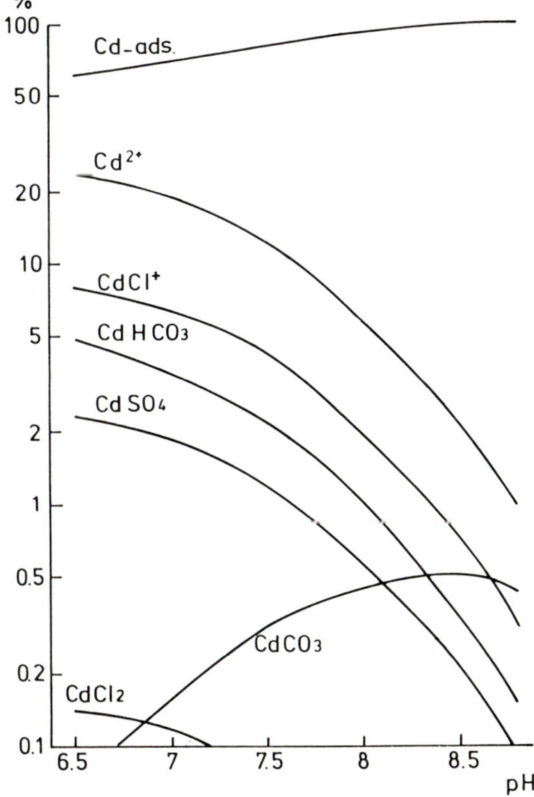

Figure 3. Calculated distribution of cadmium between particulate and dissolved phases, and the calculated speciation of cadmium in solution as a function of the pH.

Cadmium in lakes

Sources for cadmium in lakes are the atmosphere, rivers and waste discharges. In discussing processes affecting cadmium in lakes, the lake will be considered as a system with boundaries. The ratio of particulate to dissolved metals entering the lake system depends on the processes at the boundaries[17]. Three boundaries can be distinguished:
– the river-lake boundary,
– the atmosphere-lake boundary, and
– the sediment-lake boundary.
The changeover between river and lake is characterized by a decrease in flow velocity. It may be expected that the coarser particles (with low cadmium concentrations) will settle in the mouth area of the river, causing a relative increase in the particulate cadmium concentration. In general the composition of the lake differs from that of the river (e.g. pH and dissolved organic matter concentrations). As a result a redistribution of cadmium between the dissolved and particulate phases will take place. This is rather well documented for the IJsselmeer, a lake in the Netherlands which is fed by the IJssel, a distributary of the river Rhine. In this case the increase in pH from about 7.4 (river) to > 8.5 for the lake, causes an adsorption of dissolved cadmium onto the suspended matter. The dissolved Cd-concentration in the Rhine is about 0.4 µg/l; in the IJsselmeer this is reduced to 0.05 µg/l.
If the pH of the lake is lower, the reverse may take place. Therefore acidification of a lake results in a higher dissolved Cd-concentration. The distribution of cadmium between the dissolved and particulate phases as a function of the pH is shown in figure 3. Figure 3 also shows the changes in dissolved speciation as a function of the pH.

The atmosphere can be a major source for some trace metals. Its relative importance depends on the contribution from other sources and on the residence time of the water in the lake (a high residence time and a high surface-area to depth ratio causes a large relative contribution). Especially for lakes close to industrial areas, the atmosphere can be an important source. For Lake Ontario and other Great Lakes about 30–60% of the annual trace metal input is atmospheric[16].
The input from the atmosphere is from both wet and dry deposition. The speciation of trace metals in the atmospheric particulate matter will determine their reactivity at the atmosphere-lake boundary. Information on the mode of occurrence of trace metals in urban particulates has been obtained with selective leaching techniques[4, 9]. The results show a high proportion of soluble cadmium. The solubility of the trace metals in atmospheric aerosols depends on the grain size, as was shown by Lindberg and Harris[8]. However, for cadmium no relationship was observed; about 80–90% was acid-soluble. The high solubility of cadmium in atmospheric particles is a general phenomenon[18], and shows that dry deposition in lakes will result in a release of cadmium to the dissolved phase. A special role is played by the surface micro-layer[18]. The surface micro-layer in freshwater systems is, as in marine systems, enriched in cadmium and other trace metals[17]. Processes at the atmosphere-lake boundary, however, are far from being understood, and the ecological significance of the metal-enriched surface layer is not known. Little is known about the possible release of cadmium from the deposited sediments in lakes. Results of analyses of pore water in the IJsselmeer[14] show equal or slightly higher Cd-concentrations compared to those in the overlying water. Diffusion, bioturbation and erosion may result in elevated Cd-levels in the surface water, especially in shallow lakes. (Co)precipitation in the oxic top layer and readsorption will counteract this effect.
In the lake system, particulate cadmium is removed by sedimentation. In addition cadmium (and other trace metals) may be taken up by algae and removed from the water column in the same way as occurs in the ocean[21]. Although inputs in lakes are generally higher compared with ocean systems, metal concentrations are of the same order of magnitude. Therefore, apparently highly efficient removal processes are operating in lakes[21].
pH is an important parameter controlling dissolved cadmium concentrations in lakes. Seasonal changes in pH, due to algal blooms, may also cause significant changes in dissolved cadmium concentrations and in its speciation.

Cadmium in estuaries

The behavior of cadmium in the estuarine environment is extremely complicated. Not only do we have to take into account the interactions between dissolved and particulate cadmium as a function of changes in the pH and chlorinity, but other processes affect cadmium as well. Examples are the formation of iron and manganese hydroxides caused by flocculation during the process of mixing of river water with seawater, and through the oxidation of dissolved Fe(II) and Mn(II) released from the estuarine bottom sediments.
Processes affecting trace metals (and other components) in estuaries are often discussed in terms of conservative and non-conservative behavior. If cadmium behaves conservatively both its particulate and dissolved concentrations can be calculated from the mixing ratio of marine to fluvial components provided that the composition of the end-members remains constant over the flushing time of the water. Deviations from the theoretical linear mixing curve point to processes other than simple mixing affecting cadmium.
A summary of the processes causing deviations from the theoretical mixing curve is given in the table.

Processes which cause a deviation from the theoretical mixing curve between cadmium and mixing ratio of marine to fluvial end-members in an estuary (based on Salomons and Foerstner[18])

---

1) Change in the composition of the end-members
2) Flocculation of colloids
3) Release and/or uptake of dissolved cadmium by bottom sediments
4) Degradation of organic matter and subsequent release or solubilization of cadmium
5) Biological uptake of cadmium
6) Changes in adsorption/desorption equilibra due to:
   a) Change in pH
   b) Changes in salinity
   c) Changes in turbidity
   d) Formation of new particulate matter

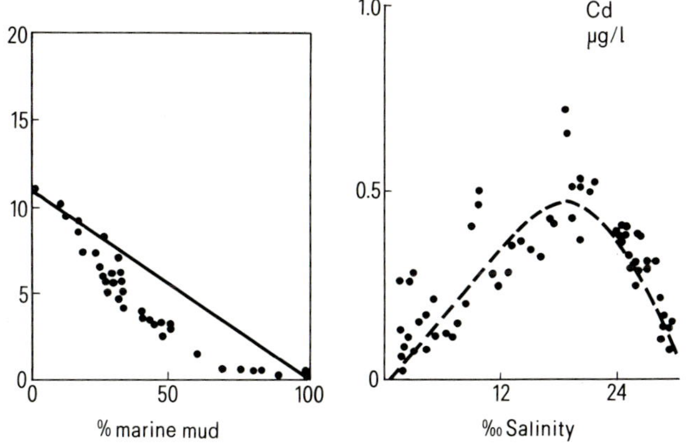

Figure 4. Positive deviations from the mixing curve for dissolved cadmium in the Scheldt estuary and the corresponding negative deviations for particulate cadmium. The tracer to determine the mixing ratio of marine to fluvial suspended matter was the isotopic composition of the carbonates.

For dissolved components the chlorinity or the chlorosity are used as tracers to determine the mixing ratio of marine to fluvial end-members. However, this tracer cannot be used to determine the mixing ratio of marine to fluvial particulate matter (and cadmium); the movement of particulates is different from that of the water. In general the residence time of the particulates will be longer than that of the water. To determine this mixing ratio a number of tracers have been developed based on difference in composition (chemical, mineralogical and isotopic) between fluvial and marine particulate matter[18, 19].

There is no common behavior of cadmium in estuaries; both positive and negative deviations from the theoretical mixing curve have been observed. An example of deviation from the theoretical mixing curve for dissolved and particulate cadmium is shown in figure 4.

An increase in chlorinity causes a decrease in the adsorption of cadmium. However, this does not imply that all cadmium will desorb from particulates; only that part which is present in the exchangeable form will be subject to a release. Changes in speciation (including the adsorbed phase) of cadmium as a function of chlorinity with a constant suspended matter concentration are shown in figure 5.

The calculations are more or less similar to those of Mantoura et al.[10], the only difference being that the adsorbed fraction is included. All model calculations show that only insignificant amounts of dissolved cadmium are associated with the dissolved organic matter. Therefore, it was not included in figure 5. A pronounced turbidity maximum is observed in a number of estuaries. If this occurs an adsorption of the riverborne cadmium may take place[13].

Other processes which enhance the removal of cadmium (and other trace metals) from the dissolved phase are the formation of new particulate matter, or of fresh coatings around particles caused by a precipitation of iron and manganese from the river water entering the estuary or by a release of iron (II) and Mn (II) from the bottom sediments. However, in all cases these increases in adsorption are counteracted by an increase in chlorinity.

Whether adsorption processes occur depends to a large extent on the hydrodynamic conditions within a particular estuary[13]. In certain parts of the estuary (high turbidity areas) the adsorption process may dominate, whereas in low turbidity areas a release of cadmium may take place.

The particulate concentrations of the trace metals are determined not only by the loss or gain from solution but in addition by the mixing of marine and fluvial suspended matter. In addition, in estuaries and coastal areas, two distinct classes of particles have been identified[3]; particles which are rather coarse, originating from the deposited sediments, and very finely grained particles (the continuous suspended fraction). Low particulate cadmium concentrations are associated with high suspended matter concentrations (contribution from eroded coarse bottom sediments is high). High particulate cadmium concentrations are associated with low suspended matter concentrations.

There is strong evidence that changes from anoxic to oxic conditions may have a strong effect on the mobility of cadmium.

Changes from anoxic to oxic conditions occur when anoxic river water enters oxygen-rich estuarine waters (the Scheldt estuary), when anoxic bottom sediments become eroded by the tidal currents and during the dumping of anoxic dredged material. There is strong evidence that under anoxic conditions cadmium is present as a sulfide[15]. Cadmium sulfide is not stable under oxic conditions and

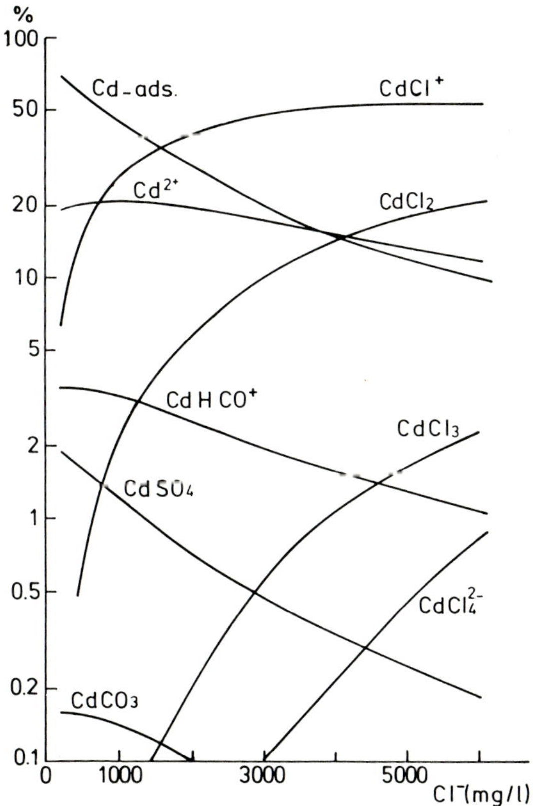

Figure 5. Calculated speciation of cadmium as a function of chlorinity.

will decompose; again, depending on the chlorinity, a (partial) readsorption will take place.

The erosion of sediments takes place in most shallow water environments; each cycle of erosion and sedimentation causes a decomposition of the sulfides, and partial readsorption on the suspended matter. This readsorption process depends to a large extent on the chlorinity of the water and on the suspended matter concentration. The net effect will be a release of dissolved cadmium into the surface water. Laboratory experiments show that under these conditions more than 50% of cadmium may be released from the resuspended sediments[12]. The amount of release depends on the suspended matter concentration, showing the importance of the readsorption process. Repeated cycles of deposition (sulfide formation) and erosion (decomposition of sulfides) might result in an extensive remobilization of cadmium in shallow water sediments. The unexpectedly low concentration in Wadden Sea sediments (North Sea), which are lower than expected on the basis discharge data, might be the net result of this process (Salomons and Kerdijk, unpublished data). In addition this process takes place during dumping of dredged material in high energetic estuarine or marine environments.

Few data are available on the retention of cadmium in estuaries and on the net amounts of cadmium reaching the oceans through river inputs. In the Netherlands only 10% of the supply of the rivers Scheldt and Rhine enters the North Sea, because of the high settling rates. The supply of dissolved cadmium is of more importance, and affects the dissolved cadmium concentration along the coastline to a large extent. Yeats and Bewers[22] estimate that about 69% of the cadmium introduced by the rivers into the coastal zone is retained and only 31% enters the world's oceans.

1 Alberts, J. J., Giesy, J. P., and Evans, D. W., Distribution of dissolved organic carbon and metal-binding capacity among ultrafiltrable fractions isolated from selected surface waters of the South Western United States. Envir. Geol. Water Sci. 6 (1984) 91–101.
2 del Castilho, P., and Salomons, W., Biological activity and its influence on metal complexation in river systems. Int. Conf. 'Environmental Contamination', London 1984.
3 Duinker, J. C., Effects of particle size and density on the transport of metals to the oceans, in: Trace Metals in Sea Water, p. 209–226. Eds C. S. Wong, E. Boyle, K. W. Bruland, J. D. Burton and E. D. Goldberg. Plenum Publishing Co., New York, London 1983.
4 Harrison, R. M., Laxen, D. P. H., and Wilson, J. S., Chemical associations of lead copper and zinc in streets dusts and roadside soils. Envir. Sci. Technol. 13 (1981) 1378–1383.
5 Hoffmann, M. R., Yost, E. C., Eisenreich, S. J., and Maler, W. J., Characterization of soluble and colloidal-phase metal complexes in river water by ultra-filtration. A mass-balance approach. Envir. Sci. Technol. 15 (1981) 655–661.
6 Kranck, K., Sedimentation processes in the sea, in: The Handbook of Environmental Chemistry, vol. 2A, pp. 61–75. Ed. O. Hutzingen. Springer, Berlin, Heidelberg, New York 1980.
7 Laegreid, M., Alstad, J., Klaveness, D., and Selp, M., Seasonal variation of cadmium toxicity toward the alga Selenastrum capricornutum Printz in two lakes with different humus content. Envir. Sci. Technol. 17 (1983) 357–361.
8 Lindberg, S. E., and Harris, R. C., Water and acid soluble trace metals in atmospheric particulates. J. geophys. Res. 88 (1983) 5091–5100.
9 Lum, K. R., The potential availability of P, Al, Cd, Co, Cr, Cu, Fe, Mn, Ni, Pb and Zn in urban particulate matter. Envir. Technol. Lett. 3 (1982) 57–62.
10 Mantoura, R. F. C., Dickson, A., and Riley, J. P., The complexation of metals with humic materials in natural waters. Estuarine coastal mar. Sci. 6 (1978) 387–408.
11 Martin, J. M., and Whitfield, M., The significance of the river input of chemical elements to the ocean, in: Trace Metals in Sea Water. pp. 265–296. Eds C. S. Wong, E. Boyle, K. W. Bruland, J. D. Burton and E. D. Goldberg. Plenum Publishing Co., New York, London 1983.
12 Salomons, W., Release of trace metals from anoxic sediments resuspended in oxygenated seawater. Delft Hydraulics Laboratory Report R 1024, 1978 (in Dutch).
13 Salomons, W., Adsorption processes and hydrodynamic conditions in estuaries. Envir. Technol. Lett. 1 (1980) 356–365.
14 Salomons, W., Trace metal cycling in a polluted lake (IJsselmeer, The Netherlands). Delft Hydraulics Laboratory Report S357/EEC contract no 199-7-1 ENV N. 1983.
15 Salomons, W., Sediments and water quality. Envir. Technol. Lett. 6 (1985) 315–326.
16 Salomons, W., Impact of atmospheric trace metals on the hydrological cycle in: Toxic Metals in the Air. Eds J. Nriagu and C. Davidson. Wiley Publishing Company, 1986.
17 Salomons, W., and Baccini, P., Chemical speciation and metal transport in Lakes. Dahlem Konferenzen 1985.
18 Salomons, W., and Foerstner, U., Metals in the Hydrological Cycle. Springer, Berlin, Heidelberg, New York 1984.
19 Salomons, W., and Mook, W. G., Trace metal concentrations in estuarine sediments: mobilization, mixing or precipitation. Neth. J. Sea Res. 11 (1977) 199–209.
20 Salomons, W., and Van Pagee, H., Prediction of NTA-levels in river systems and their effect on metal concentrations. Proc. Int. Conf. 'Heavy metals in the environment', Amsterdam 1981.
21 Sigg, L., Sturm, M., Stumm, W., Mart, L., and Nürnberg, H. W., Schwermetalle im Bodensee – Mechanismen der Konzentrationsregulierung. Naturwissenschaften 69 (1982) 546–548.
22 Yeats, P. A., and Bewers, J. M., Discharge of metals from the St.-Lawrence River. Can. J. Earth Sci. 19 (1982) 982–992.

The distribution of cadmium in the sea

by L. Mart and H. W. Nürnberg*

Institute of Applied Physical Chemistry, Nuclear Research Establishment (KFA), D–5170 Juelich (Federal Republic of Germany)

Introduction

Cadmium occurs in the sea at ultra trace levels. The concentration varies in various regions of the oceans according to the different oceanographic conditions. Among the heavy metal trace-elements with general or potential ecotoxic significance (Pb, Hg, Cu, Ni, Cr, As, Zn), Cd occurs in surface waters of the open oceans in the relatively lowest concentrations, mostly between 0.1 and 20 ng/kg. The lower section of this range corresponds to large areas in the subtropical and central gyres, whereas the upper section of the range is observed in areas of upwelling, at oceanic divergencies and at the subpolar fronts. Somewhat higher values can occur in shelf regions, due to release of Cd from the bottom sediment depot. In areas of pronounced upwelling the values in surface waters may even be raised to 50–70 ng/kg. Comparable or even higher levels are also observed as a consequence of anthropogenic input in polluted coastal waters (viz. e.g. North Sea and particularly its Southern Bight and German Bight). Even more elevated levels, between 100 and 500 ng/kg, occur in the estuaries of polluted rivers, e.g. Scheldt, Rhine, Weser, Elbe; at particular pollution locations, as e.g. the entries of big ports, the local Cd-level may reach or exceed 1000 ng/kg. A comprehensive comparative consideration of the only locally significant elevated Cd levels in estuaries and near-shore waters, controlled by many specific parameters and local influences, is beyond the scope of this paper which will focus on the Cd-distribution in the oceans. A more comprehensive treatment of the distribution and fate of heavy metals with ecotoxic significance in the sea is in preparation[49].

The biogeochemical cycle in the oceanic water column

In this context it has to be emphasized that in the oceans the Cd-levels increase significantly with depth below the euphotic zone, attaining values up to 125 ng/kg between 500 and 1000 m depth in the North Pacific and remaining there at about 80 ng/kg at depths between 1000 m and the sea bottom at about 5000 m[7,39,50]. In other oceans analogous increases of the Cd-concentration with depth have been observed, yet the maximum concentrations at 800 to 1000 m depth remained below 50 ng/kg (fig. 5). In the Pacific, also, the typical depth profiles of Cd with a distinct maximum at 800 to 1000 m depth, in the zone of oxidative decomposition of organic sinking particles loaded with Cd which are the product of planktonic productivity in the euphotic zone, are most pronounced, because modifying influences are relatively small in this ocean. The water masses below the mixing zone become much older in the Pacific, and are less affected by the deep water circulation than in the Atlantic; therefore the depth profiles of nutrient-like heavy metals, like Cd, and of the nutrients can adjust in the Pacific in a much more undisturbed way[8]. Moreover, there is horizontal transport of the water masses by currents in the surface zone and at various depths which can influence the depth profile pattern.

Thus, differences in the significance of sources and of the oceanographic parameters give rise to depth profiles of Cd and nutrients different in quantity but analogous in general pattern in Pacific and Atlantic. These depth profiles are modified even more strongly in the Eastern Arctic Ocean[15,34,35], owing to water mass stratification, and in the North Atlantic[52] and in the Norwegian Sea[36] (fig. 6) as well as in the Baltic[23] and the western Mediterranean[13] (fig. 7). As recent investigations in the Black Sea reveal, special patterns of the depth profile are to be expected in completely anoxic waters, due to the formation of soluble bisulphide complexes of Cd[16].

The typical depth profiles in oxic waters (viz. figs 5–7) originate from the fact that Cd is a typical nutrient-like metal[7,8]. These metals are taken up in surface waters by living phytoplankton and by zooplankton grazing on it. With the sinking dead phytoplankton and the fecal pellets and detritus of zooplankton the Cd is carried into the depths. Between 500 and 1000 m the organic particles are oxidatively decomposed. The accumulated Cd and other heavy metals are released as well as the nutrients phosphate and nitrate. Cd and other nutrient-like heavy metals (e.g. Ni, Zn and to a certain extent also Cu) are advected vertically together with the nutrients phosphate and nitrate (and silicate) in the water column. In the euphotic surface zone the nutrients are again consumed by the phytoplankton and the advected Cd (and other nutrient-like heavy metals) will be taken up again by the plankton. Cd can undergo such vertical cycling many times and only a certain amount of Cd will ultimately escape this vertical cycling and be transported to greater depths and to the sea bottom. The mean residence time of Cd in the water column of the oceans is estimated as about 450 years and is thus somewhat less than one half of the oceanic stirring time[63].

A consequence of vertical cycling is the existence of very strong correlations between the depth profiles of Cd and the nutrients phosphate and nitrate[7,8]. For large regions of the oceans the surface water concentration of Cd is controlled in the first place by vertical cycling. Obviously, this biologically controlled vertical cycling is the more intensive the higher the primary production is. Therefore the lowest surface water concentrations for Cd occur in the oceanic regions with depleted nutrient levels and consequently low primary productivity. These regions are the subtropical and central gyres, whereas higher Cd-levels are observed at the divergencies and subpolar fronts and particularly in the upwelling areas where deep water, rich in nutrients and Cd, is advected.

Other sources of cadmium input into surface waters

It has to be borne in mind, however, that vertical cycling is not the only parameter controlling the surface water level of Cd in the oceans. Other cadmium input sources of different significance in different regions of the oceans are riverine input and aeolian input of Cd originating from terrestrial sources, mainly of anthropogenic kind. In the more shallow shelf and coastal water regions, the release of Cd from the depots in the bottom sediments can also contribute. These other Cd-sources will have to balance at least the Cd-deficit which occurs during vertical cycling because of the transport of a certain amount of Cd to greater depths and ultimately to the sea bottom. For different oceans the significance of sources other than the vertical cycling for the Cd-level varies. In the Atlantic, depending on the region, all three other types of Cd-sources (riverine input, aeolian input and release from shelf sediments), provide not insignificant contributions to the Cd level. In general these other sources seem to be more significant in comparison with vertical cycling in the Atlantic than in the Pacific. In the Pacific, riverine input is negligible and thus the surface water levels of Cd are primarily controlled by vertical cycling and secondarily to a much smaller extent by deposition from the atmosphere.

Speciation of cadmium

In general, Cd is found preferentially in the dissolved state in the sea. The tendency for it to be bound to suspended particulate matter is moderate, except for uptake by phytoplankton. Cd forms only complexes of moderate stability with most organic ligands, and as in the oceans the overall concentration of particulate matter is usually low, even for particles with organic coatings there is no change to scavenge significant amounts of Cd by surface complexation in the suspended matter phase, unless this is phytoplankton. Exceptions do of course occur locally in coastal waters or estuaries with high levels of suspended material.

The moderate stability of Cd-complexes with organic ligands and the rather low levels of dissolved organic matter (DOM) are the reasons why in the oceans usually only inorganic speciation of dissolved Cd has significance. It has been shown that at the levels occurring in the oceans, contributions of dissolved humic substances[53,54] and amino acids[62] to the speciation of dissolved Cd will remain negligible. Exceptions could only occur at certain spots with substantially higher DOM levels in coastal waters and estuaries, or in the interstitial water of sediments. Among the inorganic ligands only chloride has significance for Cd complexation in oxic ocean waters. 97% of dissolved Cd exists as chlorocomplexes, i.e. 10% as $CdCl^+$ and 87% as the uncharged complex $CdCl_2$ [45].

Cadmium in marine organisms

Cadmium is of course also taken up by higher organisms, like molluscs, crustaceans and fishes, either via the marine food web or directly from the water via the gills. The predominant organs of accumulation are, as is common for higher organisms, the kidneys and the liver; the level of Cd in the muscle meat of fish which is the only part usually consumed, remains very low and constitutes no problem with respect to tolerance levels for human food[61]. Heavy metal burdens can be somewhat more critical for items of sea food of which the whole soft body is eaten, like mussels, oysters and clams, if the ambient water of the cultivation site is polluted. Recently the distribution of Cd and other toxic heavy metals over all significant organs of *Ostrea edulis* and *Mytilus galloprovincialis* was described in a detailed study[41,42]. For krill it has also been established that only in the peeled krill tail does the Cd level remains insignificant with respect to human consumption[60].

General analytical aspects

The overwhelming majority of accurate and oceanographically consistent data on Cd-levels in the sea has been obtained since 1975 by a still rather small group of laboratories which can master with sufficient expertise the substantial requirements for reliable ultra trace analysis. In this respect the necessary prerequisites were the development and adaptation of reliable analytical procedures and powerful determination methods, and a full awareness of the accuracy risks, mainly due to the possibilities of analytical contamination.

A most critical step with respect to exclusion of contamination was and is still collection of samples. Reliable and approved procedures for surface water sampling (usually to be performed from rubber rafts at a sufficient distance from the research vessels)[11,30] and special devices for deep water collection have been developed and successfully applied[11,36,55].

Pretreatment of collected samples has to be kept simple, and manipulations have to be performed in suitable laboratories using clean benches with a laminar flow of filtered air (class 100) and with scrupulously cleaned labware and reagents. If decomposition of DOM is necessary, this can be achieved without contamination risks by UV-irradiation of the sample[31].

For the determination of Cd and the other trace metals mentioned, a very suitable and convenient method is differential pulse voltammetry. It has, furthermore, the additional advantage that Cd, Pb and Cu can be determined simultaneously. A detailed treatment of the whole analytical procedure from sampling to the voltammetric determination stage is given in two recent reviews[46,47], while the particular potentialities of voltammetry in studies on Cd speciation are treated in a further review[48]. A reliable alternative for the analysis of Cd (and other trace metals) in sea water is the procedure based on extraction of Cd by dithiocarbamate. A 200–300-fold preconcentration is achieved. The subsequent Cd determination by graphite furnace atomic absorption spectrometry (GFAAS), an analytical approach developed for sea water analysis to ultimate perfection[11], has permitted the establishment in a series of important studies of the most significant fundamental aspects of the chemical oceanography of cadmium[7,8].

It has to be emphasized in this context that only meticulous fulfilment of the admittedly high trace analytical requirements will provide realistic data on the distribution of Cd and other heavy metal traces in the sea. This key analytical prerequisite can be still underestimated by chemical oceanographers[1].

Levels of cadmium in the oceans and European seas

The following section surveys the hitherto existing, still incomplete knowledge of the distribution of Cd in the oceans. Only data which meet the necessarily high standards of reliable ultra trace analysis and are oceanographically consistent, realistic results have been taken into account. Therefore only data obtained after 1975 by a comparatively small number of laboratories could be considered. The surface water Cd levels, typical for the oceanic regions investigated, are presented in figures 1 to 4. To make the maps for certain transects or areas clearer, representative average values have been evaluated from the individual sets of data reported in the original publications. Representative depth profiles of Cd from various parts of the oceans are given in figures 5–7.

Pacific Ocean

For a survey on the Cd-levels in surface waters see figures 1 and 2.

Investigations in the eastern North Pacific[40] yielded a Cd

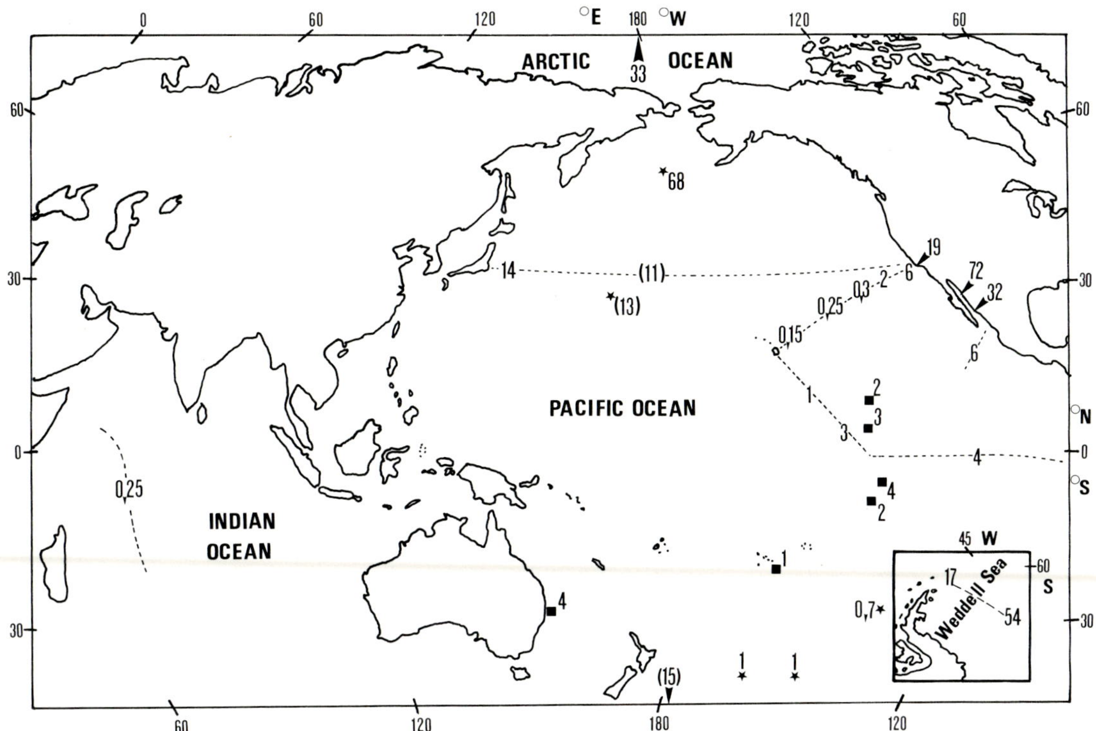

Figure 1. Cd distribution in the surface waters of the Pacific, Indian, central Arctic Ocean and in the northern Weddell Sea (inset).

range from 3 to 67 ng/kg. The survey comprised upwelling waters of the California Current and was limited to depths down to 100 m, as a pumping system was used for sampling. A rather close covariation of Cd with labile nutrients was described. The authors suggested deep sea water Cd concentrations on the order of 110 ng/kg, a value obtained by extrapolation. As more detailed investigations were carried out later by these authors, these results are not presented in figure 1.

Boyle et al.[3] determined deep water profiles in the Pacific, their 3 stations being located in regions of different oceanographic conditions. One station with 15 ng/kg was situated in the South West Pacific, south of New Zealand (see arrow at bottom of fig. 1), one in the temperate mid-Pacific east of Japan (14 ng/kg are given as mean in fig. 1) and one in the North Pacific upwelling region in the Bering Sea. With 68 ng/kg, Cd at the latter station was elevated (fig. 1). A common finding of the deep water profiles was a Cd maximum around 120–130 ng/kg in water depths of 800 to 2000 m. The authors found a distinct correlation of Cd with phosphate and an atomic ratio P: Cd of $1:3.5 \times 10^{-4}$.

A more detailed examination of Cd levels in the California Current and an area off central California as well as in the Bay of California was presented by Bruland et al.[12]. The authors realized at that time that, although their earlier findings[40] were in general agreement with results published at that time by other authors for the oceans, considerable improvements, both in sampling and analytical procedures, could be achieved. Thus, during the different cruises of this investigation, samples were collected from a raft in movement (surface water), by a pumping system (surface and intermediate depth water), with a modified Go-Flo system at a nonmetallic hydroline (deep water), with a 4-1 lucite sampler (deep water) as well as with the so-called CIT sampler by Schaule and Patterson[55]. The samples collected were not filtrated and Cd values given were considered to approximate closely 'total' concentrations.

For surface waters, which were considered at that time as 'nutrient depleted', values from 3.9 to 5.1 ng/kg were determined, whereas near-surface waters ranged from 9 to 72 ng/kg during upwelling periods. Typical values of 72 and 32 ng/kg were obtained for the Bay of California (fig. 1) as well as a mean of 6 ng/kg in the California Current off northern Mexico. 19 ng/kg are depicted in fig. 1 for the coastal upwelling area off Monterey, central California. A correlation equation for Cd and phosphate, as well as Cd and nitrate, was developed[12].

For one of the 2 stations (not depicted in figs 1 or 2), owing to upwelling within the California Current off central California[11], the surface water Cd-value, with 16 ng/kg, was rather elevated. A further increase to 61 ng/kg at 100 m and a rather broad maximum of about 110 ng/kg at 800 m were observed. The primary objective of this study was the comparison of deep water sampling devices. It could be proved that consistent results were obtainable with the different samplers tested. Using different preconcentration techniques (extraction with dithiocarbamate or ion exchange) before GFAAS determination, an excellent agreement for Cd could be attained.

The findings concerning the behavior of Cd in ocean waters have been widened by a comprehensive investigation of Bruland[7] in the North Pacific. Cd profiles, vertical and horizontal, have been determined on a transect from Hawaii to Monterey, California (fig. 1). The unfiltered samples were collected from a raft, rowing crosswind to a distance of at least 200 m from the vessel. The deep sea water samples were collected with the so-called CIT sampler[55] at one station. At 2 other stations, a 30-l teflon-coated modified Go-Flo sampler (General Oceanics) was clamped to a plastic hydroline.

The Cd was determined after a 200–300-fold preconcentration from approximately 300-ml samples by dithiocarbamate extraction prior to GFAAS. In this mission, special care was devoted to a further reduction of blanks by a factor 3 in comparison to those reported earlier[11]. Surface samples from 4 stations were processed on board the ship using a modified ion-exchange technique, offering a preconcentration factor of as much as 800:1. This methodolocigal intercomparison was mainly used to verify the extremely low surface water values on this transect, which comprises three important oceanic regimes. Within the oligotrophic central gyre of the eastern North Pacific exist warm and high salinity waters in contrast to the cool, lower salinity mixed subarctic waters of the California Current, and between the two extremes occurs a complex transition zone. In these different zones three deep water vertical profiles, representative for the respective oceanographic conditions, have been established[7].

An examination of the surface water data reveals a strong decrease in Cd from the coastal zone off Monterey in the direction of Hawaii[7] (fig. 1). At the California Current boundary station the Cd surface level has been determined to be 4.7–6.5 ng/kg (given as 6 ng/kg in fig. 1). This was the only station where the nutrient nitrate could be detected, with 0.1 µmol/kg. The salinity was 33‰. After one further station in a westerly direction, with 2 ng/kg Cd, the surface water levels decreased rapidly to values of 0.3 ng/kg and less (fig. 1), going parallel with decreasing levels of phosphate, whereas nitrate was no longer detectable. Within the central gyre waters with oligotrophic conditions also phosphate became undetectable (< 0.01 µmol/kg) and Cd values averaged 0.15 ng/kg. The lowest analytical value reported was even 0.08 ng/kg. The maximum salinity was 35‰. These results from oligotrophic central gyre waters are more than one order of magnitude lower than any other Cd data in the oceans reported previously.

In one oligotrophic gyre station (fig. 1) about half way from Hawaii to the continent, Cd is depleted at the surface to 0.25 ng/kg and increases with depth to 19 ng/kg at 185 m and a maximum of 117 ng/kg at about 1000 m (fig. 4). A moderate decrease with further depth to levels from 94–86 ng/kg in 3000–4800 m has been observed. This station is characterized by a rather sharp maximum at 1000 m depth, in contrast to the broad maximum between 500 and 1000 m at the boundary station influenced by eutrophic coastal upwelling. The latter station (not shown in fig. 1) yields a higher surface water Cd level of 18 ng/kg at 10 m depth, characteristic for upwelling coastal waters. Due to the special hydrographic conditions at this station, as well as at the next station described below, these elevated surface water values have not been included in the discussion of the horizontal profiles from Hawaii to Monterey, mentioned above. The last deep water profile within the California Current yielded findings in between the two extremes described. In general it can be concluded from the three depth profiles in contrasting oceanic regions, that at depths below

1000 m the profiles become indistinguishable down to the sea bottom.

In this investigation[7] oligotrophic oceanic regimes were included for the first time, and the detection limit of the analytical procedure had been lowered enough to determine Cd levels in extremely depleted surface waters. This provided an extension of the data range necessary for the establishment of consistent relationships along the depth profiles between Cd and nutrients, like phosphate and nitrate. Dissolved Cd is correlated with phosphate according to the equation:

$$(\text{Cd, nmol/kg}) = (0.347 \pm 0.007)\,(\text{PO}_4,\,\mu\text{mol/kg})$$
$$- (0.068 \pm 0.017)$$

with r = 0.992. Furthermore, phosphate is correlated with nitrate by the equation:

$$(\text{NO}_3,\,\mu\text{mol/kg}) = (15.2 \pm 0.2)\,(\text{PO}_4,\,\mu\text{mol/kg})$$
$$- (4.9 \pm 0.5)$$

with r = 0.991. According to these findings, for every Cd atom involved in the vertical biogeochemical cycle, roughly 300 000 atoms of organic carbon are required as carrier.

Comparing the before-mentioned results with earlier data from surface waters in some extreme upwelling zones, including the Bay of California[12] with a maximum Cd level of 72 ng/kg (fig. 1), it could be demonstrated that Cd surface water levels may be oceanographically consistent over an extremely wide range.

Results from four cruises were reported by Boyle et al.[2]. In comparison to more recent findings, the data from the 1976 cruise might be more affected by analytical contamination, and therefore for the central latitudes only a mean of 11 ng/kg and a more elevated mean of 14 ng/kg near Japan is given in fig. 1. The other results show a large scatter for individual measurements, and only the given mean values should be therefore taken into consideration. Many surface water levels were reported to be below the detection limit of 1.2 ng/kg and had a one σ precision in this order of magnitude, so that Bruland's reliable findings[7] in nutrient depleted subtropic gyre waters could not be reproduced by Boyle et al.[2].

One cruise off California is given in figure 2 with a mean of 2 ng/kg[1]. Further cruises from California to the Galapagos Islands and to Panama[1] are summarized in figure 2 for the respective areas, with 2.6 and 3 ng/kg.

Surface water Cd levels have been reported by Mart et al.[39] for different areas of the Pacific. 7 ng/kg were obtained for the Peru Basin in surface waters influenced by upwelling (not depicted in figs 1 or 2).

Four sampling areas close to the equator (indicated by squares in fig. 1) yielded means of 2, 3, 4 and 2 ng/kg in the equatorial Pacific waters. Going southwest to the Aitutaki Passage at the Cook Islands, the Cd levels became even lower with a range of 0.7 to 1.7 ng/kg, represented by a mean value of 1 ng/kg in figure 1.

The same authors[39] found in the east Australian shelf region (Tasman Sea) in surface waters 3.2–6.2 ng/kg yielding a mean of 4 ng/kg.

Data for a transect through the equatorial Pacific and north to Hawaii have been reported by Nürnberg et al.[50]. Levels from the equatorial east Pacific with a range from

2 to 5.9 ng/kg (corresponding to a mean of 4 ng/kg in fig. 1) decrease going north to about 3 ng/kg in the equatorial counter current and to 1 ng/kg in the north equatorial current south of Hawaii (fig. 1). On this transect a number of depth profiles have been determined (see e.g. data in fig. 4). They have in common a broad maximum of Cd with values between 80 and 109 ng/kg around 1000 m for stations in the region of the South Equatorial Current and the Equatorial Counter Current. Levels in bottom waters 1 m above the seafloor are 79–92 ng/kg. These bottom water samples were collected with a device attached to a grab sampler originally designed for collection of manganese nodules[58]. At stations south of the Galapagos archipelago the maximum values occur between 500 and 1000 m with levels from 90 to 114 ng/kg. The same is observed for 2 stations between the Galapagos archipelago and the coast of Ecuador. A comparison of the data obtained from this mission with deep water profiles from the investigations of Bruland[7] yields comparable levels and also shapes of the profiles for different areas of the Pacific.

Within the confidence limits, identical values for filtered and unfiltered samples have been found, due to the low content of particulate matter in these waters. The same is valid for untreated samples and water UV-irradiated in order to destroy organic compounds that might chelate trace metals, which indicates very low levels of DOM[50]. For three stations[32] (marked by an asterisk in fig. 1) in the southern Pacific, 0.7 ng/kg was found at the northern station and 1 ng/kg is given for the two stations on the transect at 42° S. On this complete transect from 111° W, the east Pacific rise, to 163° W, bottom water (3000–5700 m depth) total Cd levels from 80–108 ng/kg were found in samples from 1 m above the sea bottom. Collection was made with a special device according to the concept of Sipos et al.[58].

Weddell Sea

Results from the northern Weddell Sea[39] (fig. 1a) show that this area is influenced by upwelling waters. The southern station has 54 ng/kg and levels decrease continuously in a northerly direction to 17 ng/kg.

Indian Ocean

Only sparse information about the Indian Ocean is available. Data presented by Danielsson[14] yielded a mean of 15 ng/kg in surface waters and 64 ng/kg below 100 m. But with our present knowledge about trace metals in the adjacent oceans as well as according to our own data (fig. 1) it can be concluded that these results are at least a factor 10 too high for surface waters. In surface waters investigated in a cruise in 1983 the Cd-level had a narrow range of 0,2–0,3 ng/kg[32]. The area of the transect corresponds to a nutrient-depleted zone of very low primary productivity in the Indian Ocean, as in oligotrophic regions of the Pacific and on the Atlantic side in the Sargossa Sea, which are comparable with respect to latitude. For the *Red Sea* (fig. 2) only the value of 5 ng/kg from a single station is available[39].

Atlantic Ocean

Most surface water data are given in figure 2 together with some data from the Pacific and the Arctic Oceans. Data for the northern North Atlantic are presented together with levels in the North Sea and Baltic Sea in figure 3.

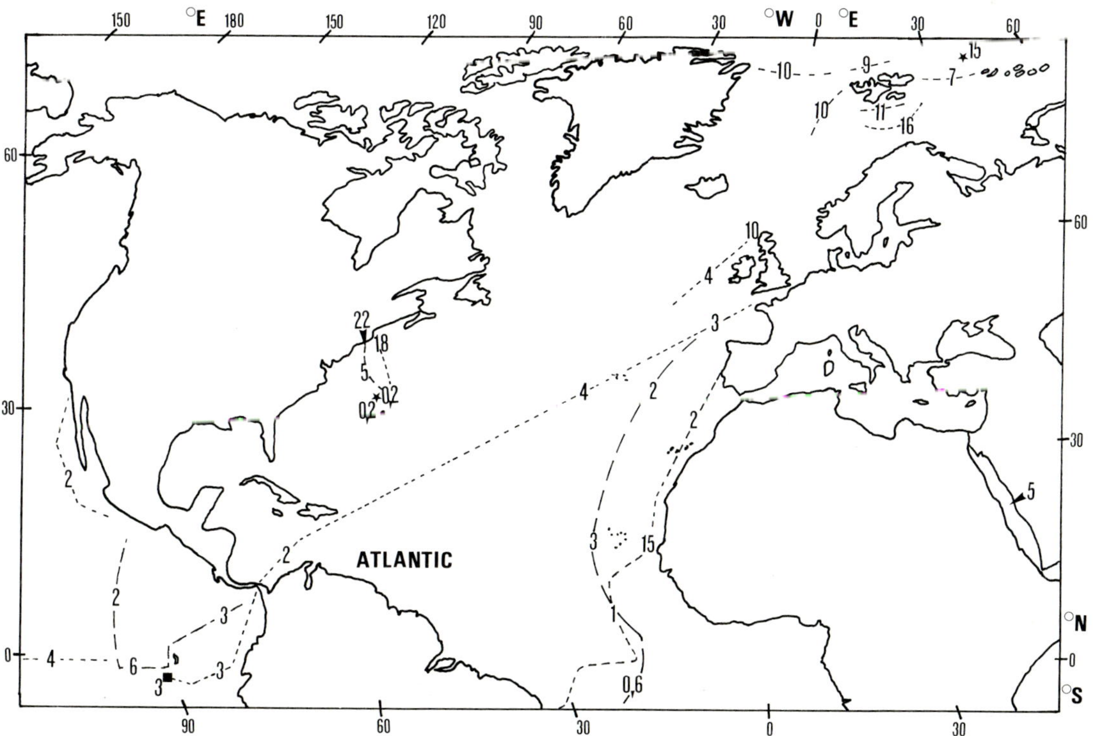

Figure 2. Cd distribution in the surface waters of the Atlantic Ocean, the eastern Arctic Ocean and a part of the eastern Pacific Ocean.

Data from two Atlantic cruises have been presented by Boyle et al.[2] and Boyle and Huested[1]. The authors conceded that they had problems with elevated blanks. Their precision, given with 1.2 ng/kg for the best 1 σ-precision, is obviously too limited for fully consistent measurements of Cd in depleted waters. This is reflected by results from repetitive sampling at the same stations, where results diverge by a factor up to 40, which cannot be easily explained by oceanography. Thus, only trends can be observed from their work. The authors state that Cd varies with the labile nutrients, but that much data on surface water Cd is below their detection limit. In figure 2 their cruise from Recife to Lisbon is summarized with means of 1 and 2 ng/kg. Only in the upwelling area off Dakar did the Cd-level increase substantially, to the order of 15 ng/kg. In a further cruise from the American east coast to the Sargasso Sea, 18 ng/kg were measured in shelf waters off the coast of New England. The lowest value obtained in the Sargasso Sea is, however, still 5 times higher than data reported by Bruland and Franks[10] which are presented in figure 2 and will be discussed later. Cd levels have been determined in the North and Central Atlantic by Mart et al.[39]. Further details about this same cruise (the final destination was Hawaii) have been reported elsewhere[50]. The Cd levels over the transect Biskaya-Panama range from 3–6 ng/kg (mean 4 ng/kg) in the Atlantic. For the Caribbean 1–3 ng/kg (mean 2 ng/kg) were obtained. A depth profile in the north equatorial drift yielded relatively low levels in the deep waters with a maximum of 34 ng/kg at 500 m (fig. 5). The values of the depth profile are low in so far that these Atlantic deep waters have only about ¼ to ⅓ of the Cd levels observed in the Pacific.

Western North Atlantic surface waters have been collected on a transect (fig. 2) from the shelf waters off New England to the Sargasso sea southeast of Bermuda[10]. A deep sea vertical profile has been investigated in the Sargasso Sea, approximately 200 km northwest of Bermuda. Results from the unfiltered samples are given by the authors as total dissolved levels. Elevated Cd values have been found in the slope and shelf waters off the New England coast. The sources seem to be riverine input and/or shelf sediments. The typical value given in figure 2 for the shelf waters is 22 ng/kg, whereas the Sargasso Sea, to be considered as the end member, yields levels of only 0.2 ng/kg. These latter results are characteristic for nutrient-depleted surface waters as in the central subtropical gyres of the North Pacific. On the transsect from the coast to the Sargasso Sea, Cd levels are reversely proportional to the salinity. A comparison of the surface waters of the North Atlantic with the North Pacific reveals that in both oceans Cd can be depleted in oligotrophic surface waters to values equal to or less than 0.2 ng/kg.

The vertical profile starts with 0.2 ng/kg in the nutrient-depleted surface waters of the Sargasso Sea increasing to a maximum of 37 ng/kg at about 1000 m depth. A slight decrease to levels around 31 ng/kg can be noted with further depth. A comparison of vertical Cd profiles from the North Atlantic and North Pacific[7] yields practically identical shapes and maxima at similar depths, but the overall levels in the deep waters are about a factor of 3 lower in the Atlantic (fig. 5). These deep water concentrations average 15% less than would be predicted by the correlation equation Cd:PO_4 found for Pacific waters[7]. Similar conclusions arise from the comparison of the depth profiles obtained at different locations in the Pacific and Atlantic by Mart et al.[39] and Nürnberg et al.[50]. Results of a transect through the central Atlantic from Recife to the Straits of Dover[32] are depicted in figure 2. The central equatorial Atlantic is characterized by a range of surface Cd levels of 0.4–0.9 ng/kg with a mean of 0.6 ng/kg. Distinctly more elevated levels are found close to the Cap Verde Islands with 2.5–4.1 ng/kg (mean 3 ng/kg), corresponding to surface waters influenced by upwelling at the West African coast. Around the Azores a range of 1.9–3.5 ng/kg (given as a mean of 2 ng/kg in fig. 2) is found. At the last station northwest of Cap Finisterre, Spain, again 3 ng/kg were found.

Cd levels around 4 ng/kg in surface waters of the North Atlantic west of Ireland (fig. 2) have been reported by Kremling[22]. The author noted a sharp increase in the nutrient and trace metal levels (including Cd) entering the shelf waters northwest of Scotland. Levels ranged from 3 to 30 ng/kg in these waters, as a result of combined physical and geochemical processes operating over the shelf area. Typical means of 10 ng/kg off the Hebrides (fig. 2) and 28 ng/kg at the northern edge of Scotland (fig. 3) are given.

European North Atlantic

The surface water levels in this section are mainly depicted in figure 3 but some data are given in figure 2.

A station southwest of Iceland[52] yielded a surface water result of 12 ng/kg. This value, corresponding to an area richer in nutrients and of higher primary productivity, fits into the overall pattern given by Kremling[22] and Mart et al.[37]. The depth profile (fig. 6) yields 35 ng/kg around 1200 m, which seems to characterize the water mass as North Atlantic water. Olafsson[52] also reported a strong correlation between nitrate and Cd.

Figure 3 shows that Cd levels cluster around 14 ng/kg in waters of higher primary productivity northeast of Scotland (Orkney Islands) and the Faroe Islands and around 15 ng/kg northeast of the Shetland Islands[36]. In general, surface levels in the northern Norwegian Sea coastal area are somewhat enhanced (20 ng/kg) and increase southward, to 22 and 25 ng/kg (fig. 3). This is explained by residual current waters from the North Sea with more elevated trace metal levels, which are carried north along the Norwegian coast up to the western Barents Sea (fig. 2), where the mean is 16 ng/kg[36]. Results in surface waters around Spitzbergen[34,35] range from 9 to 11 ng/kg, while for the transect in the Arctic Ocean from Spitzbergen to Franz-Josef-Land 7 ng/kg are found[34,35] (fig. 2). The mean of the surface water levels in the area from Greenland to Spitzbergen, which is under the influence of the East Greenland Current, is 9.6 ng/kg[34,35] (depicted as 10 ng/kg in fig. 2).

A depth profile from the Norwegian Sea at 63° N, 2° E, reported by the same authors[34,35] showed only a slight increase from 16 ng/kg at the surface to 22 ng/kg at 500 m depth. Almost identical findings to much greater depths,

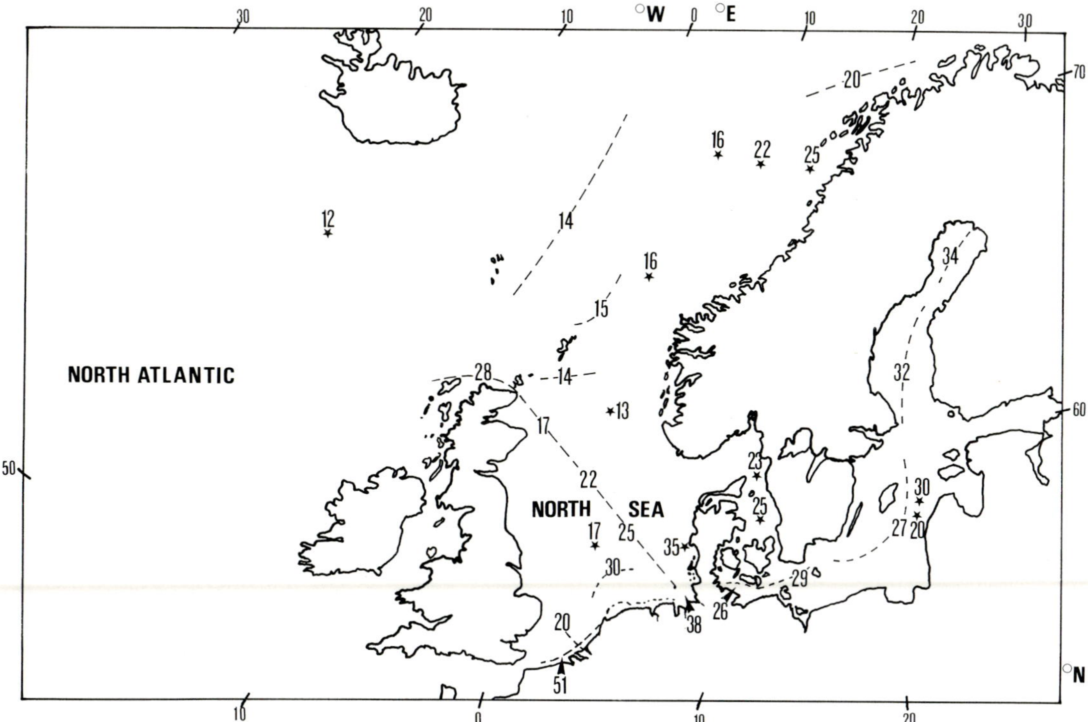

Figure 3. Cd distribution in the surface waters of the North Sea, Baltic Sea and the northern Atlantic Ocean.

i.e. a rather constant profile with no distinct variations, have been reported by Brügmann et al.[6]. Unfortunately their investigations suffer from severe methodological discrepancies, which reduces the reliability of their data. An investigation by Bruland et al.[9] has proved that such methodological deficiencies can be avoided.

Arctic Ocean

A depth profile in the central Arctic Ocean has been determined[43], indicated by an arrow in figure 1. A striking fact is the rather elevated surface water concentration of 33 ng/kg. The author relates this to a supply from underlying waters and the low biological activity, as no relevant information about river input was available. A sharp increase to 67 ng/kg can be observed around 100 m depth, which coincides with the maximum of Si levels. For greater water depths a mean of 22 ng/kg was found. These deep water results are identical to those found by Danielsson and Westerlund[15] and Mart et al.[34] in the eastern Arctic Ocean. In this area, a rather smooth increase of Cd with depth, without any distinct maximum, could be observed (fig. 6).

Danielsson and Westerlund[15] found a range of 15–22 ng/kg in surface waters in the region from North Greenland to the area north of Spitzbergen and the transsect from Spitzbergen to Franz-Josef-Land. Somewhat lower results have been obtained by Mart et al.[34,35] for the same regions, given as means of 10, 9 and 7 ng/kg in figure 2. These findings can be explained by the fact that the surface samples collected in this mission[34,35] stem from waters depleted in trace metals during the season of maximum biological productivity. Primary productivity in the open water leads of the ice cover was such that surface water had a brownish color. As far as analytical preci-

sion is concerned, Danielsson and Mart obtained identical Cd results in a methodological intercomparison[36]. Danielsson and Westerlund[15] as well as Mart et al.[35] obtained Cd: phosphate relationships about 20% lower than those reported by Bruland[7] for Pacific waters.

North Sea

The characteristic surface water data are depicted in figure 3.

The first consistent Cd results from near-shore waters of the Belgian and Dutch Sea coast were presented by Mart[30], who collected samples in 1976, by a rather simple but reliable sampling method which had been in use since 1975. Surface water collection was carried out from the bow of a smaller ship or raft, slowly progressing to windward, by immersing a polyethylene bottle attached to a fiberglass pole which could be extended to an appropriate length[29]. The author found dissolved Cd levels for the coastal areas mentioned[30] from 13–77 ng/kg with a mean of 33 ng/kg and total levels of 21–127 ng/kg with a mean of 51 ng/kg (fig. 3). Samples from port entries have been omitted in this discussion, in order to allow comparison with other authors[18] who took samples further away from the coast, thus avoiding this locally restricted land-borne pollution. Summaries of the results from the thesis of Mart[30] have been reported elsewhere[39,44]. An important finding from this work was that even in so-called 'heavily polluted' coastal areas, including the outer estuaries of the Scheldt and Rhine, Cd and other trace metal levels might be low.

The trace metal situation in the Dutch coastal area as well as in the Southern Bight has been studied by Duinker and Nolting[18]. The authors propose levels around 120 ng/kg dissolved Cd for near-shore waters and minimum values of 20–30 ng/kg for the Southern Bight. The mean coastal value of 120 ng/kg seems to be somewhat on the high side. Earlier results[17] were obviously affected by severe methodological discrepancies and problems.

After a considerable improvement in their sampling and analysis techniques, realistic data from the Belgian coast have been presented also by Gillain et al.[19]. The authors claim a range of 30–180 ng/kg for dissolved Cd.

A transect from North Scotland to the German Bight across the North Sea by Kremling[22] yielded a Cd range of 10–28 ng/kg. Characteristic means of 17, 22 and 25 ng/kg are given in figure 3. Two minima have been found, 10 ng/kg west of Scotland and 12 ng/kg in the Central North Sea. A similar low value, 17 ng/kg, has been reported by Mart et al.[36] in an adjacent area (marked by an asterisk in fig. 3), rather close to a zone with distinctly higher Cd levels, represented by a mean of 30 ng/kg. This reflects clearly the patchiness of North Sea surface waters, which might be influenced by penetrating tongues of North Atlantic waters with lower Cd levels (fig. 2) mainly entering the North Sea between the Orkneys and Shetlands.

A trace metal survey in the German Bight and its adjacent Wadden Sea, as well as in the estuaries of the major rivers Elbe and Weser, was started in 1975 and continued until 1984[33]. Preliminary results were published earlier[51]. Data about restricted areas like the Jade Bay, affected by industrial activities, ranged from only 13 to 21 ng/kg dissolved Cd, results which were confirmed in later missions. This gave us a strong line of evidence that Cd levels in the German Bight should be rather low, although anthropogenically enhanced, compared to Cd levels in the open Atlantic. Mean results presented for different areas were as follows[33]. Dissolved Cd around Helgoland was 28 ng/kg and total Cd was 46 ng/kg in 1977. In 1983 the results there were 28 and 34 ng/kg, respectively, for dissolved and total Cd. The open sea area of the German Bight yielded a mean of 28 and 34 ng/kg for dissolved and total Cd. Results for coastal areas, including the Wadden Sea, were only slightly higher. Substantially higher Cd-levels have been encountered in the polluted estuaries of Weser and Elbe[37]. The results from the open sea in the German Bight are of the same order as those reported by Schmidt et al.[57] whose open sea data cluster around 30 ng/kg. A characteristic mean of 38 ng/kg for the German Bight is given in figure 3.

The riverine input of Cd and other heavy metal traces from the Weser and Elbe is not unsignificant[37]. Patches of polluted water are carried away by the northern bound residual current along the Danish and Norwegian coast[36].

Baltic Sea

The surface water data are depicted in figure 3. For Skagerrak and Kattegat in the surface waters outflowing from the Baltic Sea, a mean of 23 ng/kg has been found. The same level is maintained in the North Sea bottom waters inflowing through the Skagerrak[28].

Realistic Cd levels of 25 ng/kg have been presented for a limited central Kattegat area in Danish coastal waters[26] (fig. 3). This proved that earlier results by Magnusson and Westerlund[27] were still influenced by contamination and therefore only their lowest earlier results from the range reported (20–150 ng/kg) were correct. The same is true for Cd-results from studies on the basis of 2 anchor stations, reported by Kremling[21]. Only the surface water level of 39 ng/kg from the 'Bosex' station (Gotland Sea) was close to the highest levels reported in 1984 from a synoptic survey on dissolved trace metal levels in Baltic

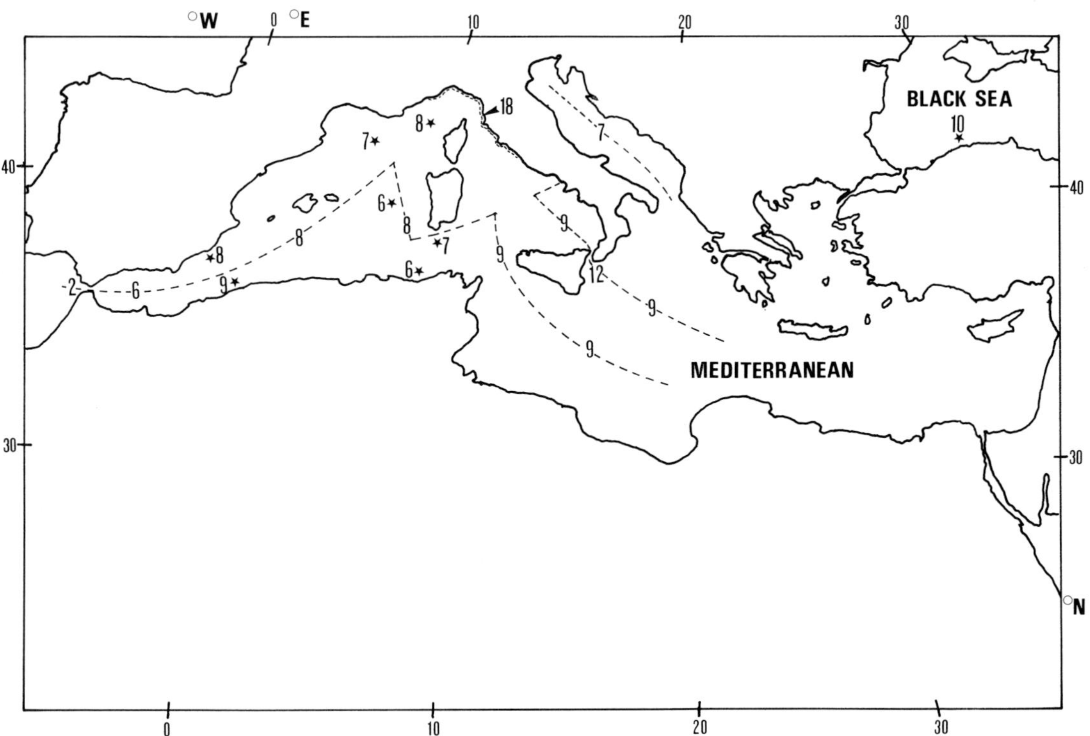

Figure 4. Cd distribution in the surface waters of the Mediterranean Sea.

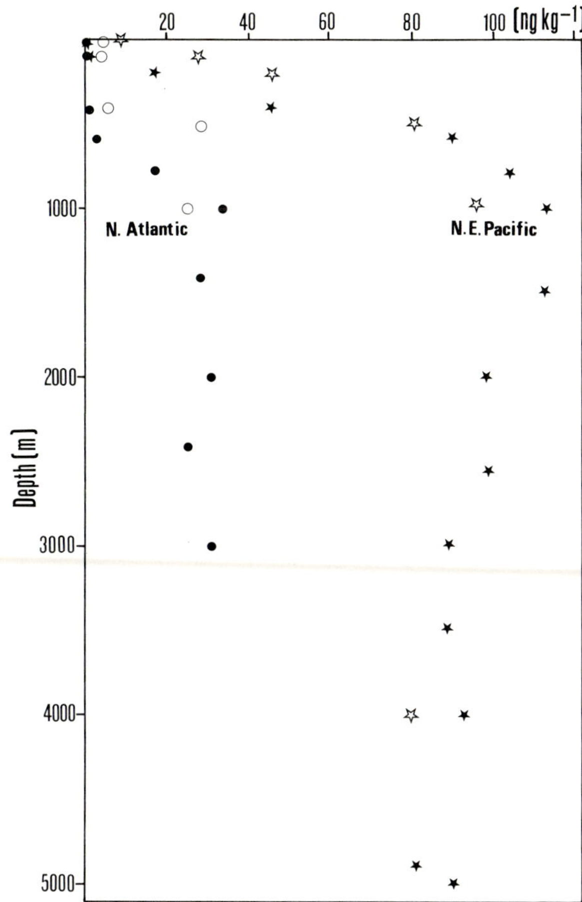

Figure 5. Depth profiles of Cd in the North Atlantic:
● 34°6′N; 66°7′W (ref. 10);
○ 24°40′N; 57°0′W (ref. 50);
and in the North East Pacific:
★ 17°32′N; 144°59′W (ref. 10);
☆ 1°9′S; 115°20′W (ref. 50).

surface waters[25]. Their results showed little variation in the Cd-levels from the Bothnian Bay to the Belt. Mean values for the different Baltic regions are given with 34, 32, 27 and 29 ng/kg, going from the Bothnian Bay to the Kiel Bight (fig. 3).

Results from 2 stations in the Gotland Sea have been reported by Kremling[23] with 20 and 30 ng/kg in the surface waters (fig. 3). The author noted a decrease of Cd at the O_2-H_2S interface (fig. 7) and his study supports the assumption that the solubility of Cd is greatly enhanced and controlled by the formation of sulfide complexes in the anoxic deeper waters.

Data by Gustavsson[20] from a cruise in 1979 through the Gotland and Bornholm Sea with a range from 29 to 73 ng/kg (not depicted in fig. 3) in the upper water layer from 0.5 to 200 m seem to be consistent, although it cannot be excluded that levels from the upper end of the reported range might be affected by analytical contamination.

Results (not depicted in fig. 3) from a cruise in the Baltic Sea covering the area between Flensburg, Travemünde and the Isle of Bornholm[55] cluster around 20 ng/kg for

dissolved Cd, which seems to be realistic, compared to recent data of other authors in figure 3. The range presented goes up to 60 ng/kg, with high-flyers up to 180 ng/kg, which reflects some analytical contamination problems, possibly during sampling.

The Kiel Bight, as well as the area around Fehmarn was investigated in 1983 (Mart and Backhaus, unpublished results) yielding for dissolved Cd a range from 8 to 44 ng/kg with a mean of 26 ng/kg (fig. 3). Even samples from the Kieler and Flensburger Förde (outer part) fit into this range.

Mediterranean Sea

All surface water data are shown in figure 4.

A survey along the Ligurian and northern Tyrrhenian shore line was undertaken in 1976[30,38,44]. It could be demonstrated that, even close to areas with heavy anthropogenic pollution, like the port of Genova, levels of Cd and also other trace metals decrease rapidly to background values. The remarkable determination limit of 0.5 ng/kg and the high contamination control and reduction of blanks introduced by Mart in 1976[51] made it possible to establish consistent coastal data. The mean reported by Mart[30] is 14 ng/kg for dissolved Cd and for total Cd 18 ng/kg, shown in figure 4 as a dotted line along the coast. It should be borne in mind that these values are representative for shore waters influenced by waste water inlets and run-off from ports. Riverine input from the Arno and Tiber remains negligible, because the major amount of the polluting heavy metal freight is trapped by sedimentation in the estuaries[5].

A conclusion that could be drawn from the shore water data was that the open Mediterranean should have Cd levels around or below 10 ng/kg[30]. Such values were presented in 1983 by Spivack et al.[59] in a synoptic survey of the surface waters of the western and central part of the Mediterranean Sea in two transects depicted in figure 4. One transect begins in the Atlantic just outside the Strait of Gibraltar with a mean of 2 ng/kg. Levels increase to 6 ng/kg already in the Strait of Gibraltar, remaining at the same level in the Alboran Sea and increasing to 8 ng/kg in the Alboran Sea and 9 ng/kg in the Strait of Sicily and the Ionian Sea. Similar levels are found in the southern Tyrrhenian Sea. The highest values, 12 ng/kg on average, were found in the Strait of Messina.

Further proof was given by results from 7 depth stations in the western Mediterranean[13]. Surface levels ranged from 6–9 ng/kg (see asterisks fig. 4) with rather constant depth profiles and levels of 5–7 ng/kg at 2000 m depth (fig. 7). In this context it is emphasized that the Mediterranean Sea is a nutrient-depleted sea with consequently rather moderate primary productivity. Similar low values in deep waters of the Mediterranean have been presented by Kremling and Petersen[24] but in general their range was distinctly larger; up to 40 ng/kg with high-flyers up to 58 ng/kg. The overall mean was 17.4 ng/kg and thus a factor 2,5 higher than the recent data presented by Copin-Montegut et al.[13], who took special care to avoid contamination during sampling.

Investigations in 1980 at 40 stations scattered over the Adriatic Sea from the northern basin to the Strait of

Otranto resulted in a mean value of 7 ng/kg for the dissolved Cd level in surface waters[4].

Black Sea

Consistent data, apart from a deep sea profile in the southwestern part of the Black Sea off the Turkish coast (42°10′ N, 32°30′ E) at Amasra have not been published to our knowledge. Surface water levels for total Cd around 10 ng/kg increase to a maximum of 30 ng/kg in 100 m depth. The subsequent pattern of the profile is further affected by solubility reactions, due to sulphide complex formation, in the completely anoxic waters existing below 150 m[16].

Concluding remarks

Inspection of figures 1–4 reveals the following general aspects. Surface levels of Cd hardly exceed 5 ng/kg in the central regions of the Pacific and Atlantic, and can even be below 1 ng/kg in the particularly nutrient-depleted oligotrophic gyres. This seems to hold also for the al-

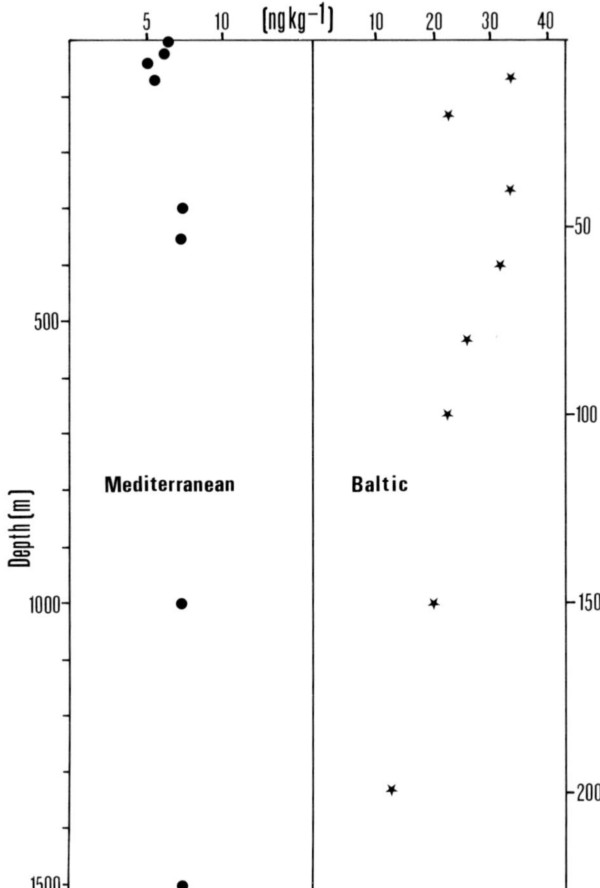

Figure 7. Depth profiles of Cd in the Mediterranean Sea:
● 37°30′N; 7°30′E (ref. 13);
in the Baltic Sea:
★ 57°4′N; 19°50′E (ref. 23).

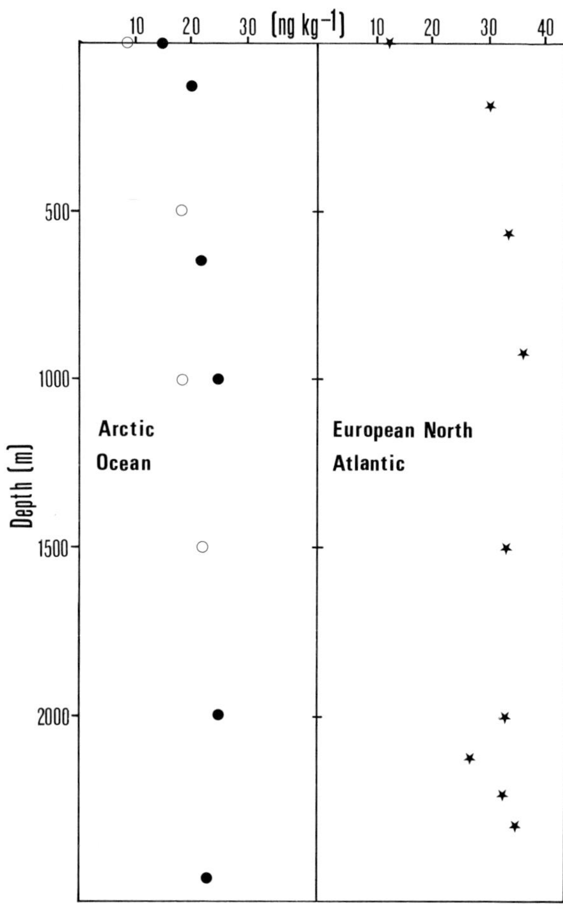

Figure 6. Depth profiles of Cd in the eastern Arctic Ocean:
● 82°31′N; 43°57′E (ref. 15);
○ 81°43′N; 8°51′W (ref. 35);
in the European North Atlantic:
★ 61°N; 17°W (ref. 52).

ready investigated transect in a nutrient-depleted region of the Indian Ocean. Somewhat higher levels exist in waters containing more nutrients with higher primary productivity at more northern latitudes in the Pacific, Atlantic and Arctic Oceans. Significantly higher levels, between 15 and 75 ng/kg, are observed in areas of upwelling in the Pacific and Atlantic oceans and also in the Weddell Sea. The comparatively elevated Cd levels in the North Sea and Baltic Sea and off the North American east coast are the consequence of anthropogenic pollution by riverine and aeolian input. A more moderate pollution from anthropogenic sources, probably in the first place by input from the atmosphere, is also the reason for the moderately elevated Cd levels in the Mediterranean Sea, compared to the central parts of the oceans. As it is a rather nutrient-depleted sea, significantly lower Cd levels would be expected to exist there if there were no anthropogenic pollution.

Although some crude contours and trends of Cd distribution, and certain findings of fundamental significance concerning the behavior of Cd and other heavy metals in the sea have emerged during the last decade, based on analytically accurate and oceanographically consistent data, the situation is still characterized by large gaps in our knowledge. Only sporadic data – if any at all – are available for large areas of the oceans. These areas are the South Atlantic, the Central and South Pacific, the Indian Ocean, the seas around Antarctica and the Central Arctic Ocean. More data are also needed for the North Atlantic and the Mediterranean Sea, particularly for the eastern part.

* Deceased, May 1985.

1 Boyle, E., and Huested, S., Aspects of the surface distribution of copper, nickel, cadmium and lead in the North Atlantic and North Pacific, in: Trace Metals in Sea Water, pp. 379–394. Eds C. S. Wong, E. Boyle, K. W. Bruland, D. Burton and E. D. Goldberg. Plenum Press, New York – London 1983.

2 Boyle, E., Huested, S., and Jones, S., On the distribution of copper, nickel and cadmium in the surface waters of the North Atlantic and North Pacific Ocean. J. geophys. Res. *86* (1981) 8048–8066.

3 Boyle, E., Sclater, F., and Edmond, J. M., On the marine geochemistry of cadmium. Nature *263* (1976) 42–44.

4 Branica, M., Peharec, Z., and Kwokal, Z., Concentrations of Zn, Pb, Cd and Cu in the surface waters of the Adriatic Sea (1980 cruise of the R/V "Andrija Mohorovičić". Rapp. Comm. int. Mer médit. (Monaco), in press.

5 Breder, R., Flucht, R., and Nürnberg, H. W., A comparative study on the toxic trace metal situation in the Thyrrhenian estuaries. Thalassia jugosl. *18* (1982) 135–171.

6 Brügmann, L., Danielsson, L. G., Magnusson, B., and Westerlund, S., Data from an expedition with R. V. "Alexander von Humboldt" on the Atlantic Ocean, June 1981, in: Report on the chemistry on sea water XXIX. University of Göteborg, Sweden, 1982.

7 Bruland, K. W., Oceanographic distributions of cadmium, zinc, nickel, and copper in the North Pacific. Earth planet. Sci. Lett. *47* (1980) 176–198.

8 Bruland, K. W., Trace elements in sea water, Chapt. 45, Chemical Oceanography, Vol. 8. Eds J. P. Riley and R. Chester. Academic Press, London 1983.

9 Bruland, K. W., Coale, K. H., and Mart, L., Analysis of sea water for dissolved cadmium, copper and lead: an intercomparison of voltammetric and atomic absorption methods. Mar. Chem. *17* (1985) 285–300.

10 Bruland, K. W., and Franks, R. P., Mn, Ni, Cu, Zn, and Cd in the western North Atlantic, pp. 395–414. Eds C. S. Wong, E. Boyle, K. W. Bruland, D. Burton and E. D. Goldberg. Plenum Press, New York-London 1983.

11 Bruland, K. W., Franks, R. P., Knauer, G. A., and Martin, J. H., Sampling and analytical methods for the determination of copper, cadmium, zinc, and nickel at the nanogram per liter level in sea water. Analyt. chim. Acta *105* (1979) 233–245.

12 Bruland, K. W., Knauer, G. A., and Martin, J. H., Cadmium in northeast Pacific waters. Limnol. Oceanogr. *23* (1978) 618–625.

13 Copin-Montegut, G., Courau, P. and Nicolas, E., Cd and Pb in Mediterranean waters. Mar. Chem., in press.

14 Danielsson, L. G., Cadmium, cobalt, copper, iron, lead, nickel and zinc in Indian Ocean water. Mar. Chem. *8* (1980) 199–215.

15 Danielsson, L. G., and Westerlund, S., Trace metals in the Arctic Ocean, in: Trace Metals in Sea Water, pp. 85–95. Eds C. S. Wong, E. Boyle, K. W. Bruland, J. D. Burton and E. D. Goldberg. Plenum Press, New York – London 1983.

16 Dorten, W., Nürnberg, H. W., Mart, L., and Valenta, P., Depth profiles of cadmium, lead and copper in the Black Sea. Sci. tot. Envir. (1986) in press.

17 Duinker, J. C., and Kramer, C. J. M., An experimental study on the speciation of dissolved Zn, Cd, Pb and Cu in River Rhine and North Sea water by differential pulse anodic stripping voltammetry. Mar. Chem. *5* (1977) 207–228.

18 Duinker, J. C., and Nolting, R. F., Dissolved copper, zink and cadmium in the Southern Bight of the North Sea. Mar. Poll. Bull. *13* (1982) 93–96.

19 Gillain, G., Duyckaerts, C., and Disteche, A., Trace metals (Zn, Cd, Pb, Cu, Sb and Bi) levels (ionic and dissolved organic complexes) in the Southern Bight (Belgian Coast). Action Interuniversitaire, Oceanologie, Service Premier Ministre, Bruxelles 1980.

20 Gustavsson, I., The concentrations of cadmium, lead and copper in filtrated Baltic sea water samples collected at two seasons 1979. Acta hydrochim. hydrobiol. *11* (1983) 309–317.

21 Kremling, K., The distribution of selected trace metals in Baltic waters; a study on the basis of 2 anchor stations. Proc. of the XI. Conf. of Baltic oceanographers, 1978, vol. 1 pp. 290–297.

22 Kremling, K., Trace metal fronts in European shelf waters. Nature *303* (1983) 225–227.

23 Kremling, K., The behaviour of Zn, Cd, Cu, Ni, Co, Fe, and Mn in anoxic Baltic waters. Mar. Chem. *13* (1983) 87–108.

24 Kremling, K., and Petersen, H., The distribution of zinc, cadmium, copper, manganese and iron in waters of the open Mediterranean Sea. 'Meteor' Forsch. Ergebn. *23* (1981) 5–14.

25 Kremling, K., and Petersen, H., Synoptic survey on dissolved trace metal levels in Baltic surface waters. Mar. Poll. Bull. *15* (1984) 329–334.

26 Magnusson, B., and Rasmussen, L., Trace metal levels in coastal sea water: Investigation of Danish waters. Mar. Poll. Bull. *13* (1982) 81–84.

27 Magnusson, B., and Westerlund, S., The determination of Cd, Cu, Fe, Ni, Pb and Zin in the Baltic sea water. Mar. Chem. *8* (1980) 231–244.

28 Magnusson, B., and Westerlund, S., Trace metal levels in sea water from the Skagerrak and the Kattegat, in: Trace Metals in Sea Water, pp. 467–473. Eds C. S. Wong, E. Boyle, K. W. Bruland, J. D. Burton and E. D. Goldberg. Plenum Press, New York – London 1983.

29 Mart, L., Collection of surface water samples, Fresenius Z. analyt. Chem. *299* (1979) 97–102.

30 Mart, L., Ermittlung und Vergleich des Pegels toxischer Spurenmetalle in nordatlantischen und mediterranen Küstengewässern. Ph. d. thesis, RWTH Aachen, FRG, 1979.

31 Mart, L., Minimization of accuracy risks in voltammetric ultra trace determination of heavy metals in natural waters. Talanta *29* (1982) 1035–1040.

32 Mart, L., Golimowski, J., Klahre, P., Rützel, H., and Nürnberg, H. W., Trace metal levels in different oceanic regions. Sci. tot. Envir. in preparation.

33 Mart, L., and Nürnberg, H. W., Cd, Pb, Cu, Ni, and Co distribution in the German Bight, Mar. Chem. *18* (1986) in press.

34 Mart, L., Nürnberg, H. W., and Dyrssen, D., Low level determination of trace metals in Arctic sea water and snow by differential pulse anodic stripping voltammetry, in: Trace Metals in Sea Water, pp. 113–130. Eds C. S. Wong, E. Boyle, K. W. Bruland, J. D. Burton and E. D. Goldberg. Plenum Press, New York – London 1983.

35 Mart, L., Nürnberg, H. W., and Dyrssen, D., Trace metal levels in the eastern Arctic Ocean. Sci. tot. Envir. *39* (1984) 1–14.

36 Mart, L., Nürnberg, H. W., and Rützel, H., Comparative studies on cadmium levels in the North Sea, Norwegian Sea, Barents Sea and the eastern Arctic Ocean. Fresenius Z. analyt. Chem. *317* (1984) 201–209.

37 Mart, L., Nürnberg, H. W., and Rützel, H., Levels of heavy metals in the tidal Elbe and its estuary and heavy metal input into the sea. Sci. tot. Envir. *44* (1985) 35–49.

38 Mart, L., Nürnberg, H. W., Valenta, P., and Stoeppler, H., Determination of levels of toxic trace metals dissolved in sea water and inland waters by differential pulse anodic stripping voltammetry. Thalassia jugosl. *14* (1978) 171–188.

39 Mart, L., Rützel, H., Klahre, P., Sipos, L., Platzek, U., Valenta, P., and Nürnberg, H. W., Comparative studies on the distribution of heavy metals in the oceans and coastal waters. Sci. tot. Envir. *26* (1982) 1–17.

40 Martin, J. H., Bruland, K. W., and Broenkow, W. W., 'Marine pollutant transfer', Eds H. L. Windom, and R. A. Duce. Health, Lexington, Mass., 1976.

41 Martincic, D., Nürnberg, H. W., Stoeppler, M., and Branica, M., Toxic metal levels in bivalves and their ambient water from the Lim Channel. Thalassia jugosl. *16* (1980) 297–315.

42 Martincic, D., Nürnberg, H. W., Stoeppler, M., and Branica, M., Bioaccumulation of heavy metals by bivalves from Lim Fjord (North Adriatic Sea). Mar. Biol. *81* (1984) 177–188.

43 Moore, R. M., Oceanographic distribution of zinc, cadmium, copper and aluminium in waters of the central Arctic. Geochim. cosmochim. Acta *45* (1981) 2475–2482.

44 Nürnberg, H. W., Potentialities and applications of advanced polarographic and voltammetric methods in aquatic and marine trace metal chemistry. Acta Univ. Upsaliensis Annum Quingentesimum Celebrantis *12* (1978) 270–287, Almquist and Wiksell Internat., Stockholm.

45 Nürnberg, H. W., Investigations on heavy metal speciation in natural waters by voltammetric procedures, Fresenius Z. analyt. Chem. *316* (1983) 557–565.

46 Nürnberg, H. W., The voltammetric approach in trace metal chemistry of natural waters and atmospheric precipitates. Analyt. chim.

Acta *164* (1984) 1–21.
47   Nürnberg, H. W., Trace analytical procedures with modern voltammetric determination methods for the investigation and monitoring of ecotoxic heavy metals in natural waters and atmospheric precipitates. Sci. tot. Envir. *37* (1984) 9–34.
48   Nürnberg, H. W., Potentialities of voltammetry for the study of physicochemical aspects of heavy metal complexation in natural waters, in: Complexation of Trace Metals in Natural Waters, pp. 95–115. Eds C. J. M. Kramer and J. C. Duinker. Martinus Nijhoff/W. Junk Publ., The Hague – Boston – Lancaster 1984.
49   Nürnberg, H. W., and Mart, L., Distribution and Fate of Heavy Metals with Ecotoxic Significance in the Sea, Topics in Current Chemistry. Springer-Verlag, Heidelberg – Berlin – New York, to appear 1986.
50   Nürnberg, H. W., Mart, L., Rützel, H., and Sipos, L., Investigations on the distribution of heavy metals in the Atlantic and Pacific oceans. Chem. Geol. *40* (1983) 97–116.
51   Nürnberg, H. W., Valenta, P., Mart, L., Raspor, B., and Sipos, L., The polarographic approach to the determination and speciation of toxic metals in the marine environment. Fresenius Z. analyt. Chem. *282* (1976) 357–367.
52   Olafsson, J., Mercury concentrations in the North Atlantic in relation to cadmium, aluminium and oceanographic parameters, in: Trace Metals in Sea Water, pp. 475–485. Eds C. S. Wong, E. Boyle, K. W. Bruland, D. Burton and E. D. Goldberg. Plenum Press, New York – London 1983.
53   Raspor, B., Nürnberg, H. W., Valenta, P., and Branica, M., Significance of dissolved humic substances for heavy metal speciation in natural waters, in: Complexation of Trace Metals in Natural Waters, pp. 317–327. Eds C. J. M. Kramer and J. C. Duinker. Martinus Nijhoff/W. Junk Publ., The Hague – Boston – Lancaster 1984.
54   Raspor, B., Nürnberg, H. W., Valenta, P., and Branica, M., Studies in sea water and lake water on interactions of trace metals with humic substances isolated from marine and estuarine sediments. Mar. Chem. *15* (1984) 217–230, 231–249.

55   Schaule, B., and Patterson, C. C., The occurrence of lead in the Northeast Pacific and effects of anthropogenic input, in: Lead in the Marine Environment, pp. 31–43. Eds M. Branica and Z. Konrad. Pergamon Press, Oxford 1980.
56   Schmidt, D., Comparison of trace heavy metals from monitoring in the German Bight and the southwestern Baltic Sea. Helgoländer Meeresunters. *33* (1980) 576–586.
57   Schmidt, D., Freimann, P., and Zehle, H., Changes of trace metal levels in waters through the coastal zone of the German Bight (North Sea). Symp. on contaminant fluxes through the coastal zone, Nantes, France, 1984.
58   Sipos, L., Rützel, H., and Thijssen, T. H. P., Performance of a new device for sampling sea water from the sea bottom. Thalassia jugosl. *16* (1980) 89–94.
59   Spivack, A. J., Huested, S. S., and Boyle, E. A., Copper, nickel, and cadmium in the surface waters of the Mediterranean, in: Trace Metals in Sea Water, pp. 505–512. Eds C. S. Wong, E. Boyle, K. W. Bruland, J. D. Burton and E. D. Goldberg. Plenum Press, New York – London 1983.
60   Stoeppler, M., and Brandt, K., Comparative studies on trace metal levels in marine biota. II. Trace metals in krill, krill products and fish from the Antarctic Scotia Sea. Z. Lebensmittelunters.-Forsch. *169* (1979) 95–98.
61   Stoeppler, M., and Nürnberg, H. W., Comparative studies on trace metal levels in marine biota, III. Typical levels and accumulation of toxic trace metals in muscle tissue and organs of marine organisms from different European seas. Ecotox. envir. Safety *3* (1979) 335–351.
62   Valenta, P., Simoes-Goncalves, M. L. S., and Sugawara, M., Voltammetric studies on the speciation of Cd and Zn by amino acids in sea water, in: Complexation of Trace Metals in Natural Waters, pp. 357–366. C. J. M. Kramer and J. C. Duinker, Martinus Nijhoff/W. Junk Publ., The Hague – Boston – Lancaster 1984.
63   Whitfield, M., The salt sea – accident or design?, New Sci. *94* (1982) 14–17.

Cadmium in sediments

by U. Förstner

Arbeitsbereich Umweltschutztechnik, Technische Universität Hamburg-Harburg, Eissendorfer Strasse 38, D–2100 Hamburg 90 (Federal Republic of Germany)

Introduction

Compared to the analyses of hydrous phases, the investigations on solid substances has only recently become a major subject of interest in the research of aquatic systems. Even so, present activity in sediment analysis is so great that the overall effort is quite comparable to the study of water and biological sample material.

The sediment approach has a number of perspectives: First, sediments are an expression of the condition of a water system[40]. They can reflect the current quality of the system as well as the historical development of certain hydrologic and chemical parameters. Comparitive analysis of the total concentrations of longitudinal profiles and sediment cores is performed to determine metal anomalies in zones of mineralization as well as from pollution sources. The study of dated sedimentary cores has proved to be especially useful as it provides a historical record of the various influences on the aquatic system by indicating both the natural background levels and the man-induced accumulation of elements over an extended period of time. Two examples are given in figure 1 from marginal seas of northern Germany; one from the German Bight, North Sea[12], the second from the Kieler Bucht, Baltic Sea[7]. In both examples the natural background in the deeper section of the cores for cadmium is approximately 0.25–0.30 mg/kg, that is, similar to the background values determined by other methods, e.g., average shale composition, fossil sediments from typical environments, and recent deposits from 'undeveloped' aquatic systems. From that 'background' the cadmium-concentrations increase about 7–8 times to reach their present levels in the upper layers of the core profiles.

Sediment analysis is used for selecting of critical sites for routine water sampling for contaminants that, upon being discharged to surface waters, do not remain soluble, since they are rapidly adsorbed by particulate matter and may thus escape detection by water analysis. Thus, sediment data play an increasing role within the framework of environmental forensic investigations[25], particularly in those cases, in which a short-term or past pollution event is not or only insufficiently traceable from water analysis. Here, however, the effects of different grain size must be considered[13,14].

In addition, sediments are increasingly recognized as a pollutant proper and as a carrier and possible source of contaminants in aquatic systems as well as with respect to the biological effects of polluted solid materials on agricultural land. Metals are not necessarily fixed permanently by the sediment, but may be recycled via biological and chemical agents, both within the sedimentary compartment and also back to the water column. This is especially true for 'dredged materials', which threaten not only organisms but also the quality of water. Human activities promote the accumulation of polluted sediments and the resulting increased maintenance dredging results in a high amount of contaminated sediments for which safe disposal sites on land or in the waters have to be found. An example of both the quantity and quality problems arising from sediment accumulations is given in figure 2 from Rotterdam Harbor[29], where the increase in cadmium concentration in Rhine sediments (100-fold in 80 years; note logarithmic scale!) is due to increased industrial use and the increase in annual dredging rates is due to harbor extension and maintenance.

Sources and distribution of cadmium in river sediments

Towards the end of the 1960's, the possibilities of applying geochemical methods of exploration in water quality assessment was recognized both in North America as well as in Europe. Investigations on suspended matter and sediments by Turekian and Scott[37] in the Susquehanna River revealed signs of the characteristic pollution sources of industrial and communal emittants. In Europe, the studies of De Groot and coworkers[5,6] and Hellmann and coworkers[17,18] centred around metal investigations on the Rhine River and its tributaries, as well as on the behavior of trace sediment-associated metals in the mixing zone of fresh water and sea water[11]. Meanwhile hundreds of similar investigations have shown the usefulness of sediment analyses in monitoring and assessment of pollution sources[14].

Table 1 compiles characteristic examples of river sediment studies on cadmium, which were carried out during the last 10 years: Strong point-source emissions of cadmium from industrial plants have been recorded in sediments of the Hudson River and in a river of the Hitachi area, northeast of Tokyo. Maximum concentrations of 3000–50,000 mg Cd/kg sediment occurred near the effluent of a nickel-cadmium battery plant at Foundry Cove near Cold Springs, New York[2]; up to 368 mg Cd/kg was measured in the sediments of the irrigation ditch of a Braun tube factory in Japan[1]. Strong contamination by cadmium was detected in sediments from European rivers draining highly industrialized areas, for example, in the tributaries and reservoirs in the vicinity of Miasteczko Slaskie, the zinc and lead work on the upper reaches of Mala Panew River (Poland), in the Voglajna River near Celje in Slovenia (Yugoslavia), in the Ginsheimer Altrhein near Mainz (FR Germany) and in the Meuse and Vesdre Rivers at Liège (Belgium). There is particularly strong cadmium pollution in the catchment area of River Tawe in the lower Swansea Valley, Wales, where coal has been mined since the 14th century, providing the basis for smelting and other industries[39]. The major source of strong accumulations of cadmium in the lower Neckar River, a tributary of the Rhine, was a pigment dye production plant that released (until February 1973) an annual total of approximately 10–20 tons of cadmium in dissolved and particulate form into the river; because of the numerous lock reservoirs in sections of the river, where highly polluted sediments settle, regeneration necessarily proceeds at a very slow rate[35].

In mineralized areas, particularly in regions of sulfidic lead-zinc mineralizations, significant accumulation of cadmium takes place in the river sediments. The area around Coeur d'Alène River in Idaho, the Tennessee River near Knoxville, many rivers in Wales, southeast England, and especially the Takahara River/Jintsu River region in Japan, in which the catastrophic Itai-Itai disease occurred are examples of the lead-zinc mining effect. The discharge of effluents from tin and

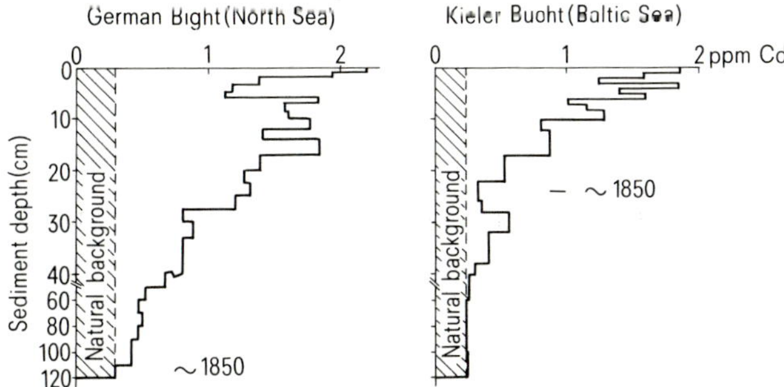

Figure 1. Chronological development of cadmium concentrations in the North Sea and Baltic Sea as derived from analyses of sediment cores from the German Bight[12] and Kieler Bucht[7] (from: Förstner[9]).

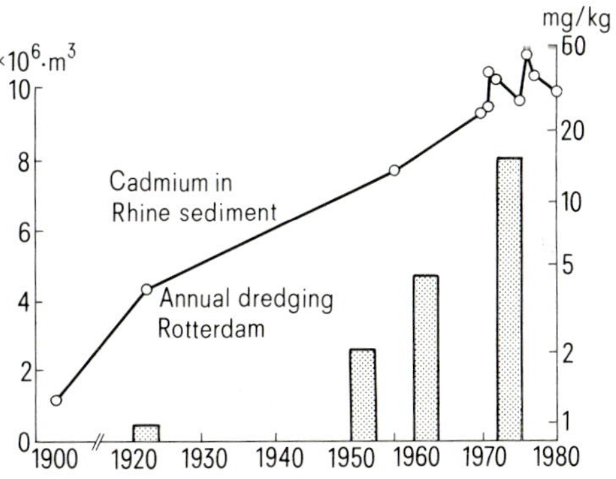

Figure 2. Increase of Cd concentrations in sediments and of dredging activities in Rotterdam Harbor[29].

Table 1. Cadmium in polluted river sediments (examples). For references see Förstner and Wittmann[14]

Sediment source	Cadmium (ppm)	Source	Reference
North America			
Susquehana River	1.68		Malo (1977)
Harrisburg, PA			
Grand River, MI	3.5	Domestic effluents	Fitchko and Hutchinson (1975)
Grand Calumet River, IN	9.7		Hess and Evans (1972)
	3.1–7.9	Domestic effluents	Romano (1976)
Murderkill River, DE	0.8–8.7		Bopp et al. (1973)
Illinois River	2.0		
	(0.2–12.1)		
Rideau River, Ont.	0.3–15	Mixed effluents	Agemian and Chau (1977)
Lake Cayuga tributaries	15.6		Kubota et al. (1974)
Saginaw River, MI	28		Hess and Evans (1972)
Coeur d'Alene River	Max. 80	Mine effluents	Maxfield et al. (1974)
Milwaukee River, WI	16.6	Industrial effluent	Fitchko and Hutchinson (1975)
	Max. 149		
Tennessee River	Max. 227	Mine effluents	Perhac (1972)
Los Angeles River, CA	860	Sewage effluent	Chen et al. (1974)
Hudson River estuary, NY	2.3	Ni-Cd-battery	Vaccaro et al. (1972)
Foundry Cove, NY	Max. 50,000	factory	Kneip et al. (1974)
South Africa, Australia, Japan			
Gold mine drainage,	0.21	Domestic and mine	Wittmann and Förstner (1976a)
South Africa	(0.05–1.0)	effluents	
Jukskei River, South Africa	0.25–4.9		Wittmann and Förstner (1976b)
Molonglo River, Australia	0.8–3.3	Mining wastes	Australian Government Technical Commission (1974)
Tamar River, Tasmania	3.6	Mine effluents	Ayling (1974)
	(< 0.1–6.0)		
South Esk River, Tasmania	Max. 153	Mine effluents	Tyler and Buckney (1973)
Tama River, Tokyo	0.7–9.8		Suzuki et al. (1975)
Jintsu River, Toyama Pref.	3.27	Mine effluents	Goto (1973)
	(0.16–5.0)		
Takahara River	121	Mine effluents	Kiba et al. (1975)
(near Kamioka mine)	(4.1–238)		
Rivers around Himeji City	0.56–10.4		Azumi and Yoneda (1975)
(W of Osaka)	Max. 129		
Rivers in the Hitachi area, northeast Tokyo	Max. 368	Braun tube factory	Asami (1974)
Israel and Europe			
Gadura River	Max. 123	Battery factory	Kronfeld and Navrot (1975)
(Bay of Haifa, Israel)			
Lake Geneva tributaries,	1.4		Vernet (1976)
Switzerland	(0.09–12.4)	Industrial effluents	Viel et al (1978)
Upper Rhône, Switzerland	0.1–73		Ribordy (1978)
Elbe, FRG	2.9–19.9		Lichtfuss & Brümmer (1977)
Sajo River, Hungary	Max. 20		Literathy and Laszlo (1977)
Blies, Saar, FRG	0.5–24.0	Industrial effluents	Becker (1976)
Bavarian rivers, FRG	< 0.05–29.2	Industrial effluents	Bayerische Landesanstalt für Wasserforschung (1977)
Main River, FRG	17–151		Schleichert and Hellmann (1977)
Ginsheimer Altrhein, FRG	2–95	Industrial effluents	Laskowski et al. (1975)
Neckar River, FRG	Max. 320	Pigment factory	Förstner and Müller (1974)
River Conway, GB	21	Mine effluents	Thornton et al. (1975)
mineralized areas	(3-95)		
River Tawe, GB	Max. 355	Metal processing	Vivian and Massie (1977)
Sava basin, Yugoslavia	Max. 66	Industrial effluents	Štern and Förstner (1976)
Voglajna River			
Stola River, Poland	Max. 116	Mine effluents	Pasternak (1974)
Meuse River, Belgium	Max. 230	Industrial effluents	Bouquiaux (1974)
Vesdre River, Belgium (near Liège)	Max. 430		Bouquiaux (1974)

tungsten mines has caused severe pollution of the South Esk River in northeastern Tasmania.

Cadmium enrichments between 5 and ca. 10 mg/kg in river sediments are often caused by communal wastewater effluents, sometimes with an industrial component, for instance from the electroplating industry. In the Milwaukee River, Wisconsin, this effect for the most part is dominant over domestic sources: effluents from brewing, tanning, incineration, the chemical industry, foundries, metal works and manufacturing are common sources of cadmium[8]. Particularly high concentrations of cadmium have been analyzed by Chen et al.[3] in suspended silts in the dry weather flows from the urbanized Los Angeles area. By analyzing sediments collected since 1922 and from polders reclaimed in the 15th and 18th centuries Salomons and De Groot[30] were able to show that pollution of the lower Rhine is still increasing for cadmium (fig.2), whereas mercury, arsenic and lead pollution is decreasing.

It is a general experience, that as a result of increasing efforts to reduce the wastewater input the metal discharges in highly polluted rivers are now decreasing after a maximum in the early 1970's. From a review of data from the Rhine River at the German-Dutch border it is shown by Malle and Müller[24] that the suspended solid contents did hardly change since 1973, while the contents of Cr, Cu, Zn, Cd, Hg and Pb are retrograde; the decrease in the Cr and Hg contents mainly occurred in the suspended solids, Zn and Cd concentrations were found to be reduced in the liquid phase. It should be noted, however, that of the total amount of metals transported by the rivers Rhine and Meuse, about $\frac{2}{3}$ accumulates in the Netherlands, which in fact acts as an effective 'treatment plant' for these river discharges[34]. This situation may be regarded as beneficial for the North Sea, but poses problems for the management of inland waters and the land-fill areas in the Netherlands (fig. 2).

Mobilisation of cadmium from sediments

Trace metals temporarily immobilized in the suspended matter and in bottom sediments of aquatic systems may be released as a result of physicochemical changes, such as a) increased salinity, b) an alteration in the redox conditions, c) lowering of pH, and d) increased input of organic chelators. In figure 3 examples are given for these parameters from laboratory experiments.

a) Increased salinity at the river/sea interface

Experiments performed by Van der Weijden et al.[38] with artificial seawater indicate partial desorption of cadmium from particulate matter, presumably by inorganic complex formation; preferential remobilization is probably due to the high stability of the cadmium-chloro-complex. Similar developments have been described by Rohatgi and Chen[27] from experiments with sewage sludge in seawater solutions; 93% of the original cadmium content of sewage particles was released after 4 weeks' treatment. The simultaneous influence of chlorinity and pH on the adsorption was studied by Salomons[28]; figure 3A shows that with an increase in chlorinity the amount of added metal (5 µg Cd/l) which adsorbs onto the suspended sediment (100 mg/l) decreases; with an increase in the pH the adsorption increases. It has been demonstrated by Salomons et al.[34] that already at chloride levels of 200 mg/l – which are 'normal' for the lower Rhine River – the adsorption of cadmium on the suspended matter is affected. In figure 3B the combined influence of chlorinity and suspended matter on the adsorption of cadmium is shown; already a small increase in the chlorinity causes a large decrease in the amount of metals removed from solution.
These sorption effects obviously are not fully reversible, as has been demonstrated by Salomons[28] from desorption experiments (table 2).

Table 2. Percentage of cadmium not released from suspended matter in river water after treatment with NaCl or ammonium acetate

Adsorption period	1 day	3 days	8 days	24 days	60 days
NaCl (35‰)	24%	30%	33%	37%	40%
1 N ammonium acetate	31%	36%	38%	45%	52%

Apparently, the adsorbed cadmium becomes more strongly bound to the sediment with increase in time.

b) Redox changes

A change in redox conditions is usually caused by an increased input of nutrients (reducing effects) or by the land disposal of strongly polluted sediments (oxidizing effects). Oxygen deficiency in sediments leads to an initial dissolution of manganese oxides, followed by that of hydrous iron oxides. Since these metals are readily soluble in their divalent states, any coprecipitates with metallic coatings become partially remobilized. On the other hand, it was found from dredge experiments that larger concentrations of cadmium were released to the water column under oxygen-rich than under oxygen-deficient conditions; under oxidized conditions larger releases of cadmium were measured as salinity increased. The first effect is attributed to the release, upon oxidation, of trace metals bound to sulfide phases; the second observation can be explained by the enhanced formation of soluble inorganic complexes (see section a)). In figure 3C the effects are described of the redox conditions on the cadmium concentrations in equilibrium solutions from marine mud samples under

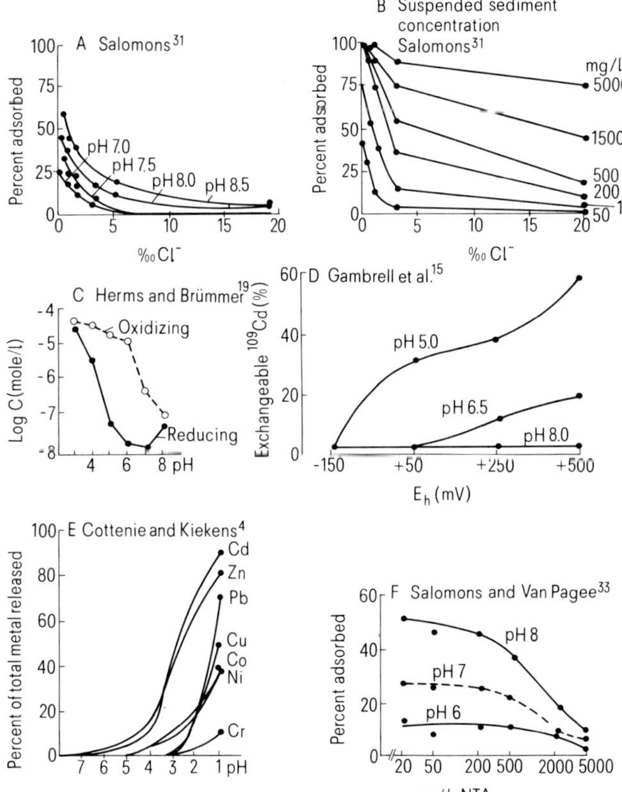

Figure 3. Factors affecting the remobilization of cadmium from solid matter (explanations see text).

different pH-values[18]. In the range between pH 4 and pH 7 the solubility of Cd in the oxidizing, sulfide-free milieu is strongly increased (at pH 6 approximately 1000-fold compared to the reducing, sulfidic conditions). Isotope studies performed by Gambrell et al.[15] with Mississippi River sediments indicated that exchangeable ^{109}Cd levels are significantly increased as cadmium-contaminated sediment is transported from a near-neutral pH, reducing environment to a moderately acid, oxidizing environment (fig. 3D). Under these conditions, Cd levels of subsurface drainage water from upland disposal of dredged materials may be increased, and cadmium availability to plants growing on the material enhanced[15].

c) Acidic waters

A lowering of pH leads to the dissolution of carbonate and hydroxide minerals and (as a result of hydrogen ion competition) to an increased desorption of metal cations. Long-term changes in pH conditions have been observed in waters poor in bicarbonate ions, which are affected by atmospheric SO_2 emissions. From a compilation of Haines[16] it is suggested that lakes with a pH of 4.1–5.3 exhibit 5- to 10-fold higher Cd concentrations in the water phase than lakes with a pH of 6.0–7.8. Cadmium enrichment in acidic mine effluents is significant, but less than metals such as Fe, Mn, Ni and Co. On the other hand the experimental data for step-wise acidification of soil samples (fig. 3E) indicate a higher mobility of Cd and Zn compared to Pb, Cu, and Cr in the pH-range of 3.5–1.5[4].

d) Organic complexing

Interaction with organic substances plays an increasingly important role in the transport of heavy metals in both surface water and groundwater, since the amount of organic complexing material increases further because of secondary sewage treatment effluents. An even more serious impact on heavy metal remobilization from polluted sediments may result (a matter of considerable controvercy during the last few years) from the growing use of synthetic complexing agents (e.g., nitrilotriacetic acid, NTA) in detergents to replace polyphosphates. The data in figure 3F (from Salomons and Van Pagee[33]) indicate that the effect of reduced adsorption of cadmium is even more problematic than the possible (active) remobilization of cadmium, since already at NTA-concentrations of 200 µg/l and at higher pH-values (see pH 8 curve) the normal elimination of dissolved cadmium by fixation upon solid phases is significantly lowered. This could imply a potential danger for drinking water obtained from bank filtration or artificial recharge[36].

Estimation of environmental impact of Cd-polluted sediments

Of the various processes, products, and substrates of metal enrichment in aquatic solids, three major types should be distinguished for the interpretation and assessment of their sources, distribution and environmental effects:

– A rise in pH and oxygen content promotes the formation of metal hydroxides, carbonates and other metal precipitates. Hydrous Fe and Mn oxides constitute significant 'sinks' for heavy metals through the effects of sorption/coprecipitation;
– in waters rich in organic matter, minerals may be solubilized by the combined processes of complexation and reduction; reincorporation of metals into the sediment involves the mechanisms of adsorption, flocculation, polymerization, and precipitation[21];
– metals are transported and deposited as major, minor, or trace constituents in the detrital minerals derived from rocks and soils, in organic residues, and in solid waste material.

Only part of the metals present – mainly the former types – may take part in short-term geochemical processes and/or are bioavailable. For the differentiation of the relative bonding strength of metals in different phases, sequential extraction procedures have been developed; while the determination of the relative binding strength seems to pose basically operational problems, the correlation with the biological uptake mechanisms is as yet not satisfactory. A promising approach to bridge the gap between the chemical speciation data of the solid matter and of the biological uptake values is the application of pore water analyses[32].

Examples for the solid phase partition of Cd have been given in table 3, according to an extraction sequence developed by Förstner and Calmano[10]. Compared to other metals, such as Pb and Cu, cadmium is characteristically enriched in the more mobile fractions 'cation exchange' and 'easily reducible phases', and is, therefore, more mobile than most of the other heavy metals. The data from table 3 also suggest a general decrease in the residual bonding form, that is, the predominantly inertly fixed cadmium content, as the anthropogenic metal enrichment increases. Such species differentiations can be used for the estimations on the remobilization of metals under changing environmental conditions and on the potential uptake by biota:

In the estuarine environment the 'exchangeable fraction' might be affected in particular; however, changes of pH and redox potential could also influence other easily extractable phases, e.g., carbonates and manganese oxides. Lowering of pH will affect, according to its strength, the 'exchangeable', then the 'easily reducible' and in case parts the 'moderately reducible fraction', the latter consisting of Fe-oxyhydroxides in less crystallized forms. Redox changes concern under 'postoxic' conditions, during reduction of, for example, nitrate and manganese oxide, the 'easily reducible fraction', i.e., metals associated with manganese oxides and partly amorpous Fe-oxyhydrates; in strongly reducing environments, e.g., in highly polluted sediments, the 'moderately reducible fraction' is affected too, especially the iron compounds present as coatings[20]. The effects on organically bound metals are more complicated; however, it has been argued that this fraction is highly susceptible to environmental changes, especially dur-

Table 3. Chemical forms of cadmium in aquatic solids

Sediment fraction[a–e]	Weser (1)	Estuary (2)	Rhine River (3)	Rotterdam Harbor (4)	Neckar River (5)
Cation exchangeable[a]	8%	18%	27%	12%	13%
Easily reducible[b]	32%	49%	42%	61%	64%
Moderately reducible[c]	45%	14%	2%	13%	10%
Organic/sulfidic[d]	12%	17%	17%	12%	12%
Residual fraction[e]	3%	2%	12%	2%	1%
Total Cd (mg/kg)	3.0	4.3	9.0	18.1	33.6

(1) Dredged sediment; (2) suspended matter; (3) station Wesel (Dutch/German border); (4) Broekpolder; (5) lock reservoir Lauffen.
[a] 1 M ammonium acetate, 1:20 solid/solution ratio, 2 h shaking time; [b] 0.1 M hydroxylamine hydrochloride +0.01 M nitric acid, 1:100, 24 h; [c] 0.2 M ammonium oxalate +0.2 M oxalic acid, pH 3, 1:100, 24 h; [d] 30% hydrogen peroxide +1 M ammonium acetate, pH 2.5; 1:100, 24 h; [e] hydrofluoric/perchloric acid digestion.

ing early diagenetic reaction, where recycling of mineralized organic matter and pore-fluid transfer processes are controlling the dynamics of pollutants and nutrients in sediments.

The still somethwat unsatisfactory situation with respect to the assessment of the quantitative extent of bioavailable element concentrations from solid speciation data has been explained by Luoma and Bryan[22,23] as being due mainly to the effect of competition between more or less strong adsorption sites in the solid substrate and selective mechanisms of the metal uptake by the different organisms which cannot simply be drived from the chemical extraction data. On the other hand, the usefulness of a differentiated approach on the interactive processes between water/ biota and – even if only operationally – defined solid phases has been clearly evidenced. The possible environmental implications, e.g., during dredging operations, after land disposal of waste material, from acid precipitation, for redox changes in the subsoil, and organic complexing near sewage outfalls, which may lead to a partial remobilization of the particle-bound cadmium and other metals, can be qualitatively estimated, particularly when the hydrochemical conditions are known. With respect to the more important environmental parameters, such as pH and redox potential[26] a weak acid-reducing extractant offers a relatively good indication of the mobile and potentially bioavailable fraction of transition metals, equivalent to the non-apatite inorganic phosphorous fraction of the solid P-compounds[31]. The results of these extractions as well as the methods used in soil science, e.g. treatment with dilute EDTA or DTPA solutions, show a good correlation with the data from pot experiments, i.e., that cadmium is more readily accessible for plant uptake than most of the other potentially toxic trace metals.

1 Asami, T., Environmental pollution by cadmium and zinc discharged from a braun tube factory. Ibaraki Daigaku Nogaukubu Gakujutsu Hokaku 22 (1974) 19–23.
2 Bower, P.M., Simpson, H.J., Williams, S.C., and Li, Y.H., Heavy metals in the sediments of Foundry Cove, Cold Spring, New York. Envir. Sci. Technol. 12 (1978) 683–687.
3 Chen, K.Y., Young, T.K., Jan, T.K., and Rohatgi, N., Trace metals in waste water effluents, J. Water Pollut. Control Fedn 46 (1974) 2663–2675.
4 Cottenie, A., and Kiekens, L., Beweglichkeit von Schwermetallen in mit Schlamm angereicherten Böden. Korresp. Abwasser 4 (1981) 206–210.
5 De Groot, A.J., Mobility of trace elements in deltas. Trans. Comm. II and IV. Int. Soc. Soil Sci., Aberdeen 1966, pp. 267–297.
6 De Groot, A.J., Goeij, J.J.M., and Zegers, C., Contents and behaviour of mercury as compared with other metals in sediments from the rivers Rhine and Ems. Geol. Mijnbow 50 (1971) 393–398.
7 Erlenkeuser, H., Suess, E., and Willkomm, H., Industrialization affects heavy metal and carbon isotope concentration in recent Baltic Sea sediments. Geochim. cosmochim. Acta 38 (1974) 823–842.
8 Fitchko, J., and Hutchinson, T.C., A comparitive study of heavy metal concentrations in river mouth sediments around the Great Lakes. Great Lakes Res. 1 (1975) 46–78.
9 Förstner, U., Cadmium in polluted sediments, in: Cadmium in the Environment, vol.1, pp. 305–363. Ed. J.O. Nriagu Wiley-Interscience, New York 1980.
10 Förstner, U., and Calmano, W., Bindungsformen von Schwermetallen in Baggerschlämmen. Vom Wasser 59 (1982) 83–92.
11 Förstner, U., and Müller, G., Schwermetalle in Flüssen und Seen als Ausdruck der Umweltverschmutzung, 225 p. Springer Verlag, Berlin 1974.
12 Förstner, U., and Reineck, H.-E., Die Anreicherung von Spurenelementen in den rezenten Sedimenten eines Profilkernes aus der Deutschen Bucht. Senckenberg. Marit. 6 (1974) 175–184.
13 Förstner, U., and Salomons, W., Trace metal analysis on polluted sediments. I. Assessment of sources and intensities. Envir. Technol. Lett. 1 (1980) 494–505.
14 Förstner, U., and Wittmann, G., Metal Pollution in the Aquatic Environment, 486 p. Springer Verlag, Berlin 1979.
15 Gambrell, R.P., Khalid, R.A., Verloo, M.G., and Patrick, W.H., Transformations of heavy metals and plant nutrients in dredged sediments as affected by oxidation reduction potential and pH. Dredged Material Research Program, U.S. Army Corps of Engineers, Vicksburg, MS, Rept. D-77-4 1977; 309 p.
16 Haines, T.A., Acidic precipitation and its consequences for aquatic ecosystems: A review. Trans. Am. Fish Soc. 110 (1981) 669–707.
17 Hellmann, H., Die Charakterisierung von Sedimenten auf Grund ihres Gehaltes an Spurenelementen. Dt. gewässerk. Mitt. 14 (1970) 160–164.
18 Hellmann, H., and Griffatong, A., Herkunft der Sinkstoffablagerungen in Gewässern (1. Mitt.: Chemische Untersuchungen der Schwermetalle). Dt. gewässerk. Mitt. 16 (1972) 14–18.
19 Herms, U., and Brümmer, G., Einfluss der Redoxbedingungen auf die Löslichkeit von Schwermetallen in Böden und Sedimenten. Mitt. Dt. bodenk. Ges. 29 (1979) 533–544.
20 Jenne, E.A., Trace element sorption by sediments and soils – sites and processes, in: Symposium on Molybdenum, vol.2, pp. 425–553. W. Chapell and K. Petersen (eds.). Marcel Dekker, New York 1976.
21 Jonasson, I.R., Geochemistry of sediment/water interactions of metals, including observations on availability, in: The Fluvial Transport of Sediment-Associated Nutrients and Contaminants, pp. 255–271. Eds H. Shear and A.E. Watson. IJC/ PLUARG, Windsor/Ont. 1977.
22 Luoma, S.N., and Bryan, G.W., Factors controlling the avail-

ability of sediment-bound lead to the estuarine bivalve *Scobicularia plana*. J. mar. Biol. Assoc. U.K. *58* (1978) 793–802.
23 Luoma, S.N., and Bryan, G.W., Trace metal bioavailability: Modeling chemical and biological interactions of sediment-bound zinc, in: Chemical Modeling in Aqueous Systems. Am. chem. Soc. Symp. Ser. *93* (1979) 577–609.
24 Malle, K.-G., and Müller, G., Metallgehalt und Schwebstoffgehalt im Rhein. Z. Wasser Abwasser Forsch. *15* (1982) 11–15.
25 Meiggs, T.O., The use of sediment analysis in forensic investigations and procedural requirements for such studies, in: Contaminants and Sediments, vol. 1, pp. 297–308. Ed. R.A., Baker. Ann Arbor Sci. Publ., Ann Arbor, Mich. 1980.
26 Pickering, W.F., Selective chemical extraction of soil components and bound metal species. CRC Critical Reviews in Analytical Chemistry, Nov. 1981, pp. 233–266.
27 Rohatgi, N.K., and Chen, K.Y., Fate of metals in wastewater discharge to the ocean. J. Envir. Enging Div. ASCE *102* (1976) 675–685.
28 Salomons, W., Adsorption processes and hydrodynamic conditions in estuaries. Envir. Technol. Lett. *1* (1980) 356–365.
29 Salomons, W., and Eysink, W.D., Pathways of mud and particulate metals from rivers to the Southern North Sea. Proc. Conf. Holocene Marine Sedimentation in the North Sea. Blackwell Sci. Publ. 1981.
30 Salomons, W., and De Groot, A.J., Pollution history of trace metals in sediments, as affected by the Rhine River, in: Environmental Biogeochemistry, vol. 1, pp. 149–162. Ed. W.E. Krumbein. Ann Arbor Sci. Publ., Ann Arbor, Mich. 1978.
31 Salomons, W., and Förstner, U., Trace metal analysis on polluted sediments. II. Evaluation of environmental impact. Envir. Technol. Lett. *1* (1980) 506–517.
32 Salomons, W., and Förstner, U., Metals in the Hydrocycle. Springer Verlag, Berlin 1984, in press.
33 Salomons, W., and Van Pagee, J.A., Prediction of NTA levels in river systems and their effect on metal concentrations. Proc. Intern. Conf. Heavy Metals in the Environment, Amsterdam 1981; pp. 694–697.
34 Salomons, W., Van Driel, W., Kerdijk, H., and Boxma, R., Help! Holland is plated by the Rhine. Proc. Exeter Symp. IAHS, Effects of Waste Disposal on Groundwater, No 139, 1982, pp. 255–269.
35 Schoer, J., and Förstner, U., Die Entwicklung der Schwermetall-Verschmutzung im mittleren Neckar. Dt. gewässerk. Mitt. *24* (1980) 153–158.
36 Schöttler, U., Das Verhalten von Schwermetallen bei der Langsamsandfiltration. Z. dt. geol. Ges. *126* (1975) 373–384.
37 Turekian, K.K., and Scott, M., Concentration of Cr, Ag, Mo, Ni, Co, and Mn in suspended material in streams. Envir. Sci. Technol. *1* (1967) 940–942.
38 Van der Weijden, C.H., Arnoldus, M.J.H.L., and Meurs, C.T., Desorption of metals from suspended material in the Rhine Estuary. Neth. J. Sea Res. *11* (1977) 130–145.
39 Vivian, C.M.G., and Massie, K.S., Trace metals in waters and sediments of the River Tawe, South Wales, in relation to local sources. Envir. Pollut. *14* (1977) 47–61.
40 Züllig, H., Sedimente als Ausdruck des Zustandes eines Gewässers. Schweiz. Z. Hydrol. *18* (1956) 7–143.

Removal of cadmium from wastewaters

by C. P. C. Poon

University of Rhode Island, Kingston (Rhode Island 02881, USA)

Introduction

There has been increasing concern over the discharge of heavy metals into the environment as evidenced by actions taken by pollution regulating agencies in various industrialized nations to impose severe discharge limits for certain heavy metals. This paper describes the source and quantity of cadmium in waste waters. The effluent limitation of cadmium is presented, followed by discussions of various treatment methods including chemical precipitation and flotation, ion exchange, adsorption osmosis as well as other effluent polishing techniques applied to both industrial discharges and municipal waste waters. Finally, several methods of cadmium recovery from industrial discharges are discussed.

I. *Sources of wastewater and cadmium concentration*

According to Stubbs[50], a total of 11,444 metric tons of cadmium were processed by 5 major industrial nations (France, the Federal Republic of Germany, Japan, the United Kingdom, and the United States) in 1974; the end products were as follows: plating 3830 t (34%), pigments 2975 t (26%), stabilizers 1676 t (15%), batteries 1646 t (14%), others including alloys 1317 t (11%). Cadmium is also used to some extent in the production of television tube phosphorus, golf courses

fungicides, rubber curing agents and nuclear reactor sheilds and rods, according to an Environmental Science report 1971[13]. High concentrations of cadmium in lead mine drainage have also been reported by McKee et al.[36].

The quality and quantity of cadmium-bearing industrial discharges depend upon the industrial source and

Table 1. Cadmium concentrations reported for industrial and municipal wastewaters

Sources and process	Cd concentration (mg/l)	Reference
Automobile heating control manufacturing	14–22	Gard et al.[20]
Plating rinse waters Automatic barrel Zn and Cd	10–15	Lowe[34]
Mixed manual barrel and rack	7–12	
Plating rinse waters (large installations)	15 average 50 maximum	Pinkerton[44]
Plating rinse waters 0.5 gph dragout	48	Nemerow[37]
2.5 gph dragout	240	
Bright dip and passivation baths	2000 5000	Lowe[34]
Lead mine acid drainage	1000	McKee et al.[36]
Municipal raw wastewater	0.008 average 0.002–0.016	Culp et al.[10]

the degree of treatment imposed. Patterson[40,43] reported cadmium concentrations in various waste waters ranging from 0 to 1000 mg/1, as summarized in table 1. Also included in table 1 is the cadmium concentration found in municipal waste water reported by Culp et al.[10].

Cadmium exists in wastewater in many forms, including soluble, insoluble, inorganic, metal organic, reduced, oxidized, free metal, precipitated, adsorbed, and complexed forms. Treatment processes for cadmium removal must be selected to remove the existing form, or cadmium must be converted to a suitable form compatible with the removal process. The following sections discuss the effluent limitations and methods available for the removal and/or recovery of cadmium from wastewaters along with the treatment efficiency and treatment cost, when such data are available.

II. *Effluent limitations*

The maximum allowable discharge concentration for cadmium was at one time set a 10 µg/1 according to Federal Register 1975[14]. However, guidelines have since been developed for effluent cadmium concentrations for different categories of industrial discharges[42] as shown in table 2. These guidelines were developed from performance data obtained from various industrial wastewater treatment plants. Variations of wastewater characteristics from industrial categories or subcategories cause differences in cadmium's susceptibility to the applied treatment technology; therefore, limitation guidelines are not all alike.

Table 2. Proposed and promulgated BATEA effluent limitation guidelines – 30-day average*

Industry category	U.S. EPA development document	Cadmium (mg/l)
Inorganic chemicals	440/1-79/007	
Chlor-alkali mercury cells		0.05
Titanium dioxide-sulfate process		0.15
Titanium dioxide-chloride		
ilmenite process		0.10
Chrome pigments		0.19
Copper sulfate		0.05
Iron and steel	440/81-80/024b	0.10
Steam electric power	440/1-80/029b	1.00
Range: Minimum		0.05
Maximum		0.19
Median value		0.10

* Adapted from Patterson[43].

Pretreatment discharge limitations were also developed for wastewater discharged into publicly owned wastewater treatment works. The maximum allowable concentration for cadmium was 2.0 mg/1 in 1976, regardless of industrial categories, according to Olson[38]. Pretreatment standards for heavy metals were later developed for different industrial categories. As an example, the cadimum pretreatment standards for the electroplating industry[15] are given below:

	Daily maximum Cd concentration	4-day average Cd concentration
Process water, flow less than 10,000 gallons/day (gpd)	1.2 mg/l	0.7 mg/l
Process water, flows equal to or greater than 10,000 gallons/day (gpd)	1.2 mg/l	0.7 mg/l

Interest in wastewater reuse is prevalent in many countries. The Orange County Water District of the state of California, USA, has conducted studies in wastewater reclamation and ground water recharge through injection wells since 1965. Public health considerations are the ultimate concern in reuse of wastewater. Stringent standards have been set on the injection water by both the California Department of Public Health and the California Regional Water Quality Control Board[3]. The cadmium concentration is not to exceed 0.01 mg/1 in the injection water. This effluent limitation is consistent with the U.S. Public Health Service Drinking Water Standards.

As seen previously, the effluent limitations for cadmium varies widely from 0.01 to 1.2 mg/1 depending on industrial category, where the effluent is discharged to, and whether or not the treated effluent is reused. Nevertheless, all effluent limitations can be met by the use of a treatment system employing one or more of the currently available treatment processes.

III. *Chemical precipitation*

a) *Hydroxide precipitation*

Metal hydroxide precipitation with coagulants is known to be effective for metal removal, including cadmium. The solubility of the metal hydroxide is determined by solubility products (K_s) and by equations relating the metal hydroxide solid species in equilibrium with soluble free metal ion or metal-hydroxide species. Solubility product constants for a number of metals have been published by Feitknecht[16]. However, because of precipitate aging, insufficient time for the soluble metal species to establish equilibrium with the precipitated metal, incomplete solids separation, or coprecipitation and adsorption effects in wastewater, solubility products provide only a general guide to metal residuals to be expected in practice. The following cadmium hydroxide solubility products constants are given by Feitknecht[17] and Bard[4].

	Aged precipitate	Fresh precipitate
$K_{s0} = -[Cd^{2+}][OH^-]^2$	Log $K_{s0} = -14.4$	-13.7
$K_{s1} = -[Cd(OH)^+][OH^-]$	Log $K_{s1} = -11.2$	-9.5
$K_{s2} = -[Cd(OH)_2^0]$	Log $K_{s2} = -6.1$	-5.3
$K_{s3} = -[Cd(OH)_3^-]/[OH^-]$	Log $K_{s3} = -5.3$	-4.6
$K_{s4} = -[Cd(OH)_4^{2-}]/[OH^-]^2$	Log $K_{s4} = -5.6$	-4.9

pH of minimum cadmium hydroxide solubility is 11.5.

Jenkins[28] reports that freshly precipitated cadmium hydroxide leaves approximately 1 mg/l of cadmium ion in solution at pH 8, but only 0.1 mg/l at pH 11. Similarly, Anderson[1] reports an effective removal of cadmium at pH 9.7. In a study of cadmium precipitation using hydrated lime for the treatment of an electric arc furnace blowdown at a steel mill, Brantner[5] reports that lime precipitation at pH 9.9 and clarification removed total cadmium from 1.66 mg/l to 1.15 mg/l while soluble cadmium was removed from 1.23 mg/l to 1.12 mg/l. Nearly all of the residual cadmium in both clarifier and filter effluents was soluble, indicating that the residual level was limited by the solubility rather than the effectiveness of solid-liquid separation.

Co-precipitation with other metal hydroxides can be more effective and affect the pH at which the minimum cadmium solubility occurs. Both Anderson[1] and Weiner[53] indicate co-precipitation with iron hydroxide at pH 8.5 resulting in near complete removal of cadmium. It was found by El-Gohary[12] that in alkaline precipitation using sodium hydroxide, co-precipitation with 100 mg/l of ferric chloride at pH 9.0 removed 97% of cadmium while co-precipitation with 100 mg/l of aluminum sulphate at pH 9.0 removed only 91.5% of cadmium. Using lime in alkaline precipitation, the same report indicates that a co-precipitation with 100 mg/l of either ferric chloride or aluminum sulphate resulted in better than 98% removal of cadmium. With coagulant aid addition (Separan NP-10) at 1–2 mg/l to lime precipitation of a plating waste, Hanson[22] reported an effluent with pH 9.0 and cadmium concentration of 0.54 mg/l. Since the effluent also contained 2 mg/l of suspended solids, it was not known how much of the residual cadmium was in the particulate form and how much in the dissolved form. The impact of solids removal on effluent concentration has been reported by Chalmers[7]. The effluent cadmium concentration was 0.7 mg/l after cyanide-chromium treatment plus neutralization and sand filtration. The effluent cadmium concentration was further reduced to 0.08 mg/l using 10-h settling followed by paper press filtration.

Cadmium precipitation is inhibited in the presence of complexing agents such as cyanide. Cyanide removal is, therefore, imperative prior to cadmium precipitation from plating baths and plating rinse waters containing cyanide. In this regard, the development of a cyanide-free cadmium plating bath with acceptable product quality can be an answer to simplify the treatment needs.

It was recently found by Patterson[43] that carbonate alkalinity initially present in the wastewater, or induced into the wastewater as a result of treatment, could have a significant effect on alkaline precipitation of cadmium. A reduction in cadmium solubility with increasing carbonate concentration was found at treatment pH values below pH 11. Cadmium solubili-8.4, soluble cadmium was 1.7 mg/l at total carbonate (C_T) of 6.7 mg/l, versus 0.6 mg/l at C_T of 9.8 mg/l. At pH 7.4–7.6, soluble cadmium ranged from 62 to 2.6 mg/l as C_T varied from 3.7 to 5.7 mg/l. Carbonate, therefore, adds to a long list of interfering substances and circumstances affecting the cadmium precipitation process. The interfering characteristic may reflect site-specific or geographical factors as well as factors associated with a particular industrial category. The nature and degree of interference should be fully assessed in each specific case before any effective treatment system can be designed using the hydroxide precipitation process for cadmium removal.

b) Carbonate precipitation

Carbonate precipitation of cadmium can be successful. A laboratory study by Patterson[41] shows that by adding sodium carbonate, treatment nearly equivalent to that for cadmium hydroxide was obtained with cadmium carbonate ($C_T = 10^{-1.33}$ mole/l, or 2800 mg/l) at pH 8.4 which yielded 1.2 mg/l cadmium compared to 0.2 mg/l for hydroxide treatment at pH 10.4 and 126 mg/l at pH 8.6. Consequently, treatment at lower pH can be achieved with the carbonate system. In addition, lower volumes of denser sludge can be obtained with relative filtration rates approximately twice that of the cadmium hydroxide sludge.
Barntner's study in 1981[5] compared pilot plant tests of carbonate precipitation with hydroxide precipitation of cadmium from an electric arc furnace blowdown. At a Na_2CO_3 dosage of 650 mg/l and a pH ranging from 8.3 to 8.7, the total cadmium concentration was reduced from 1.37 mg/l down to 0.14 mg/l in the clarified effluent. Soluble cadmium concentration was reduced from 0.79 mg/l down to 0.1 mg/l and in another test from 1.2 mg/l down to 0.04 mg/l. The clarified effluent (350 gpd/ft^2 or 14.3 m^3/day-m^2) was polished by filtering through a dual media filter (5 gpm/ft^2 or 204 l/min-m^2) which further improved the effluent quality (0.02–0.06 mg/l total cadmium) comparable to that of the hydroxide precipitation-clarification-filtration system. El-Gohary[12] reported 98% cadmium removal at pH 7.5 using 3500 mg/l of calcium carbonate. Despite the little attention given to the carbonate precipitation process, it is apparent that the process can be equally successful in cadmium removal.

c) Sulfide precipitation

There are many attractive features associated with the sulfide precipitation process in metal removal including cadmium. A high degree of metal removal can be attained over a broad pH range. Cadmium can be effectively precipitated at very low pH because of the solubility of CdS is very low at pH below 4. Figure 1 shows the calculated solubilities of metal sulfides and metal hydroxides. The high reactivity of sulfides shortens the detention time required for the process.

According to Whang[54]. metal sulfide sludge exhibits better thickening properties and dewaterability than metal hydroxide sludge from the hydroxide precipitation process. In addition, metal sulfide sludge is three times less subject to leaching at pH 5 as compared to hydroxide sludge and, therefore, final disposal is easier and safer. Although metal sulfide precipitates are very fine, anionic floculant can be used to facilitate liquid-solid separation.

Scrubber wastes containing high concentrations of heavy metals including cadmium from 10.5 mg/l to 15.0 mg/l were treated with a hydroxide-sulfide precipitation process reported by Schwitzgebel[48]. Lime was added to the wastes to raise the pH to 5.0–6.0. The sulfide precipitation reaction pH was varied between 7.5 and 8.5. The maximum metal separations in the water were obtained with 0.6 times of the stoichiometric quantity of sulfide required to react with $Cd+Cu+Zn+Fe+Pb$ and at a final pH of 8.0. Both sulfide and hydroxide precipitates of metals were formed. In all cases, 98–99.6% of all heavy metals were removed with the effluent (precipitation-clarification – multi media filtration) containing only 0.05–0.08 mg/l of cadmium.

Full scale testing of the sulfide precipitation process was also reported by Schwitzgebel[48] at the Boliden Metall Corporation, Skelleftehamn, Sweden. The wastewater was partially neutralized to pH 2.5–3.0 with NaOH, and then Na_2S was added. The removal of all heavy metals with the exception of zinc was complete. Cadmium concentration after sulfide precipitation-polymer sedimentation-multi media filtration was reduced from 4 mg/l down to 0.01 mg/l.

The parallel testing of 3 precipitation processes to remove metals from an electric arc furnace blowdown as reported by Barntner[5] also indicates the successful application of a hydroxide-sulfide process for metal removal. Ferrous sulfide (FeS) was applied to maintain a dosage of 130 mg/l FeS and pH was maintained at 8.0–8.5 by addition of lime. The FeS dosage of 130 mg/l was selected to be 1.5 times the calculated stoichiometric requirement for precipitation of all the metals. At pH 8.2, total cadmium was reduced from 3.3 mg/l to 0.18 mg/l (clarified effluent) and to 0.06 mg/l (filtered effluent) while soluble cadmium was reduced from 0.4 mg/l to 0.02 mg/l (both classified and filtered effluents). The removal was comparable or better than the hydroxide precipitation and the carbonate precipitation processes.

d) Other chemical processes

Sodium borohydride precipitation. The use of sodium borohydride ($NaBH_4$) as a reducing agent is a cost-effective alternative to remove cadmium from wastewater. The $NaBH_4$ process is irreversible, extremely rapid, and results in nearly total removal of the dissolved metal even at low initial concentration. It also offers the potential to recycle a significant portion of the metal since the metal is chemically reduced to the elemental state, usually as a compact precipitate.

Sodium borohydride may be used as the primary treatment procedure or in conjunction with or following alkaline neutralization. The stoichiometric quantity of $NaBH_4$ required is $\frac{1}{12}$ the weight of cadmium reduced. In actual practice the borohydride to cadmium ratio is usually somewhat higher, since other reducible compounds may react with the borohydride. When a low initial level of dissolved cadmium (2–15 mg/l) is present in the wastewater, thorough mixing of the $NaBH_4$ is essential. Sodium borohydride is generally used as an aqueous caustic solution which is stored and handled in the same manner as 50% caustic soda.

Cook[9] reports successful cases of $NaBH_4$ application in metal removal and/or recovery. One of these cases concerns the removal of silver and cadmium complexes in a wastewater from a lithographic film coating process in Glen Cove, New York. This process water is first pretreated to remove photographic gelatin and dyes as they may interfere with the borohydride reduction and solid-liquid separation steps. After pretreatment, the process water contains 10–120 mg/l of silver and 5–60 mg/l of cadmium. The silver is present in the form of finely divided insoluble silver halides with some soluble silver species present. Ferric chloride is introduced into a 7500-gal (28.4 m^3) reactor followed by caustic soda to adjust to pH 11 prior to borohydride addition. The reactor is stored continuously. At pH 11, substantial amounts of cadmium crystallize and ferric hydroxide also occur, improving the subsequent solid-liquid separation. Sodium borohydride is then added over a 15-min

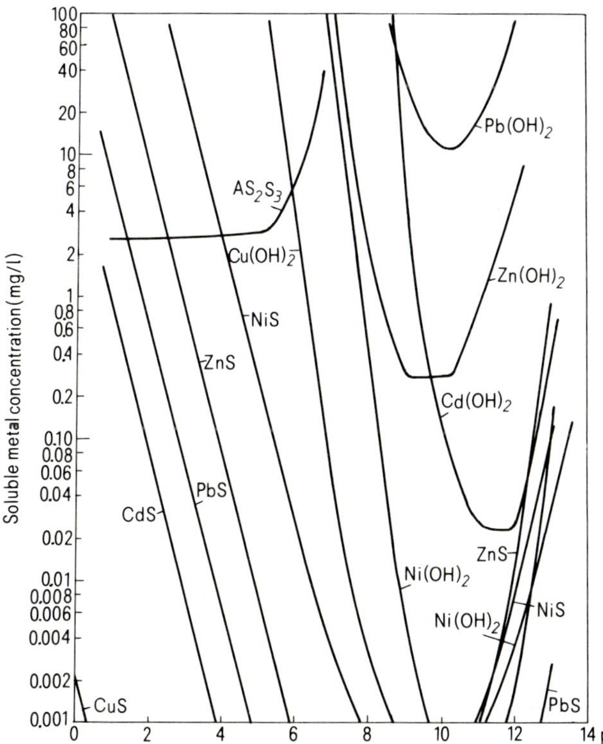

Adapted from M. Blythe et al., Characterization of Boliden sulfide-lime precipitation system. USEPA report, 600/S2-81-081, August 1981.

period followed by mild agitation for 30 min to improve the setting characteristics of the precipitate. After 60 min of setting, the supernatent is filtered through a precoated filter. The filtrate has averaged a consistent 0.09 mg/l for both silver and cadmium in the final discharge. The alkaline-NaBH$_4$ process in this case was evaluated versus alkaline precipitation sulfide precipitation, and carbon absorption. The alkaline-NaBH$_4$ process gave the best overall result based on effluent silver and cadmium concentrations, cost and suitability of the precipitated silver for further recovery.

Hydrogen peroxide oxidation-precipitation. This process which simultaneously oxidizes cyanides and cadmium to form cadmium oxide precipitate is called Kastone process[2]. It is claimed that cadmium oxide precipitates better than cadmium hydroxide resulting in better solid-liquid separation and therefore better overall removal. This proprietary process is reported as suitable for small plating operations but more detailed information on dosage requirement and effluent cadmium concentration is lacking.

Ozonation. Metal complexes in solution disappear upon ozonation. Krause[33] conducted tests on ozonation of organically complexed heavy metals. Experimental studies indicated that the complexing ability of chelates was destroyed during ozonation of Cd-EDTA, Zn-EDTA, and Ni-EDTA. Shambaugh[51] reported that in general an ozone contact time on the order of 10 min would destroy metal-EDTA complexes. A much shorter contact time would be required to reduce uncomplexed metal solubilities when the metal has a higher oxidation state. Uncomplexed metals such as zinc and cadmium which do not possess higher oxidation states do not exhibit a solubility reduction upon ozonation. Data and information leading to a successful design of a full scale treatment plant removing cadmium by ozonation do not exist to date.

Electroflotation. The successful removal of cadmium and cyanide simultaneously has been demonstrated by Poon[45], using an electroflotation process. The specially designed reactor consists of a stainless steel screen cathode placing above a 5-cm layer of sea water, and a platinated columbium screen serving as the anode below the cathode. The plating wastewater is placed on top of the cathode. Hypochlorite and chlorine gas generated from the sea water by the anode oxidize cyanide and liberate cadmium from the cyanide-cadmium complex preceeding alkaline precipitation. Free oxygen and ozone produced at the anode also assist in destroying the cyanide-cadmium complex. Simultaneously cathodic generation of hydroxides raise the pH resulting in alkaline precipitation of cadmium which is carried up to the surface by the rising gas bubbles. With 20 mg/l of cadmium and 45 mg/l of cyanide, it takes 53 min or longer to reduce to 0.2 mg/l cadmium and 0.5 mg/l cyanide with a power consumption of 0.23–0.40 kwh/g of cadmium removal.

A performance equation was developed for cadmium removal by electroflotation as in the following:

$$R = 4.45 \ P^{-0.36} \ D^{0.37} \ C^{0.66} \ S^{0.077} \ H^{-0.91}$$ where R is Cd

removal in g/kwh, P is power input in W, D is wastewater depth in the reactor in cm, C is initial Cd concentration, S is electrode spacing in cm, and H is initial pH of the wastewater. The equation indicates that a combination of lower power input and deeper wastewater column make more effective treatment. For operation, the residual chlorine concentration in the solution can be monitored as a ocntrol. The treatment process can be terminated 3–5 min after the residual chlorine has reached 1.0 mg/l, with both cadmium and total cyanide meeting the effluent quality standards.

The treatment system is compact and easy to operate. Operation requires no skill greater than that necessary for electroplating. A cost analysis shows a total treatment cost of US$0.76/m^3 (1982 dollar value). The cost is competitive with that of alkaline precipitation with lime followed by clarification-filtration.

IV. *Physical treatment*

a) *Ion exchange*

Ion exchange is capable of removing heavy metals including cadmium. The cations cadmium are exchanged for H$_2$ or Na. The cation exchange resins are mostly synthetic polymers containing an active ion group such as HSO$_3$. When all the ionized exchange sites are replaced, the resin can be regenerated by passing a concentrated solution of ions through the bed which reverses the chemical equilibrium. The ion exchange system may be put in series with more than one bed in parallel. Mixed bed systems have also been marketed for heavy metal removal where high regeneration frequency or large ion exchange volumes are required, a moving bed ion exchanger can be used which enables continuous regeneration and eliminates a large parallel system.

Na-type zeolite can be used successfully for cadmium removal. According to Onodera[39], natural zeolite of 10 mesh size was dipped in a 10-fold solution of boiling 1 N NaCl solution to obtain Na-type zeolite. Wastewater containing 10 mg/l Cd^{2+} was adjusted to pH 4.5, mixed with the Na-type zeolite 800 g/m^3 wastewater, stirred for 1 h and settled. The Cd^{2+} removal was about 90%, compared to about 30% with natural zeolite. Another resin was also found to be very effective in removing heavy metals including cadmium from an industrial wastewater containing Pb, Zn, Hg, Cd, Cu, Ni, Al, Ca, Cr^{3+}, Fe, and Ag[8]. It was reported that an industrial wastewater containing Hg 1.2, Cd 0.72, Cu 0.1, Pb 0.2, and Ca 202 mg/l respectively was treated with a thiocarbanic acid group-containing phenolic resin to remove Hg and then with an iminopropionic acid salt group-containing epoxy resin. The Hg, Cd, Cu and Pb concentrations were respectively 0.1 μg/l, and 0.05, 0.1, and 0.05 mg/l in the effluent, after 5000 l of wastewater/l resin.

Another promising exchanger material for effective removal of Cd^{2+} and other heavy metal is chromium ferrocyanide gel, reported by Srivastava[49]. The inorganic gel is prepared by adding chromium chloride gradually to potassium ferrocyanide. The produce with an exchange capacity as great as 3.65 meq/g is obtained with $Cr^{3+} = 0.071$ M, $Fe^{2+} = 0.218$ M, $H^+ = 0.01$ M and mixing ration of $Cr^{3+}/Fe^{2+} = 0.327$. The exchanger shows a great selectivity for Ag^+, Cu^{2+}, Tl^+, Zn^{2+}, Co^{2+}, Mn^{2+} and Fe^{3+}. Wastewater from an industry containing Cd^{2+}, Zn^{2+}, and Mn^{2+} equivalent to 150, 200, and 250 mg/l was allowed to pass through a column of chromium ferrocyanide gel (15 cm $\times 0.85$ cm^2) at a flow rate of 0.8 ml/min. The effluent was found to be completely free of these ions and also did not show the presence of Cr or Fe ions even in traces. For batch operation, a stirring time of 1 h is adequate for equilibration with this exchanger material. The chromium ferrocyanide gel is stable in salt and acid solutions up to a concentration of 2 M, but is unstable in alkali solutions. However, only partial recovery of cadmium (25%) from the exchanger material is possible.

b) Adsorption

Aluminosilicate adsorption. Several commercially available aluminosilicate adsorbents are able to remove Cd^{2+} from wastewaters. The type tested by Huang[26], Zeolite A, Blazer, Mordenite, Zeolex 23, Zeo 49, Zeosyl 100, were able to adsorb Cd^{2+} successfully at pH 9. Mordenite, a crystalline solid exhibited the greatest Cd^{2+} adsorption capacity at pH 5 and performed as well as other species. All of them could achieve near 100% removal of Cd^{2+} at a dose of 3–5 g absorbent/l. It is therefore possible to remove Cd^{2+} from plating wastewaters to 0.1 mg/l. At pH 3.0, the Mordenite species exhibited decreasing Cd^{2+} adsorption. This is a feature promising for regeneration of Mordenite with strong acid.

Starch xanthate adsorption. Starch xanthate and a cationic polymer such as polyethylenimine forming a complex can precipitate heavy metal from solutions. The process is expensive because almost all cationic polymers are expensive. However making starch xanthate from corn starch made insoluble by crosslinking eliminates the need for the expensive polymer. The crosslinked corn starch is commercially available. It is combined with sodium and sulfur by reaction with carbon disulfide and lye (NaOH), yielding insoluble sodium starch xanthate. In metal solutions it acts like a resin, exchanging the sodium for the metal.

According to Wing[55], cadmium concentration is reduced from 56 mg/l to 9 μg/l with an initial pH of 3.0. The required solid starch xanthate is in the range of 0.037 to 0.0859 g/50 ml for such efficient removal. If the initial pH is greater than 5, stoichiometric quantity (one xanthate group per metal ion) of xanthate is sufficient. Since the solid starch xanthate is basic, the pH of the treated solution will increase, and effective removal is always achieved in the pH range of acceptable discharge. The process is effective over a pH range from 3 to 11 and in the presence of up to 10% salt. Only a few minutes would be required to complete the removal. The cost of 500 g of dry insoluble starch xanthate containing 8% sulfur was US\$0.20–0.22, 1975 dollars. The product has a metal binding capacity of 1.1 moles metal/g.

Carbon adsorption. Certain commercially available carbons can be very effective in removing Cd^{2+} from electroplating wastewaters. Nuchar S-A and Nuchar S-N powdered carbons are found by Huang[24] to have a Cd^{2+} adsorption capacity far superior to that of granular carbons such as Filtrasorb 400 and Darco HD 3000. Pore size is probably not a critical parameter for equilibrium absorption capacity. Instead the superiority of powdered carbons over the granular carbons is due to differences in the chemical reactivity of their surfaces. According to Huang[23], powdered activated has low pH_{zpc} (pH at zero electrophoretic mobility) values and excellent adsorption capacity for cationic metal ions while granular activated carbon, having high pH_{zpc} values, is relatively poor for metal ion adsorption.

The kinetics of adsorption in Huang's study[26] are favorable for both the fuoroborate and cyanide systems. The maximum adsorption level is reached within the first 10 min of contact of the carbon with the wastewater. This rapid rate of removal is not affected by variations in pH, ionic strength, carbon/Cd^{2+} concentration, temperature and mixing rate. The fast adsorption rate is considered a significant advantage over the chemical precipitation process which is often a slower process.

For a given carbon dose, the adsorption density increases for increasing Cd^{2+} concentration in solution. However the percent of Cd^{2+} removed decreases with higher initial adsorbate concentration. This phenomenon is typical of the carbon adsorption process. The same phenomenon is not as noticeable at pH values greater than 8 due to precipitation effects in the fluoroborate bath and cadmium cyanide complex formation in the cyanide system. At pH 7 and a carbon/Cd ration of 0.25 g/mg, as much as 85% Cd^{2+} removal is obtained in the fluoroborate system.

For both the fluoroborate and cyanide wastewaters, the adsorption density is approaching its maximum level in the neutral pH range. This constitutes an advantage over many chemical precipitation techniques requiring pH adjustment to 10–11 for maximum removal. Because of the high molar ratio of cyanide/Cd^{2+} of 10:1 in the typical cyanide system, resulting in competition between the two ions for available carbon sites, Cd^{2+} adsorption is not as rapid as in the fluoroborate system. The adsorption rate can be enhanced by either increasing the carbon dose or by reducing the initial cyanide concentration in the wastewater.

For continuous flow treatment application, Huang[25] uses a packed column in which the powdered carbon Nuchar S-A is aggregated into beads of a size of 0.6 mm in diameter using polyvinylalcohol and glutaraldehyde. The aggregated powdered carbon exhibits similar adsorption capacity as the original activated carbon (1.1337 g Cd^{2+}/g carbon); although the kinetics of adsorption is slower. The study shows a column packed with a gross mass of 41 g of aggregated carbon can be used for 4 consecutive runs, with acid regeneration following each run, and treat 173 l (or 1587 bed volume) of Cd^{2+} containing tetrafluoborate plating wastewater before losing its adsorption capacity. Chemical regeneration using strong acids such as HCl, $HClO_4$, and H_2SO_4 gives good regeneration results.

c) Foam flotation

This technique relies upon a surfactant that causes a nonsurface-active material to become surface active forming a product that is then removed by bubbling a gas through the bulk solution to form a foam. Foam flotation can be divided into 3 catagories, namely, ion flotation, precipitate flotation, and adsorbing colloid flotation.

Ion flotation is a process that involves the removal of a nonsurface-active ion (a colligend) by the addition of a surfactant (a collector) which yields an insoluble product. Kobayashi[29] reports the effects of anionic and cationic surfactants on the ion flotation of Cd^{2+}. Sodium a-sulfolaurate is determined to be most effective for the flotation of Cd^{2+} ion achieving as high as 97% removal. About 97% of Cd^{2+} ions could be floated at a pH of 11.3 when cationic surfactant is used with bentonite. Ferguson[18], however, reports that increased ionic strength, Ca^{2+}, and phosphate interference make ion flotation impractical for Cd^{2+} removal.

Precipitate flotation is the process in which the nonsurface-active material forms a precipitate with something other than the collector, which in turn makes surface active by the collector. Adsorbing colloid flotation is defined as the removal of dissolved material by adsorption on colloidal particles followed by the removal of colloid particle, dissolved material, and the collector by flotation. These 2 flotation processes, unlike the ion flotation process, do not require stoichiometric amounts of the collector since only electrical attraction exists between the collector and the precipitate but sufficient surfactant must be used to form stable foam. Kolayashi[30] also investigates the removal of Cd^{2+} ions from an aqueous solution by adsorbing particle flotation using bentonite and a cationic surfactant, hexa-decyltrimethylammonium chloride. It is found that the most suitable method is to conduct flotation in the region of the coagulation flotation of bentonite.

Huang[27] studies both precipitate flotation and absorping colloid flotation of Hg^{2+} and Cd^{2+} from aqueous systems. Sodium lauryl sulfate and hexadecyltrimethylammonium bromide (HTA) are used as collectors. $Fe(OH)_3$, $Al(OH)_3$, FeS, and CuS are used as adsorbing colloids. Floc foam flotation of both metals with CuS and HTA is very effective resulting in residual Hg^{2+} concentration as low as 5 ppb and Cd^{2+} as low as 20 ppb. Floc foam flotation of Cd^{2+} with FeS and HTA yields residual Cd^{2+} as low as 10 ppb.

V. Cadmium recovery

Various techniques are available to recover cadmium from wastewaters, among which evaporation is an established method and reverse osmosis is a very promising one.

a) Evaporation

Single-stage or two-stage (double effect) recovery unit can be used for cadmium recovery. The single-stage unit is more common but a two-stage unit saves steam (energy). The use of atmospheric evaporative towers for recycling reusable chemicals in the electroplating industry is reported by Kolesar[31]. Atmospheric evaporative recovery is economical for rinses bearing the dragout from processes not self destructive by nature (e.g. Zn, Cd, Cu, Ni, and Cr plating baths). Schrantz[47] reports evaporation as a technique to separate Cr, Ag, and Cd from rinse water for subsequent reuse in the plating baths. Three 2.4-m (8-ft) by 3.0-m (10-ft), by 5.6-m (15-ft) high evaporations handle 189 1/h (50 gal/h) of rinse water each and form a closed loop with the Cr, Ag, and Cd plating baths. No sludges are produced, and there is no connection to sewers. A 20-month payback period is projected for these evaporators. A single-effect, 378 1/h (100 gal/h) evaporator is installed on a 5-barrel, 3-m^3 (800 gal) bath cadmium plating line in a Connecticut plant, according to Gallo[19].

Both capital and operational costs for evaporative recovery systems are high. Chemical recovery and water reuse values must offset these costs for evaporative recovery to become economically feasible. A closed loop system design for rinse and plating is most attractive.

b) Reverse osmosis

Reverse osmosis with cadmium cyanide bath is less developed, but suitable membranes are now available to make it practical. There is also promise that reverse osmosis will be cheaper and less energy-consuming than evaporative recovery. Donnelly[11] examines reverse osmosis treatment of plating bath rinse waters, emphasizing closed loop operation with water recycle for rinsing and concentrated chemical recycle for plating baths. Three commercially available membrane configurations, i.e., tubular (cellulose acetate), spiral wound (cellulose acetate), and hollow fiber (polyamide) are tested with diluted plating bath wastewaters. Watts-type nickel, nickel sulfamate, copper sulfate, copper pyrophosphate, nickel fluoborate, zinc chloride, and copper, zinc, and cadmium cyanide are all found attractive systems for treatment by reverse osmosis. Kosrrek[32] reports Cd^{2+} removal amounts to 95–99% by reverse osmosis.

c) Other cadmium recovery systems

Other promising cadmium recovery systems include electrochemical ion exchange and adsorption. Roof[46], is successful in recovering cadmium from plating solutions by the use of an electrochemical cell with graphite anodes. Cadmium is deposited on a carbon-fiber cathode while cyanide is removed at the anode. Recovery of cadmium is either by electrowinning or dissolution into a heated cyanide strip solution. Full-time use of this system is expected to yield a high degree of cadmium recovery alone with an effluent Cd^{2+} concentration of 0.4–0.6 mg/l.

Treatment of wastewater containing cadmium by ion exchange and adsorption has been discussed previously. Ion exchange can be used as a polishing treatment or recovery process. The concentrated solution obtained upon regeneration of the ion exchange resin is, in general, more suitable for economical recovery procedures than the initial dilute cadmium waste streams. The same general rule applies to cadmium recovery from adsorbents. Wing[55] reports successful recovery of cadmium and other metals by washing the sludge with nitric acid in an insoluble starch xanthate process. Similarly strong acids such as HCl, $HClO_4$, and H_2SO_4 can be used to regenerate activated carbon for cadmium adsorption. A concentration factor of 30 can be obtained using $1M\ H_2SO_4$ solution according to Huang[24]. A wastewater containing $10^{-4}\ M\ Cd^{2+}$ therefore can be concentrated to approximately 3.3×10^{-3} M. In the case of cyanide baths there may be concern that acid-stripping produces hazardous cyanide gas. However, it is known that once adsorbed, cyanide ions are oxidized to less hazardous CNO^- species on the carbon surfaces.

to treat large quantities of wastewater to reduce cadmium to such a low level. Applying treatment at the source without discharging into a sewer leading to a municipal treatment plant is a most cost-effective approach.

VII. *Cost of cadmium removal*

In this section, the cost comparison of cadmium treatment and cadmium recovery systems are given. Four treatment systems are compared, namely, alkaline neutralization precipitation with lime; sulfide precipitation using insoluble FeS; ion exchange with cation exchange resin; and activated carbon adsorption. The alkaline precipitation process includes coagulation-clarification-sand filtration and discharge to sewer line. Sludge is thickened and dewatered before final disposal. The sulfide precipitation process uses two-stage neutralization while FeS and polymer are added simultaneously in a third precipitation tank. Sludge filter is used to dewater CdS and $Fe(OH)_2$ solids without thickening. The total annual costs provided by Huang[24] are estimated or taken from the

VI. *Cadmium removal from domestic wastewater treatment*

Although most publicly owned treatment works are using secondary or biological treatment processes not designed for heavy metal removal, some heavy metal, including cadmium can be removed by biological uptake and by sludge adsorption. Olson[38] reports the 1979 annual average removal of cadmium from a trickling filter treatment plant to be 77.5%, producing an effluent of 0.04 mg/l cadmium on the average. The range of removal for that year is 58–92%.

In a pilot plant study by Maruyama[35], 5 mg/l of Cd^{2+} as $CdCl_2$ is added to raw municipal wastewater which then undergoes chemica coagulation, settling, dual media filtration, and activated carbon adsorption treatment. Chemical coagulants are iron (ferric sulfate, 45 mg/l Fe), low lime (260 mg/l lime and 20 mg/l Fe as ferric sulfate), and high lime (600 mg/l lime). With the high lime system, the respective accumulated removals for coagulation-setting, filtration and adsorption are 97, 98 and 99.7%. Both the low lime and the ion systems are less effective but a total removal of 98–99% still can be achieved.

In an extensive literature review and technical evaluation of various treatment processes in removing heavy metals from municipal wastewater treatment plants, Culp al.[10] summarize the cadmium removal in the following two tables.

It appears that secondary treatment alone or secondary treatment followed by nitrification-dentrification cannot remove cadmium from municipal wastewater by more than 71%. Physical and/or chemical method of polishing the effluent is required to reduce cadmium to a concentration safe to be discharged into the natural environment. It would be very expensive

Table 3. Literature and actual plant survey data (Cd concentration in raw wastewater, average 0.008 mg/l: range 0.002–0.016 mg/l)

Treatment process	Range of removal (%)	Average removal (%)
Activated sludge	30–64	54
Trickling filtration	0–5	0
Coagulation-sedimentation (ferric addition)	61–75	68
Coagulation-sedimentation (lime)	0–95	30
Coagulation-sedimentation (alum)	68–76	72
Filtration after biological treatment		32
Filtration after physical-chemical treatment		38
Activated carbon adsorption	0–23	0
Ground injection (infil-percolation)	12–13	

Table 4. Potential of cadmium removal from wastewater (based on plant operating data and results of various test programs)

Treatment process	Average removal (%)	Average effluent concentration (mg/l)
Activated sludge	71	0.002
Trickling filter	38	0.005
Rotating biological contactor	57	0.003
2-Stage nitrification	71	0.002
2-Stage nitrification-denitrification	71	0.002
2nd effluent + filtration	71	0.002
Alum addition to aeration basin	92	0.001
$FeCl_3$ addition to primary treatment	91	0.001
Tertiary lime treatment of nitrified effluent	98	0.0002
Tertiary lime treatment followed by ion exchange	98	0.0002
Carbon adsorption of 2nd effluent	71	0.002
Carbon adsorption of 3rd lime effluent	98	0.00002
Carbon adsorption of 3rd lime, nitrified	98	0.0002
Carbon adsorption of 3rd lime, ion exchange	98	0.0002
Physical-chemical, with lime	38	0.005
Physical-chemical, with $FeCl_3$	80	0.002

54

Centec report 1979[6] for alkaline precipitation, and from a U.S. EPA report 1980[52] for sulfide precipitation, as well as Zanitsch's[56] work for ion exchange. Cost comparisons are listed in table 5.

Assuming that sludge treatment and disposal is operable, alkaline precipitation is by far the most cost-effective treatment process. However ion exchange and carbon adsorption processes offer an opportunity to recover cadmium and water for reuse which could offset some of the costs.

A cost comparison for 2 cadmium recovery systems is given by Gurklis[21]. The evaporative recovery system uses a 2-stage 50-gallon evaporative recovery unit. The associated costs are presented in table 6. A reverse osmosis recovery unit handling the same rate of flow at 50 gal/h (187.4 l/h) is less costly resulting in a lower operating cost as shown in table 7. It should be noted also that some saving in chemical cost is also realized in both recovery systems which would otherwise be spent in a treatment system. The saving, approximately $3,300/year is therefore a credit against the treatment system.

Table 5. Cost comparisons of cadmium treatment

Treatment process	30 gpm	40	60	80	100
Alkaline precipitation	50	55	62.5	70	80
Sulfide precipitation	110	175	–	–	160
Ion exchange	60	75	110	136	160
Carbon adsorption column	–	77	95	110	125
complete mix	–	90	102	110	120
Total annual cost	$1,000				

Table 6. Costs for evaporate recovery of cadmium cyanide solution double-effect 50 gal/h unit

Investment costs	
Purchased cost	$30,000
Cost of installation and auxiliary items	10,000
Operating costs	
Interest/year (10%)	2,000
Depreciation/year (20%)	8,000
Steam at $4.00/1000 lb., 800,000 lb.	3,200
Electricity at 4¢/kwh, 42,000 kwh	1,680
Labor at $12.00/h, 240 h	2,900
Total operating costs	$17,780
Area plated/year (240 days)	480,000 ft^2 (44,600 m^2)
Recovery unit cost	$0.37/ft^2 ($4.0/m^2)

Table 7. Costs for recovery of cadmium cyanide plating solution by reverse osmosis

Investment Cost	
Operating costs	
Interest/year (10%)	1,000
Depreciation/year (20%)	4,000
Electric (50 kwh/day)	500
Maintenance (5%)	1,000
Membranes	1,500
Total operating costs	$8,000
Area plated/year	480,000 ft^2 (44,600 m^2)
Unit recovery cost	$0.17/ft^2 ($1.8/m^2)

1 Anderson, J.S., Phillips, J.M., and Schriver, C.B., Water contamination by certain heavy metals. Paper presented at the 44th Annual Meeting, Wat. Poll. Control Fed., 1971.

2 Anonymous, New process detoxifies cyanide wastes. Envir. Sci. Technol. 5 (1971) 496–497.

3 Argo, D.G., and Culp, G.L., Heavy metals revomal in wastewater treatment processes: part I, Water and sewage works, vol. 119, No. 8, pp. 62–65.

4 Bard, A.J., Chemical Equilibrium. Harper and Row, New York 1966.

5 Brantner, K.A., and Cichon, E.J., Heavy metals removal: Comparison of alternative precipitation processes. Ind. Waste Proc., 13th Mid. Atlantic Conf., Ann Arbor Science 1981, p. 43–50.

6 Centec Co., Environmental pollution control alternatives: Economics of wastewater treatment alternative for the electroplating industry. U.S.E.P.A. Report No. 625/5-79-016; 1979.

7 Chalmers, R.K., Pretreatment of toxic wastes, Water Poll. Control, London 1970, pp. 281–291.

8 Chemical Abstract, Heavy metal removal from wastewater, Japan Patent 80 54, 084, 1980, Chemical Abstract 93 (1980), 155451.

9 Cook, M.M., and Lander, J.A., Sodium borohydride controls heavy metal discharge. Pollut Engng 13 (1981) No. 12.

10 Culp, Wesner, and Culp. Water reuses and recycling, vol. 2, Evaluation of treatment technology. US Department of Interior, Office of Water Research and Technology, OWRT/RU-79/2, 1979.

11 Donnely, R.G., Reverse osmosis treatment of electroplating wastes. Plating 61 (1974) 432.

12 El-Gohary, F. Lasheen, M.R., and Abdel-Shafy, H.I., Trace metal removal from wastewater via chemical treatment, Proc. 1st int: Conf. Management and Control of Heavy Metals in the Environment, London 1979.

13 Environmental Science and Technology report. Metals focus shifts to cadmium. 5 (1971) 754–755.

14 Federal Register, 1975, vol. 40, No. 248, December 24, p. 59570.

15 Federal Register, 1981, vol. 46, No. 18, January 28, p. 9469.

16 Feitknecht, W., and Schindler, P., Solubility constants of metal oxides, metal hydroxides and metal hydroxide salts in aqueous solution. Pure appl. Chem. 5 (1962) 130.

17 Feitknecht, W., and Schindler, P., Pure appl. Chem. 6 (1963) 130.

18 Ferguson, B.B., Hinkle, C., and Wilson, D.J., Foam separation of lead (II) and cadmium (II) from wastewater. Separation Sci. 9 (1974) 125.

19 Gallo, B.R., and Culotta, J.M., Save on plating waste system. Wat. Wastes Engng 9 (1972) A18–A19.

20 Gard, S.M., Snavely, C.A., and Lemon, D.J., Design and operation of a metal wastes treatment plant. Sew. ind. Wastes 23 (1959) 1429–1438.

21 Gurklis, J.A., and Vaeler, L.E., Cadmium plating with environmental restriction, Proc. Workshop on Alternatives for Cadmium Electroplating in Metal Finishing NTIS PB-298 841, p. 104–112, 1979.

22 Hansen, N.H., and Zabban, W., Design and operation problems of a continuous atumatic plating waste treatment plant at data processing division, IBM, Rochester, Minnesota. Proc. 14th Purdue Industrial Waste Conf. 1959, p. 227–249.

23 Huang, C.P., and Ostovic, F.B., The removal of Cd (II) from dilute aqueous solutions by activated carbon adsorption. J. envir. Engng Div., ASCE 104 (1978) 863–878.

24 Huang, C.P., and Smith, E.H., Removal of Cd (II) from plating waste by an activated carbon process, Ch. 17, Chemistry in Water Rinse, vol. 2, p. 355–399. Ed. W.J. Cooper. Ann Arbor Science 1981.

25 Huang, C.P., and Wirth, P.K. Activated carbon for the treatment of cadmium wastewater. Paper presented at the National Conf. Hazardous and Toxic Wastes Management, June, 1980, Newark, N.J.

26 Huang C.P., and Wirth, P.K., Treatability of cadmium (II) plating wastewater by aluminosilicate adsorption, p. 87–94. Ind. Waste Proc., 13th Mid. Atlantic Conf., Ann Arbor Science 1981.

27 Huang, S., and Wilson, D.J., Foam separation of mercury (II) and cadmium (II) from aqueous systems. Separation Sci. 11 (1976) 215.

28 Jenkins, S.H., Knight, D.G., and Humphreys, R.E., The solubility of heavy metal hydroxides in water, sewage and

sewage sludge; I. The solubility of some metal hydroxides. Int. Air Water Pollut. *8* (1964) 537–556.

29   Kobayashi, K., Effects of anionic and cationic surfactants on the ion flotation of Cd^{2+}. Bull. chem. Soc. Japan *48* (1975) 1745.

30   Kobayashi, K., Studies of the removal of Cd^{2+} ions by adsorbing particle flotation, Bull. chem. Soc. Japan *48* (1975) 1750.

31   Kolesar, T.J., The use of atmospheric evaporative towers for recycle and waste elimination. Plating *61* (1974) 571.

32   Kosrrek, L.J., Removal of various toxic heavy metal and cyanide from water by membrane processes, Ch. 12, Chemistry in water reuse, vol. 1, pp. 261–280. Ed. W.J. Cooper, Ann Arbor Science 1981.

33   Krause, H., Research on the destruction of complexing and chelating agents in radioactive wastes by oxidation. Kernforschungszentrum, Karlsruhe, AEC Accession No. 32047, Rept. No. KFK-287.

34   Lowe, W., The origin and characteristics of toxic wastes with particular reference to the metals industries. Water Pollut. Control, London (1970) 270–280.

35   Maruyama, T., Hannah, S.A., and Cohen, J.M., Metal removal by physical and chemical treatment processes. J. Water Poll. Control Fedn *47* (1975) 962.

36   McKee, J.E., and Wolf, H.W., Water Quality Criteria, 2nd ed. California State Water Quality Control Board, Publication No. 3-A, 1963.

37   Nemerow, N.L., Theories and practices of industrial waste treatment. Addison Wesley, Reading, Mass. 1963.

38   Olson, J.L., and Eick, R.W., POTW removal credits and revised categorial pretreatment standards, 3rd Conference on Advanced Pollution Control for the Metal finishing Industry, EPA-600/2-81-028, p. 11–15. U.S. EPA Publication, 1981.

39   Onodera, Y., Removal of heavy metal ion from industrial wastewater. Agency of industrial services and technology, Japan, Kokai *549* (1977) 1734, Chem. Abstract *87* (1977) 43802 b.

40   Patterson, J.W., and Minear R.A. Wastewater treatment technology, 2nd ed. National Technical Information Service, PB 216-162, 1973.

41   Patterson, J.W., Allen, H.E., and Scala, J.J., Carbonate precipitation for heavy metals pollutants. J. Wat. Pollut. Control Fedn *49* (1977) 2397–2410.

42   Patterson, J.W., Guidance for BAT-equivalent control of selected toxic pollutants. Report to the U.S. Environmental Protection Agency, Region V, Chicago, Ill, 1981.

43   Patterson, J.W., Effect of carbonate ion opn precipitation treatment of cadmium, copper, lead and zinc. Proc. 36th industrial waste conference, pp. 579–602. Purdue University, 1981.

44   Pinkerton, H.L., Waste disposal, inorganic wastes, electroplating engineering handbook, 2nd edn. Ed. A. Kenneth Graham. Reinhold Publ. Co., New York, 1962.

45   Poon, C.P.C., Removal of copper, nickel, zinc, cadmium, and cyanide from plating wastewater by electroflotation, Proc. 1st Int. Conf. Management Control of Heavy Metals in the Environment, London, 1979.

46   Roof, E., Electrochemical reactor and associated in-plant changes at Varland metal services. Paper presented at the 3rd EPA/AES pp. 16–22. Conf. on Adv. Pollution Control for the Metal Finishing Industry, 1980.

47   Schrantz, J., Pan Am. Recycles Vis. Evaporation, Ind. Finishing *50* (1974) 30.

48   Schwitzgebel, M.B.K., Terry, J.C., Sund-Hagelberg, C., and Bhattacharyya, D., Characterization of boliden's sulfide-lime precipitation system. Project Summary, EPA-600/S2-81-081, 1981.

49   Srivastava, S.K., Studies on the use of inorganic gels in the removal of heavy metals. Water Res. *14* (1980) 113–115.

50   Stubbs, R.L., Cadmium – The metal of benign neglect. Cadmium Association, London 1975.

51   Sambaugh, R.L., and Melnyk, P.B., Removal of heavy metals via ozonation. J. Water Pollut. Control Fedn *50* (1978) 113–121.

52   U.S. EPA, Control and treatment technology for the metal finishing industry-sulfide precipitation. U.S. EPA Report No. 625/8-80-003, 1980.

53   Weiner, R.F., Acute problems in effluent treatment. Plating *54* (1967) 1354–1356.

54   Whang, J.S., Young, D., and Pressman, M., Design of soluble sulfide precipitation system for heavy metals removal. pp. 63–71. Ind. Waste Proc., 13th Mid. Atlantic Conf., Ann Arbor Science, 1981.

55   Wing, R.E., Corn starch compound recovers metals from water. Ind. Wastes Jan./Feb. 1975, 26–27.

56   Zanitsch, R.H., and Stensel, M.H., Economics of granular activated carbon water and wastewater treatment systems. Carbon Adsorption Handbook, p. 215. Ed. Cheremisinoff et al., Ann Arbor Science, 1978.

Part II: Bioaccumulation of Cadmium

Cadmium in sludges used as fertilizer

by R.D. Davis

Water Research Centre, Environmental Protection Directorate, Stevenage, SG1 1TH (Great Britain)

Summary. In intensively populated countries efficient sewage treatment is essential to protect river quality. An inevitable by-product is sewage sludge which has to be disposed of safely and economically. Utilisation of sludge as a fertilizer of agricultural land is the most economic disposal route for inland sewage-treatment works and also benefits farmers by providing a cheap manure. Much of the cadmium in wastewater is concentrated into sludge which consequently contains higher concentrations of cadmium than soil does. It is impracticable to reduce cadmium concentrations in sludge below certain levels. When sludge is used on farmland rates of application must be controlled so that cadmium concentrations in soil never reach levels that could significantly contaminate food crops. Cadmium is a principal factor limiting the use of sludge on land. Nevertheless, it is a local problem since agricultural land in general receives more cadmium from aerial deposition and phosphatic fertilizers. The significance of accumulations of cadmium in soil depends mainly on its availability for crop uptake. Investigations are described which have attempted to identify and to determine the availability of forms of cadmium in soil. There is considerable research interest in cadmium in soil solution which is likely to be directly available for crop uptake. Another area of interest is the apparent disappearance of cadmium from sludge-treated soil. Soil analysis often cannot fully account for the cadmium added in sludge. Apart from the effect of soil conditions, especially pH value, crop uptake varies according to the particular crop examined. Highest concentrations of cadmium occur in tobacco, lettuce, spinach and other leafy vegetables. Using crop uptake data from field trials it is possible to relate potential human dietary intake of cadmium, on which hazard depends, to soil concentrations of cadmium, which can be controlled by regulating applications of sludge. This provides an objective basis for limits for cadmium concentrations in soils receiving sludge. Transfer of cadmium via farm animals to meat and dairy products for human consumption is thought to be minimal, even allowing for some direct ingestion of sludge-treated soil by the animals. Evidence from these and other investigations suggests that a loading rate limit of 5 kg Cd/ha (equivalent to a soil concentration of about 3.5 mg Cd/kg) affords adequate protection to the foodchain where sludge is used on agricultural land. More research work is needed to provide a basis for predicting the long-term availability of cadmium introduced to the soil in sludge.

Cadmium in sludge and sewage treatment

Water re-use and the protection of river quality depend on efficient sewage treatment. But in intensively populated countries this inevitably leads to the production of large quantities of sewage sludge requiring safe and economic disposal. Sludge treatment and disposal account for about half the total cost of sewage treatment. The size of the problem can be illustrated by some statistics for the UK. Here, over 95% of households and most industrial premises are connected to the public sewer. There are 7800 sewage works serving 82% of the population and producing an estimated 40 million tonnes of sewage sludge (or 1.3 million t dry sludge solids) each year. The annual cost of sludge treatment and disposal, including capital charges and administration, approached £200 million – or approximately £4 per head – in 1981/82. In the EEC countries together, annual production of sludge probably exceeds 5×10^6 t dry sludge solids, and is increasing.

What does all this have to do with cadmium in sludges used as fertilizer? The answer is that utilization on agricultural land is the most economic sludge disposal option for inland sewage treatment works. In the UK agricultural land receives nearly 50% of the sewage sludge produced annually. From the conservation point of view, agricultural use is the only current means of disposal which has any benefits, namely that in soil the plant nutrients and organic matter in sludge improve conditions for plant growth. Utilization on land should therefore be of benefit to farmers and to the community at large since it is the least-cost disposal option. But the contaminants in sludge are a problem. Thus cadmium and other metals usually occur in sludge in higher concentrations than in soil, and metals added in sludge accumulate in top soil. When sludge is used on land, rates of application must be controlled so that concentrations of contaminants in soil never reach levels which could be harmful to crops or the animals which eat them. Cadmium is a principal factor limiting the use of sludge on land. The environmental and economic consequences of this observation are such that the agricultural effects of cadmium in sludges used as fertiliser have received considerable research interest over the last 10 years.

The purpose of sewage treatment is essentially to produce from raw sewage (wastewater) a final effluent conforming to standards of purity which allow its safe return to rivers. Treatment of municipal sewage involves separation of settleable solids (primary treatment) and conversion of dissolved and colloidal solids by a biological process (secondary treatment) to metabolites such as carbon dioxide, sulphate and water and also to microbial cells and residues in a flocculent form (humus or surplus activated sludge) which can be separated by a secondary stage of settlement. The end-products are a comparatively clean effluent, which can be returned to the river either directly or after some further (tertiary) treatment, and a sludge containing much of the polluting load of the original raw sewage. The volume of sludge produced will be approximately 1% of the volume of raw sewage treated. Sewage sludge is a putrescible, thin mud typically containing about 98% water. The solids content is 70–80% organic matter and the inorganic fraction will include appreciable concentrations of various contaminants reflecting the quality of the raw sewage from which it was derived. Sludges from the primary and secondary treatment processes are usually combined and thickened or treated before disposal, the ratio of primary sludge solids to secondary sludge solids often being about 2:1 by weight.

About 90% (range 75–100) of the cadmium in raw sewage is separated into sludge. Usually, 70% of the metal is removed during primary sedimentation, and secondary treatment by the activated-sludge process removes approximately 60–70% of the remainder[31]. A survey of 183 UK sludges found that cadmium concentrations ranged from 0 to 180 mg/kg dry solids (ds) with a median value of 17 mg/kg ds and a mean of 29 mg/kg ds[16]. A survey of over 200 sludge samples from 8 states in the USA found cadmium concentrations of 3–3410 mg/kg ds with a median value of 16 mg/kg[38]. In the UK survey, the mean concentration of cadmium in sludges from sewage works receiving no industrial effluent was 7.5 mg/kg ds whilst sludge prepared at the Water Research Centre (WRC) using sewage derived from a residential area, contains about 2 mg Cd/kg ds. It is worth mentioning that anaerobic digestion of sludge can influence concentrations of cadmium and other contaminants. This is a popular treatment process which incubates the sludge anaerobically for about a month at 30–35 °C. During this time putrescible organic matter in the sludge is converted to carbon dioxide and methane by bacterial action, resulting in a loss of 30–50% of the original organic content of the sludge. This increases the ratio of inorganic:organic matter in the sludge and can effectively double the concentration of cadmium expressed as mg/kg dry solids. Even so, this would produce only about 4 mg Cd/kg ds in a digested sludge of residential origin or perhaps 8–10 mg Cd/kg ds in what is commonly called a 'domestic' sludge, being derived from a small sewage-treatment works with a predominantly residential catchment. Much of the camium in sludge therefore results from industrial discharges to sewers. Industry uses cadmium in a variety of ways including electroplating, electrical contacts, plastics stabilizers, ceramics, pigments, alloys, fluxes, glasses, lubricant-additives, nickel-cadmium batteries, solar cells and catalysts. In the UK, attempts to restrict industrial discharges of cadmium to sewers, and hence to control cadmium levels in sludge, have met with some success. Thus, a stricter policy on trade effluent control reduced cadmium concentrations at one large sewage-treatment works in the London conurbation from 145 mg/kg ds in 1972 to 18 mg/kg in 1977[51]. This example is indicative of a general trend towards declining cadmium concentrations in sludge. However, whilst attention to point sources of industrial discharge can result in substantial reductions where cadmium levels of sludge are high, it is not easy for large works to reduce the cadmium concentration of sludge below the range 20–40 mg Cd/kg ds. This is because as little as 0.008–0.01 mg Cd per liter of sewage will give rise to a concentration in sludge of 20 mg Cd/kg ds, assuming there is 85% removal of cadmium from sewage during treatment and that 350 mg of primary and secondary sludge solids are generated per 1 of sewage treated. It is difficult to isolate and control the many diffuse sources in the catchment of a large sewage treatment works which may be contributing to a cadmium load in sewage of < 0.01 mg/l. Poon's article in this review[39] deals with cadmium removal from effluents. Methods for the removal of cadmium from sludge itself are only at the developmental stage at the moment and their cost is likely to exceed the value of the separated cadmium and the decontaminated sludge. As to the forms of cadmium which may occur in sludge it appears that sludges contain a wide variety of sites at which metals may be held by mechanisms which include ion exchange, sorption, chelation, and precipitation. The form of the metal will be influenced by the type of treatment given to the sludge. Working with digested sludge, Stover et al.[45] used a sequential extraction technique to rank the forms of cadmium as follows: carbonate > sulphide > organically bound > adsorbed = exchangeable. The phosphate and hydroxide are also likely to occur. The organically bound fraction may be more pre-

valent in undigested sludge. It is well known that acidification of sludge will release metals including cadmium, a marked effect occurring as the pH value falls below 5.0. Whilst forms of cadmium in sludge are of interest and have some bearing on potential uptake by crops from sludge-treated soil, this will depend more directly on the fate of cadmium in sludge when it is mixed with soil. Stover et al.[45] reported that about 80% of the cadmium in sludge is present in forms which would require conversion to water-soluble, exchangeable or adsorbed forms in soil before uptake by plants could proceed. Of particular interest is microbial oxidation in soil which could result in the solubilization of carbonate, the conversion of sulphide to sulphate and the breakdown of sludge organic matter. Any of these processes could dramatically alter the availability of sludge-borne cadmium for uptake by plants.

Impact of sludge on soil levels of cadmium

Background soils normally contain 0.1–1.0 mg Cd/kg. A survey of 600 samples[1] of agricultural soil in the UK found that 50% of samples contained < 1.0 mg/kg although the range was 0.08–10 mg/kg and the mean < 1.23 mg/kg. High values occur rarely in soils derived from certain carboniferous shales or as a result of contamination by mining or smelting activities.

As described above, sludge often contains 20 mg Cd/kg or more and repeated, heavy applications could substantially increase soil concentrations as the added metal accumulates in top soil. In most countries where sludge is applied to land there are guidelines specifying maximum permissible levels of elements in soils. In the UK, the limit is 3.5 mg/kg[18] which is equivalent to an addition to the soil of 5 kg Cd/ha in sludge if it is assumed that the top soil is 20 cm deep with a density of 1 g/ml and that the background soil already contains 1 mg Cd/kg. Usually, guidelines suggest also that sludge should be applied to land in accordance with crop requirements for plant nutrients. Liquid digested sludge is widely utilized and is particularly rich in nitrogen. A dressing of 100 m³/ha would probably be appropriate to meet the nutrient requirements of many crops. Typically, such sludge has a dry solids content of about 5% so 100 m³ would contain 5 t ds and would add to the land 100 g Cd/ha if the cadmium concentration of the sludge was 20 mg/kg ds. This would increase the concentration of cadmium in the top soil by 0.05 mg/kg, an amount unlikely to be detected by routine soil analytical procedures. Over the years, 250 t ds/ha of this sludge could be applied to the same field to reach the maximum permissible soil concentration of cadmium. For fields receiving it, sludge represents the main input of cadmium. But in the UK only about 1.3% of the available agricultural land receives sludge each year and as shown above a normal annual application of sludge may be expected to increase the soil concentration of cadmium by about 0.05 mg/kg. The other main sources of cadmium for agricultural soils are aerial deposition and phosphatic fertilizer. Surveys in different parts of the world have shown that the latter contain 1–160 mg Cd/kg according to the origin of the rock phosphate from which they were prepared[13]. Whilst only a small percentage of agricultural land receives sludge, inputs from aerial deposition and from phosphatic fertilizers apply to all agricultural land. Hansen and Tjell[22] have assessed cadmium inputs to agricultural land in Denmark on a national and a local basis. Nationally, sludge contributed 5% of the cadmium; the remainder resulted from aerial deposition (70%) and inorganic fertilizers (25%). Locally, assuming that sludge containing 7 mg Cd/kg was applied at a rate of 5 t ds/ha yr, the contribution from sludge accounted for 90% of the total cadmium input, compared with 8% from aerial deposition and 2% from inorganic fertilizer. More recently, the input of cadmium to background soils in Denmark has been estimated at 5.1 g/ha yr[49]. The figure was composed of 3 g/ha yr from fertilizers, 2 g/ha yr from atmospheric deposition and just 0.1 g/ha yr from sewage sludge. Taking the EC countries together Hutton[28] has estimated the total cadmium input to arable soils in areas away from localized contamination to be 8 g/ha yr. Phosphatic fertilizer was thought to contribute 5 g/ha yr and atmospheric deposition 3 g/ha yr. The contribution from sewage sludge applications was considered to be too small on a national or regional basis to warrant inclusion. This helps to put concern about sludge into perspective.

The significance of cadmium accumulations in soils

The significance or hazard of cadmium accumulations in soils depends largely on how much of the cadmium is available for uptake by crops. The possibility of hazard resulting from leaching of cadmium through the soil profile into groundwater is likely to be minimal in agricultural soils treated with sludge, because these will normally be limed to a pH value of at least 5.5 and probably 6. Plant-available cadmium in soil can be assessed directly by means of plant analysis, indirectly by means of soil analysis or more desirably by a combination of the two. Novel suggestions for assessing the bioavailability of cadmium in soil include the use of earthworms which accumulate cadmium in sludge-treated soil[25].

Soil analysis

The 'total' amount of cadmium in soil is conventionally determined by strong acid extraction, often using nitric acid on its own or in admixture with hydrochloric or perchloric acid, before analysis of the digest by atomic absorption spectrophotometry (AAS). Instrumental methods such as neutron activation analysis and or X-ray fluorescence spectrophotometry (XRFS) can also be used. These methods involve minimal sample preparation and should in theory give a truer estimate of the total amount of cadmium present although they may be less accurate for measuring concentrations below 1 mg Cd/kg in soil. XRFS in particular has a higher limit of detection

than AAS. Total concentrations are essential for assessing the extent of contamination by cadmium and guidelines for the use of sludge on land normally seek to control the total concentration of cadmium in soil. But however it is measured, the total concentration gives no insight into the significance of the observed contamination. This depends instead on the potential solubility of the cadmium which in turn determines its availability for crop uptake. Soil scientists have therefore used a variety of mild extractants as rapid chemical tests for assessing availability. Unfortunately, because conventional extractants tend to be insensitive to the soil properties which control the availability of metals to plant roots, for example pH value, it is often as difficult to relate extractable concentrations of metals in soil to probable plant uptake as it is total concentrations. Improved correlations between extractable Cd, and Cd in plant tissue, are obtained when results from different soils are grouped together on the basis of their chemical properties. Symeonides and McRae[46] have drawn attention to the frequent lack of theoretical justification in the selection of extractants and have emphasized the importance of testing extractants against plant performance before advocating their use. These authors recommend the use of ammonium nitrate (1 M) for assessing plant-available cadmium in soil. Other favored extractants include neutral ammonium acetate (1 M) and the chelating agent DTPA which was originally developed to identify soils where crops were likely to suffer from zinc deficiency[32]. Sauerbeck and Rietz[42] have recently evaluated 25 extracting solutions and concluded that the best results could be obtained by weakly acidic but well buffered extractants.

The most important development in this area has been the recognition that extractants are best used not to assess availability directly but to identify the major forms of cadmium in soil. Sequential extraction techniques are proving increasingly popular for this purpose. The original procedure of McClaren and Crawford[33] for identifying the main forms of copper in soil was modified by Stover et al.[45] to examine forms of various metals in sludge (see above). Following studies with model trace metal compounds, the procedure has now been further modified at the University of California, Riverside, to give data concerning the fractions of trace metals in exchangeable, sorbed, organic, carbonate, and sulphide forms. These solid-phase chemical forms are determined by sequential extraction with the following reagents respectively; potassium nitrate (0.5 M) 16 h, deionized water (extract 3 times and combine data), sodium hydroxide (0.5 M) 16 h, disodium salt of EDTA (0.05 M) 6 h, nitric acid (4 M) 16 h at 80 °C. Results so far obtained suggest that much of the cadmium in sludge occurs as carbonate and in neutral soils it persists in this form[44]. There is some evidence that cadmium introduced to the soil in sludge gradually reverts towards residual forms extracted by nitric acid but not by EDTA[19]. Sequential extraction studies on soils have usually found low levels of cadmium occurring in the exchangeable and adsorbed forms. For example, Emmerich et al.[19] found less than 3% and Sposito et al.[44] found 1.1–3.7%. This is of particular interest if, as seems likely, these forms represent the fraction which is most available for crop uptake.

It is now widely accepted that in order for uptake into roots to occur, soluble forms of cadmium must exist adjacent to the root membrane for some finite period[5]. This observation has led to greater efforts to understand the soil properties which influence concentrations of cadmium in soil solution. These properties are pH value, redox potential, texture, mineral composition (content of clays and oxides of iron and manganese) and profile characteristics, cation-exchange capacity, amount and type of organic compounds in the soil and soil solution, presence of other heavy metals (which may compete for adsorption sites, etc.), soil temperature and moisture content, and other factors which affect microbial activity[23]. These soil properties function interactively to control the solubility of cadmium in soil so the system is inevitably complex. The one generalization that can be made with some confidence is that a decrease in soil pH value will increase the solubility of cadmium which in turn increases crop uptake of the element. Guidelines for sludge utiliziation on land normally recommend that the soil pH value should be maintained at ≥ 6. A detailed discussion of this area was presented by Davis and Coker[13].

Advances in analytical techniques have permitted more detailed examination of the forms of cadmium in soil solution and their relationship with the solid phase. Using High Performance Liquid Chromatography (HPLC), Tills and Alloway[48] investigated the main cadmium containing species present in soil solution. They found that all the cadmium occurred in a peak associated with low molecular weight organic molecules which would also contain inorganic compounds and Cd^{2+} ions. They concluded that most of the cadmium in the soil solution of sludged soils is cationic, confirming the observations of Bingham[3] and Bolt and Bruggenwert[4]. Apart from directly measuring the species of cadmium present in soil solution

these can be predicted with computer models which make use of known chemical equilibria and properties of the soils concerned. Using the GEOCHEM programme, Emmerich et al.[20] calculated that 50–60% of the total cadmium in soil solution occurred in the free ionic form. A similar model was developed at the Water Research Centre to predict the speciation and solubility of lead in drinking water[29].

A practical problem which has so far defied a convincing theoretical explanation concerns the apparent disappearance of some of the cadmium added to soil in sewage sludge. Many researchers have been unable to account fully for the cadmium introduced to soil in field and even laboratory trials. For instance, Chang et al.[8] recently reported investigations into the cadmium content of soil from a field trial 4 years after sludge applications ceased. They could account for only 43–60% of the cadmium originally added although there was no apparent movement beyond the depth of the soil profile examined (0–60 cm). Figure 1 shows another example from a field trial reported by Coker et al.[9]. More than a year after application of the sludge the soil concentrations of cadmium were lower than predicted although all the sludge was confined to the sampled area, worms were absent from the soil and leaching is unlikely to account for the loss since the soil in question was a calcareous boulder clay. Furthermore, crop uptake removed less than 1% of the cadmium added to the soil in sludge. Since the effect is always negative, random errors in sampling seem unlikely to explain the loss and in the example above changes in soil bulk density were taken into account. It is conceivable that the cadmium introduced to the soil reverts to forms which are not solubilized during the nitric acid digestion often used to determine total cadmium concentrations in soil but this explanation seems unlikely. This is an interesting problem with important implications which requires thorough investigation.

Crop uptake of cadmium

Apart from the influence of soil conditions described above, crop uptake varies according to the particular crop concerned. There are wide differences between species and also between cultivars of the same species. Figure 2 illustrates concentrations of cadmium found in the leaves and edible parts of a range of different crops grown to maturity in a pot trial with a soil containing 69 mg Cd/kg. Data of this type permitted Davis and Carlton-Smith[12] to develop a league table of crop sensitivity based on cadmium concentrations found in the edible parts. The most efficient assimilators of cadmium from soil were 3 cultivars of lettuce which scored 73–100 on the comparative scale used, followed by spinach (58), celery (47) and cabbage and kale (38–40). All the other crops tested were in the range (1–15) with 2 legumes (French bean and pea) at the bottom of the scale. An ornamental cultivar of tobacco was included in these uptake tests and it is of interest that this plant achieved a comparative rating of 310 based on the cadmium content of its leaves.

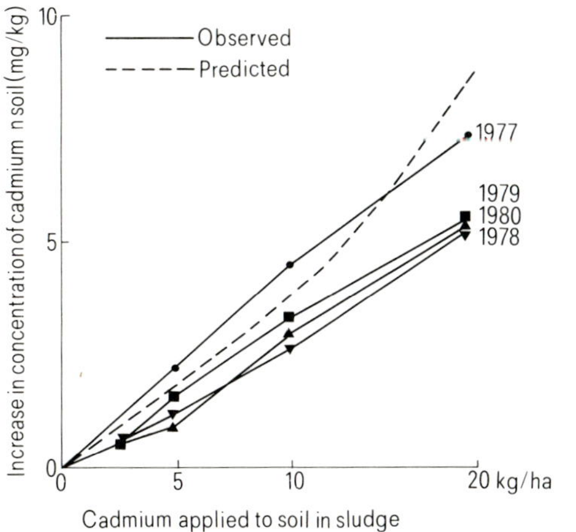

Figure 1. Observed and predicted concentrations of cadmium in soil in relation to cadmium applied in sludge (Coker et al.[9]).

Tobacco is an important source of human intake of cadmium; smoking a pack of 20 cigarettes is equivalent to 25 µg of cadmium in the diet[37]. Amongst crop plants, highest concentrations of cadmium may be expected to occur in the leaves of vegetables such as lettuce, spinach and other beets, and cabbages. Spinach is a member of the beet family (Chenopodiaceae) and this group are also able to assimilate other sludge-borne metals (nickel, copper and zinc) from soil with comparative ease. These plants tend to be rich in oxalate which has an affinity for metals and might explain the uptake ability of the beet family. This is speculation by Chaney[6] has recently reviewed plant physiological aspects of metal uptake. He observes that more research needs doing on the next step, this being the absorption by man of cadmium in ingested food. Jarvis et al.[30] examined the distribution of cadmium between the roots and shoots of 23 plant species after exposure to a nutrient solution containing 0.01 Cd/l. In all except 3 species (lettuce, kale and watercress) more than 50 per cent of the cadmium taken up was retained in the roots. The concentration in the roots was always greater than in the shoots, and in the fibrous roots of fodder beet, parsnip, carrot and radish it was greater than in the swollen storage roots. In the case of ryegrass approximately 88% of cadmium was retained in the roots. These authors concluded that although the roots of several species can

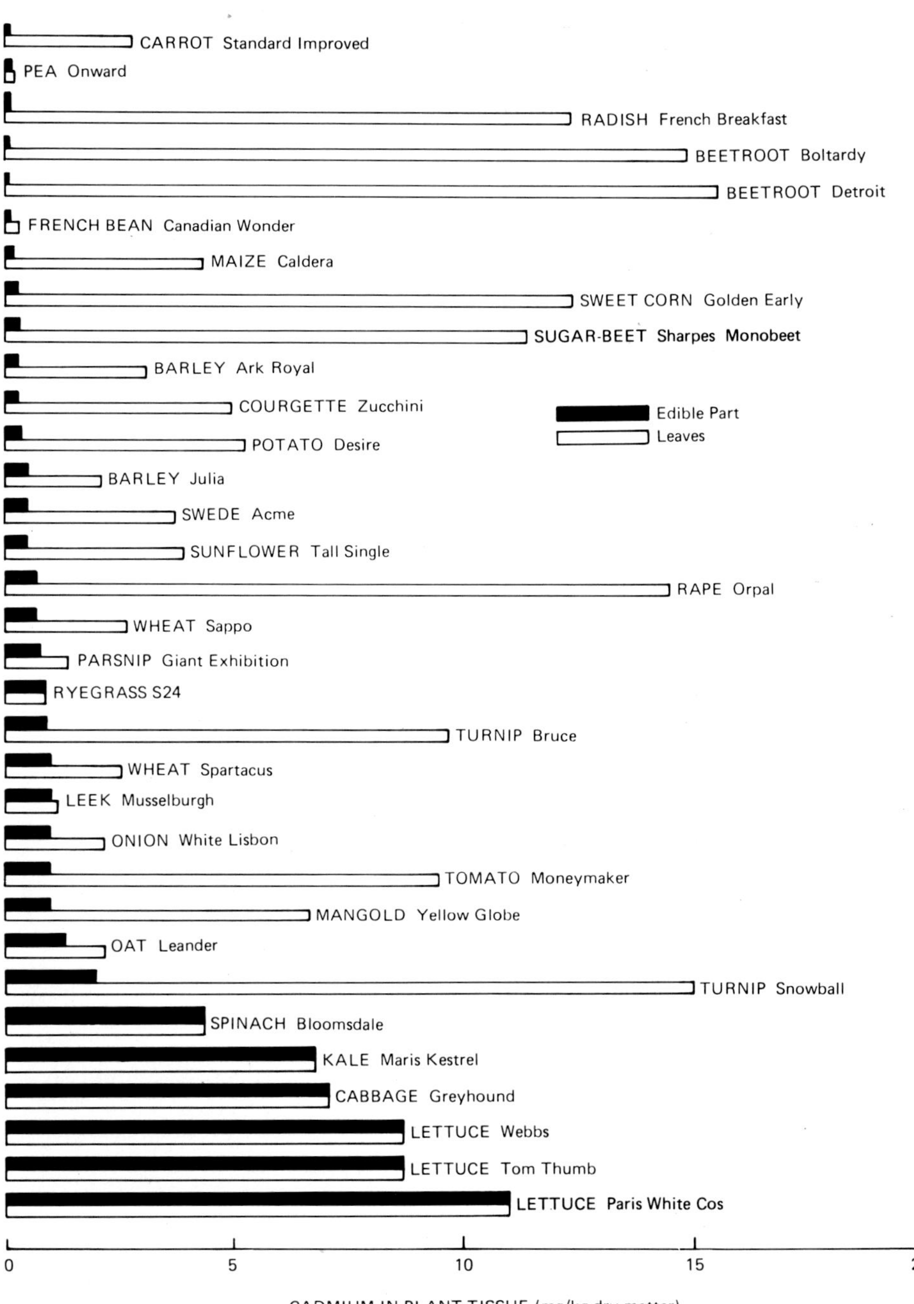

Figure 2. Cadmium concentrations in different crops all grown in the same soil (Davis and Carlton-Smith[12]).

take up large quantities of cadmium from solution, there are mechanisms (possibly associated with the solubility of cadmium phosphate) which may restrict the movement of cadmium through plants, and thus to animals. Although Chaney[6] cites exceptions, it appears in general that cadmium concentrations in plant parts decrease in the order fibrous roots > leaves > seeds = storage organs. Field sampled plants usuallly contain < 0.02–1.0 mg Cd/kg dry matter with a mean concentration of about 0.30 mg/kg; leafy crops will tend to contain cadmium concentrations towards the higher end of the range[13].

Concern about enhanced concentrations of cadmium in crop plants relates mainly to implications for the human foodchain. This is because cadmium is principally a zootoxic element; concentrations of cadmium in crops which are potentially harmful to man precede the concentrations that damage the crop itself. The foodchain could therefore be significantly contaminated by apparently healthy crops. For the principally phytotoxic elements (for instance copper, nickel and zinc) the opposite is true and the crops will display symptoms of toxicity before they contain concentrations which could damage human or animal health. In the case of cadmium, crop growth is unlikely to be affected until tissue concentrations exceed 10 mg Cd/kg dry matter, but concentrations of much less than this could lead to potentially harmful human dietary intake of the element.

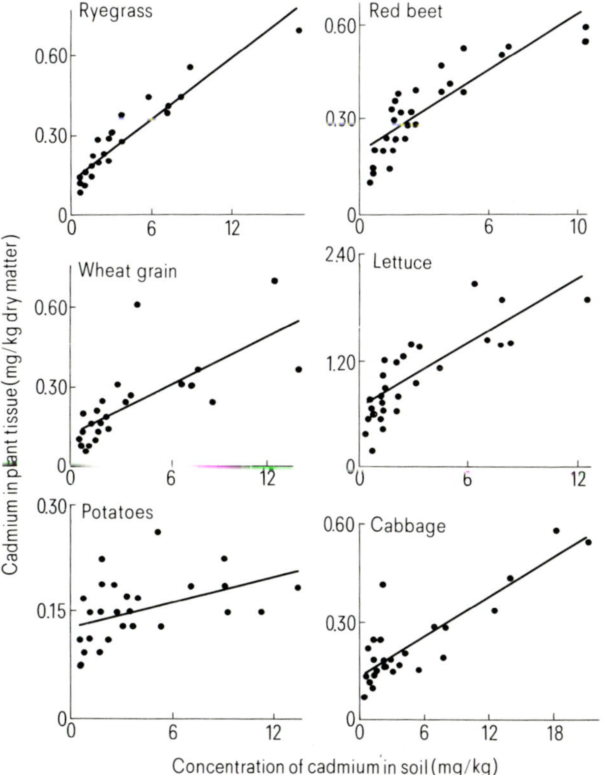

Figure 3. Results of a field trial showing the relationship between cadmium concentrations in soil and in six principal crops (Davis, Stark and Carlton-Smith[15]).

Implications for the human foodchain of cadmium in sludge applied to land

Cadmium is a cumulative poison so subtle increases in the diet sustained over long periods could conceivably lead to toxicity problems in man. Proper assessment of the problem requires that soil concentrations of cadmium can be related to potential increases in dietary intake of the element. In this way applications of sludge to agricultural land can be adjusted to keep soil concentrations of cadmium below levels likely to generate potentially a toxic dietary intake. For assessing potential dietary intake, reliable crop uptake data obtained in field conditions are needed from trials using sewage sludge. Much useful information can be obtained from agricultural investigations at sites dedicated for sludge disposal where cadmium levels have been built up gradually following a long history of sludge deposition (see Rundle et al.[40] for example). Whilst data from pot trials or even from experiments using inorganic cadmium salts are useful for some investigations these data must be eschewed for dietary modelling purposes. The drawbacks associated with such data were described recently by Chaney[6].

Data from a field trial with sludge are shown in figure 3. The details of this experiment were described by Davis and Stark[14] and Davis et al.[15]. There is an approximately linear and statistically significant relationship between the cadmium concentrations in the soil and cadmium concentrations in all 6 crops tested. As expected, highest concentrations of cadmium occurred in lettuce but significant increases occurred in all the other crops tested which included wheat grain and potato tubers. These latter two crops are the major plant components of the human diet, together supplying 74% of the plant part of the food intake of the average UK consumer[34]. Using charts of standard dietary intake in conjunction with the crop uptake data of figure 3, it has been possible to produce a preliminary model (fig. 4) relating concentrations of cadmium in sludge-treated soil to potential dietary intake of cadmium for an average consumer taking all his crops from sludge-treated soil. On this basis it is clear that small increases in the cadmium content of staple food such as potatoes and particularly wheat have a substantially more profound effect on potential dietary intake of cadmium than larger increases in concentration of the element in crops like lettuce. This would apply even to those who eat several times the normal amount of lettuce. In view of the importance of wheat (cereal products supply 40% of the plant part of the diet for the average UK consumer) it is of interest to take account of milling. In this case it was found that the flour contained only 57% of the cadmium in the whole wheat grain from which it was made. Thus potential dietary intake is reduced if an individual takes his wheat as flour instead of whole grains. No account is taken in the model of the influence of enhanced soil concentrations of cadmium on meat and dairy products for human consumption. It was felt that these are insensitive to increases in the cadmium content of soil except perhaps for offal (liver and kidney) which composes only about 0.5% of the diet on a fresh weight basis[34].

Figure 4 shows that there is a linear relationship between cadmium in soil and potential dietary intake of cadmium. The gradient of the slope is greater if cereals are eaten as whole grain rather than flour.

One basis for an objective soil concentration limit for cadmium would be that concentration which produces a potential dietary intake of 70 µg/day, the value calculated independently by both the World Health Organisation[51] and the United States Environmental Protection Agency[21] as being the maximum acceptable daily intake of cadmium. For the whole grain diet this value is generated by a soil concentration of 6.24 mg Cd/kg and for the flour diet the soil concentration is 12.90 mg Cd/kg. Tolerable accuracy of the model is suggested by the finding that unsludged soil could generate a dietary intake of about 20 µg/day, close to the estimated value for UK citizens[17]. On this basis, it is suggested that a soil concentration in the range 6.0–12.0 mg Cd/kg (equivalent to a loading rate of 10–22 kg Cd/ha) is acceptable for calcareous soils (pH value 7–8) receiving sludge. It is intended to improve and consolidate the model used to arrive at this value and to include data from acidic soils, but it has the following built-in safety factors:

a) Only 1.27% of agricultural land in the UK receives sludge each year. The average dressing increases the soil concentration of cadmium by about 0.05 mg/kg/yr. It is almost inconceivable that any individual would live on crops taken wholly from sludge-treated soil.

b) Whilst the model is based on an average consumer (who, in reality, does not exist), an unusual diet consisting for instance of 5 times the normal intake of lettuce would have little effect on dietary intake of cadmium. Wheat is much more important but few people grow their own supplies. This crop is usually collected and stored centrally so any contaminated sample would be diluted out. The diet of most individuals would include at least some refined grain (flour) of lower cadmium content.

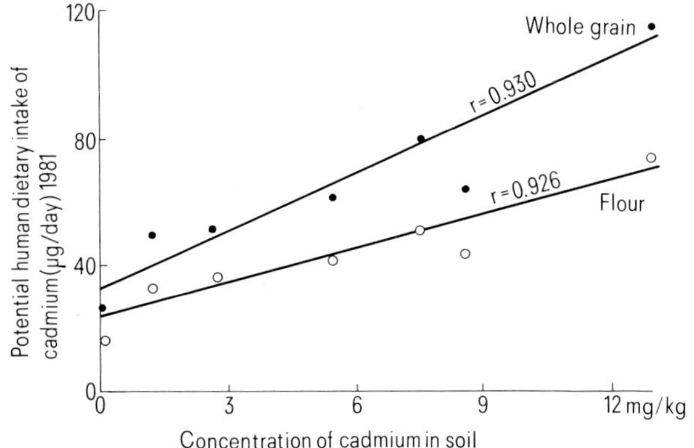

Figure 4. The relationship between cadmium concentrations in soil and potential dietary intake of cadmium for an average consumer taking all his/her crops from sludge-treated soil (Davis, Stark and Carlton-Smith[15]).

c) The WHO/EPA maximum acceptable daily intake of cadmium of 70 µg/day is not a toxic threshold but has its own safety factor. It is estimated that 200 µg/day of cadmium sustained over a 50-year period would be needed to produce kidney damage (of doubtful clinical significance) in the most sensitive individual. (See Davis and Coker[13], and Naylor and Loehr[35]).

The other possible pathway for the transfer of cadmium into the human foodchain is in meat and dairy products from animals fed with fodder crops grown on sludge-treated soil. Farm animals probably absorb < 5% of the cadmium they ingest. Ryan et al.[41] have summed up the findings of numerous investigations into the effects of elevated cadmium intake on the cadmium concentration of the muscle tissue of experimental animals. They reported almost unanimous agreement that cadmium does not seek muscle tissue and no significant differences are found between experimental animals and control animals. This finding was independent of whether the animals graze on grass grown on sludge-amended soil or are given an inorganic cadmium source in their food. Exceptions to this rule are associated with very high levels of cadmium in the animals diet which would be most unlikely to occur in practice.

In addition to muscle tissue, it has also been found that cadmium does not accumulate in other important animal products. Thus Crossmann and Seifert[11] report that recommended limits (in Germany) for cadmium in milk (0.0025 ppm of cadmium) and eggs (0.05 ppm cadmium) will only be exceeded if animal feed exceeds 50 ppm and 13 ppm of cadmium respectively. Sharma et al.[43] concluded that cadmium concentrations in animal feed of up to 10 mg/kg would cause no appreciable increase in the cadmium content of meat, milk or eggs. Constraints on sludge utilization on land are such that cadmium originating from sludge could never approach these levels in animal feed. Since cereal grains are poor assimilators of cadmium from soil (fig. 2) and farm animals do not accumulate cadmium in their muscle tissue, problems due to cadmium in sludge used on land can be minimized if the land is used to grow grain for animal feed. According to Ryan et al.[41] there would in these circumstances be no need for a cadmium soil limit to protect public health. Land use of this kind may therefore be the best option if farming is to take place on soils highly contaminated with cadmium. The average human diet includes only small amounts (see above) of offal (liver and kidney) which may accumulate cadmium and will only pose a problem for unusual consumers eating large quantities of it. Normally, dilution through the marketing process would reduce the risk considerably. Also farm animals for human consumption are short-lived so the question of long-term accumulation in the visceral organs does not arise.

One further aspect of animal nutrition must be mentioned and this relates to the inadvertent ingestion of soil and sludge by animals grazing sludgetreated grassland. Grass grazed by cattle during the months

November-March may contain 3–6% soil on a dry matter basis[47], for sheep this may well be doubled[24]. Also, sludge adheres to grass and can be ingested as the grass is grazed[6]. However, a 'no-grazing period' is usually recommended between applications of sludge to pasture land and grazing by cattle. The principal purpose of this recommendation is to avoid problems of pathogen transmission but it also serves to minimize direct ingestion of sludge. The latter can also be reduced by keeping animals off sludge-treated pasture in the winter months when grass growth is slow and the soil is soft. Baxter et al.[2] have demonstrated the potential importance of direct ingestion in an animal experiment in which cattle were fed a diet containing 12% sludge (cadmium content 88 mg/kg ds) for a 9-month period. In these experimental conditions increases in the cadmium content of kidney, liver and even muscle tissues occurred.

In practice, problems due to cadmium in sludge applied to grassland seem minimal. Nelmes et al.[36] investigated a farm with a long history of sludge deposition carried out on a sacrificial basis where soil concentrations of cadmium ranged up to 29 mg/kg and the cadmium content of herbage was 5–10 mg/kg dry matter in autumn and winter and 4–5 mg/kg in summer. These levels are much higher than is usually permitted for agricultural land receiving sludge. Nevertheless, Nelmes et al.[36] found that the health of grazing animals on this farm was unaffected by cadmium and that the transfer of cadmium to man in foodstuffs from dairy or beef cattle on the farm was minimal. Direct ingestion is avoided if the sludge is injected below the surface of grassland.

Control of contamination problems due to cadmium in sludge used as fertilizer

In most countries where sludge is used in agriculture there are guidelines which restrict soil concentrations of cadmium to a level judged to be safe. In the UK, the maximum recommended soil concentration of cadmium where sludge is used on land is 3.5 mg/kg to be approached gradually over a 30-year period[18]. It is shown above that the average annual application of sludge designed to meet crop requirements for nutrients would increase the soil concentration of cadmium by perhaps 0.05 mg/kg so about 250 t ds/ha of a sludge containing 20 mg Cd/kg ds could be added over the 30-year period before the permissible maximum was reached. The limit of 3.5 mg/kg in soil therefore allows enough sludge to be applied to the land for the operation to be beneficial to farmers and to permit economic sludge disposal. First and foremost, it is essential that the limit is environmentally acceptable in providing adequate protection to the human foodchain. The model described above which relates potential dietary intake of cadmium to soil concentrations of cadmium suggests that the 3.5 mg Cd/kg limit in soil is acceptable. The Commission of the European Communities[10] has proposed a limit of 3.0 mg Cd/kg. Webber and Monks[50] have suggested an international loading rate limit of 5 kg Cd/ha which is equivalent to a soil concentration limit of about 3.5 mg Cd/kg. They consider that such a loading rate represents little if any hazard to the foodchain. In the USA the Environmental Protection Agency[21] has recommended a maximum loading rate of 5 kg Cd/ha for soils of low cation exchange capacity (< 5 meq/100 g). There is increasing agreement that a limit of about 3.5 mg Cd/kg is environmentally acceptable for soils receiving sludge. Provided that rates of application of sludge to land can be controlled and a soil limit observed, there is no need for a concentration limit for cadmium in sludge itself. Such a limit is needed, however, if sludge is to be supplied to small farmers or gardeners in circumstances where the rate of application of sludge will not be supervised. UK guidelines[18] recommend a limit of 20 mg Cd/kg ds in sludges supplied to the general public but no limit is considered necessary in sludges for agricultural use.

A better understanding of the chemistry and bioavailability of cadmium in soils should lead to a position where soil concentration limits can be modified according to soil conditions. In developing this approach the EPA[21] has devised a system in which limits for cadmium vary according to the pH value and cation exchange capacity of the receiving soil. Certainly there would appear to be a case for a less stringent limit for calcareous soils (pH value > 7 and containing more than 5% calcium carbonate), where the soil pH value is likely to remain permanently near neu-

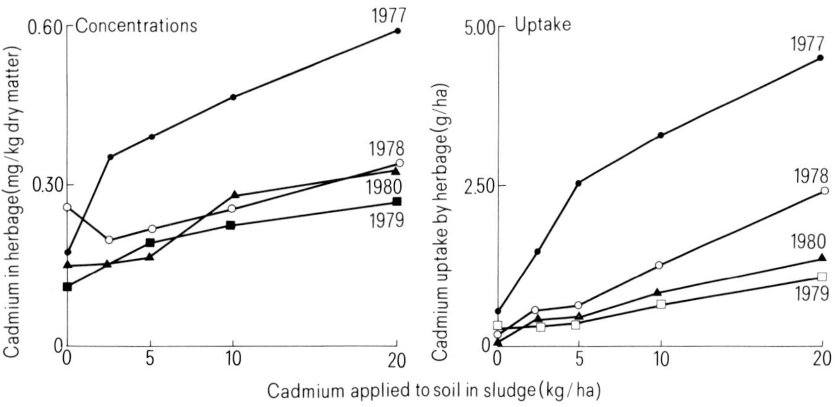

Figure 5. The effect of time on the availability to herbage (mainly ryegrass *Lolium perenne* cv. S23) of cadmium in sludge-treated soil. Sludge was applied to soil in 1976 (Coker et al.[9]).

tral. Also, there is a need to continue the development of management practices, such as growing grain for animal feed only, which minimise cadmium contamination of the foodchain and permit farming to continue on historically contaminated land.

A challenging problem from the research point of view is to predict the availability in the long-term of cadmium introduced to the soil in sludge. This question has important environmental implications. There would be less concern about cadmium in sludge if it could be confirmed that uptake of cadmium from sludge-treated soil decreases with time after sludge is added to the soil even after sludge applications stop, and that cadmium accumulations in soil cannot suddenly be released for plant uptake following a change in management practice, bearing in mind that future land use cannot be predicted. In near-neutral soils it appears that cadmium originating from sludge does gradually revert to less available forms. Figure 5 presents some data to support this observation from Coker et al.[9]. Hinesly et al.[26,27] have produced similar data for maize which led them to conclude that plant uptake depended principally on the amount of cadmium applied immediately before planting rather than the total amount applied in previous years. It is quite possible that once equilibrated with soils of near neutral pH value, cadmium accumulations remain immobilized in forms permanently unavailable for plant uptake. Experimental evidence is still awaited to support the widely held view that organic matter in sludge keeps cadmium in forms unavailable for plant uptake, and that this cadmium may be released when sludge organic matter decays in soil (the 'time-bomb' effect). Nevertheless, current knowledge about the long-term availability of cadmium is incomplete, especially with respect to non-calcareous soils (see Chaney et al.[7]).

1 Archer, F.C., Trace elements in soils in England and Wales, in: Inorganic Pollution and Agriculture. Ministry of Agriculture, Fisheries and Food (UK), Reference Book RB 326. Her Majesty's Stationery Office, London 1980.

2 Baxter, J.C., Barry, B., Johnson, D.E., and Kienholz, E.W., Heavy metal retention in cattle tissues from ingestion of sewage sludge. J. envir. Qual. *11* (1982) 616–620.

3 Bingham, F.T., Bioavailability of cadmium to food crops in relation to heavy metal content of sludge-amended soil. Envir. Hlth Perspect. *28* (1979) 39–43.

4 Bolt, G.H., and Bruggenwert, M.G., Soil Chemistry. A) Basic Elements, Cadmium, pp. 222–223. Elsevier Scientific Publishing Company, Amsterdam 1976.

5 Cataldo, D.A., and Wildung, R.E., Soil and plant factors influencing the accumulation of heavy metals by plants. Envir. Hlth Perspect. *27* (1978) 149–159.

6 Chaney, R.L., Fate of toxic substances in sludge applied to cropland. Proceedings of an International Symposium on Land Application of Sewage Sludge, October 13–15, 1982, Tokyo, Japan, in press.

7 Chaney, R.L., Hundemann, P.T., Palmer, W.T., Small, R.J., White, M.C., and Decker, A.M., Plant accumulation of heavy metals and phytotoxicity resulting from utilization of sewage sludge and sludge composts on cropland, in: Proceedings 1977 National Conference on Composting of Municipal Residues and Sludges, pp. 86–97. Information Transfer Inc., Rockville, Md 1978.

8 Chang, A.C., Page, A.L., and Bingham, F.T., Heavy metal adsorption by winter wheat following termination of cropland sludge applications. J. envir. Qual. *11* (1982) 705–708.

9 Coker, E.G., Davis, R.D., Hall, J.E., and Carlton-Smith, C.H., The use of sewage sludge in land reclamation. Water Research Centre Technical Report TR 183, 1982.

10 Commission of the European Communities. Proposal for a Council Directive on the use of sewage sludge in agriculture. Official Journal of the European Communities No. C 264/3–264/7, 8 October 1982.

11 Crossmann, G., and Seifert, D., Contamination of soil in agricultural areas by sewage sludge and aspects of transfer and carry-over to plants and farm animals. Proceedings of the Third International Cadmium Conference, Miami 1981, pp. 82–87. Metal Bulletin, London.

12 Davis, R.D., and Carlton-Smith, C.H., Crops as indicators of the significance of soil contamination by heavy metals. Water Research Centre Technical Report TR 140, 1980.

13 Davis, R.D., and Coker, E.G., Cadmium in agriculture, with special reference to the utilisation of sewage sludge on land. Water Research Centre Technical Report TR 139, 1980.

14 Davis, R.D., and Stark, J.H., Effects of sewage sludge on the heavy metal content of soils and crops, in: Second European Symposium on Characterization, Treatment and Use of Sludge, Commission of the European Communities Conference, Vienna, October 1980, pp. 687–698. Eds P. L'Hermite and H. Ott. D. Reidel, Dordrecht, Holland, 1981.

15 Davis, R.D., Stark, J.H., and Carlton-Smith, C.H., Cadmium in sludgetreated soil in relation to potential dietary intake of cadmium. Paper presented to a Seminar of Working Party 5 (Environmental Effects) of the Commission of the European Communities Concerted Action (Cost ter) on the Treatment and Use of Sludge held at the Water Research Centre, Stevenage, UK, May 26, 1982. D. Reidel, Dordrecht 1983.

16 Department of the Environment (UK). Survey of sewage sludge composition 1978 (unpublished report).

17 Department of the Environment (UK). Cadmium in the environment and its significance to man. Pollution Paper No. 17. Her Majesty's Stationery Office, London 1980.

18 Department of the Environment (UK). Report of the Subcommittee on the Disposal of Sewage Sludge to Land. Department of the Environment/National Water Council Standing Technical Committee Report 20. National Water Council, Queen Anne's Gate, London, SW 1, 1981.

19 Emmerich, W.E., Lund, L.J., Page, A.L., and Chang, A.C., Solid phase forms of heavy metals in sludge-treated soils. J. envir. Qual. *11* (1982) 178–181.

20 Emmerich, W.E., Lund, L.J., Page, A.L., and Chang, A.C., Predicted solution phase forms of heavy metals in sewage sludge-treated soils. J. envir. Qual. *11* (1982) 182–186.

21 Environmental Protection Agency (USA). Criteria for classification of solid waste disposal facilities and practices. Federal Register *44* (1979) 53438–53468.

22 Hansen, J.A., and Tjell, J.C., Guidelines and sludge utilisation practice in Scandinavia. Paper 20 in WRC Conference, Utilisation of Sewage Sludge on Land, Oxord 1978. Water Research Centre, 1979.

23 Harmsen, K., Behaviour of heavy metals in soils. Agricultural Research Report 866. Centre for Agricultural Publishing and Documentation, Wageningen 1977.

24 Healy, W.B., Ingested soil as source of elements to grazing animals, in: Trace Element Metabolism in Animals 2, pp. 448–450. Ed. W.G. Hoekstra. University Park Press, Baltimore, USA, 1974.

25 Helmke, P.A., Robarge, W.P., Korotev, R.L., and Schomberg, P.J., Effects of soil-applied sewage sludge on concentrations of elements in earthworms. J. envir. Qual. *8* (1979) 322–327.

26 Hinesly, T.D., Jones, R.L., Ziegler, E.L., and Tyler, J.J., Effects of annual and accumulative applications of sewage sludge on assimilation of zinc and cadmium by corn. Envir. Sci. Techmol. *11* (1977) 182–188.

27 Hinesly, T.D., Ziegler, E.L., and Barrett, G.L., Residual effects of irrigating corn with digested sewage sludge. J. envir. Qual. *8* (1979) 35–38.

28 Hutton, M., Cadmium in the European Community. Monitor-

ing and Assessment Research Centre Technical Report 26. MARC, Chelsea College, University of London 1982.

29 Jackson, P.J., and Sheiham, I., Calculation of lead solubility in water. Water Research Centre Technical Report, TR152, 1980.

30 Jarvis, S.C., Jones, L.H., and Hopper, L.J., Cadmium uptake from solution and its transport from roots to shoots. Plant Soil *44* (1976) 179–191.

31 Lester, J.N., Harrison, R.M., and Perry, R., The balance of heavy metals through a sewage treatment works. Sci. total Envir. *12* (1979) 13–23.

32 Lindsay, W.L., and Norvell, W.A., Development of a DTPA soil test for zinc, iron, manganese and copper. J. Soil Sci. Soc. Am. *42* (1978) 421–428.

33 McLaren, R.G., and Crawford, D.V., Studies on soil copper – the fractionation of copper in soils. J. Soil Sci. *24* (1973) 172–181.

34 Ministry of Agriculture, Fisheries and Food (UK). Household food consumption and expenditure 1979. Annual Report of the National Food Survey Committee. Her Majesty's Stationery Office, London 1981.

35 Naylor, L.M., and Loehr, R.C., Increase in dietary cadmium as a result of application of sewage sludge to agricultural land. Envir. Sci. Technol. *15* (1981) 881–886.

36 Nelmes, A.J., Buxton, R.J., Fairweather, F.A., and Martin, A.E., The implications of the transfer of trace metals from sewage sludge to man, in: Trace Substances in Environmental Health, vol.8, pp.145–153. Ed. D.B. Hemphill. University of Missouri, 1974.

37 Nordberg, G.F., and Kjellstrom, T., Metabolic model for cadmium in man. Envir. Hlth Perspect. *28* (1979) 211–217.

38 Parr, J.F., Epstein, E., Chaney, R.L., and Willson, G.B., Impacts of the disposal of heavy metals in residues on land and crops. Proceedings of 1977 National Conference on Treatment and Disposal of Industrial Wastewaters and Residues. Houston, Texas, 1977.

39 Poon, C.P.C., Removal of cadmium from wastewaters. Cadmium as a Complex Environmental Problem. Experientia *40* (1984) 127–136.

40 Rundle, H., Calcroft, M., and Holt, C., Agricultural disposal of sludges on a historic sludge disposal site. Water Pollut. Control *81* (1982) 619–632.

41 Ryan, J.A., Pahren, H.R., and Lucas, J.B., Controlling cadmium in the human foodchain: a review and rationale based on health effects. Envir. Res. *28* (1982) 251–302.

42 Sauerbeck, D.R., and Rietz, E., Soil-chemical evaluation of different extractants for heavy metals in soils. Paper presented to a Seminar of Working Party 5 (Environmental Effects) of the Commission of the European Communities Concerted

Action (Cost ter) on the Treatment and Use of Sewage Sludge, held at The Water Research Centre, Stevenage, UK, May 26, 1982. D. Reidel, Dordrecht 1983.

43 Sharma, R.P., Street, J.G., Verna, M.P., and Shupe, J.L., Cadmium uptake from feed and its distribution to food products of livestock. Envir. Hlth Perspect. *28* (1979) 59–66.

44 Sposito, G., Lund, L.J., and Chang, A.C., Trace metal chemistry in aridzone field soils amended with sewage sludge. J. Soil Sci. Soc. Am. *46* (1982) 260–264.

45 Stover, R.C., Sommers, L.E., and Silviera, D.J., Evaluation of metals in wastewater sludge. J. Water Pollut. Control Fedn *48* (1976) 2165–2175.

46 Symeonides, C., and McRae, S.G., The assessment of plant-available cadmium in soils. J. envir. Qual. *6* (1977) 120–123.

47 Thornton, I., Biogeochemical and soil ingestion studies in relation to the trace-element nutrition of livestock, in: Trace Element Metabolism in Animals, vol.2, pp.416–454. Ed. W.G. Hoekstra. University Park Press, Baltimore, USA, 1974.

48 Tills, A.R., and Alloway, B.J., Cadmium speciation in soil solutions of sewage sludge amended soils. Paper presented to a seminar of Working Party 5 (Environmental Effects) of the Commission of the European Communities' Concerted Action (Cost ter) on the Treatment and Use of Sewage Sludge, held at the Water Research Centre, Stevenage, UK, May 26, 1982. D. Reidel, Dordrecht 1983.

49 Tjell, J.C., Hansen, J.A., Christensen, T.H., and Hovmand, M.H., Prediction of cadmium concentrations in Danish soils in The Second European Symposium of Characterisation, Treatment and Use of Sewage Sludge, Vienna, October 1980, pp.652–664. Eds P. L'Hermite and H. Ott. D. Reidel, Dordrecht, Holland, 1981.

50 Webber, M.D., and Monks, T.L., Cadmium concentrations in field and vegetable crops – a recommended maximum cadmium loading to agricultural soils. Paper presented to a Seminar of Working Party 5 (Environmental Effects) of the Commission of the European Communities' Concerted Action (Cost ter) on the Treatment and Use of Sewage Sludge, held at the Water Research Centre, Stevenage, UK, May 26, 1982. D. Reidel, Dordrecht 1983.

51 Wood, L.B., King, R.P., and Norris, P.E.E., Some investigations into sludge – amended soils and associated crops and the implications for trade effluent control. Paper 16 in WRC Conference, Utilisation of Sewage Sludge on Land, Oxford, 1978. Water Research Centre, 1979.

52 World Health Organisation. Evaluation of certain food additives and the contaminants mercury, lead and cadmium. WHO Technical Report Series No.505. WHO, Geneva 1972.

Bioaccumulation of cadmium in marine organisms

by S. Ray

Department of Fisheries and Oceans, Fisheries and Environmental Sciences, Biological Station, St. Andrews (N.B. EOG 2XO, Canada)

Introduction

Cadmium (Cd) occurs in very low concentrations in open ocean water, averaging about 40 ng/l in unpolluted surface waters[57]. Similarly, Eaton[43] and Bewers et al.[10] suggested the background concentration of Cd for North Atlantic surface waters to be 40–60 ng/l. The level for Pacific oceanic water is 36 ng/l[19]. Increased concentrations have been observed in the Mediterranean, Baltic, and North Sea, where circulation and water mass turnover are limited. Cadmium level in coastal and estuarine water normally is higher, primarily due to weathering and anthropogenic inputs; levels higher by several orders of magnitude have been reported[1,13,31,68,82,133].

Cadmium bioaccumulation by marine organisms has been the subject of considerable interest in recent years because of serious concern that high levels of Cd may have detrimental effects on the marine organisms and may create problems in relation to their suitability as food for humans.

Marine geochemistry of Cd has been discussed by Eaton[43] and Boyle et al.[19]. It is well established that, although concentration of Cd in surface water may be less than 10 ng/l, it increases to a maximum of about 125 ng/l at about 1000-m depth and then decreases slightly at lower depths[8,19,22,78,89,117]. Nriagu[100] estimated that atmospheric input of Cd from the lithosphere to the oceans is 2.4×10^9 g Cd/yr, while the annual input via stream runoff is 7.5×10^9 g, and that most of the net gain of Cd to the oceans is due to human activities.

Cadmium residence time[11,15,19,100] in ocean water has been estimated at 0.7×10^4 to 25×10^4 yr; for an estuarine coastal system it has been estimated[142] to be only 2 yr; in particulate matter[100] it is also very low and estimated to be 1.3 yr.

1. Chemical form in marine environment

Cadmium may be present in one or all three phases of the marine environment: water, particulate matter, and sediment, and may be in equilibrium with each other. The transfer rates between phases will vary, depending upon local conditions.

Bioaccumulation of Cd by marine organisms is governed in part by chemical form of Cd. Progress has recently been made by using electrochemical techniques to speciate the dissolved Cd in filtered sea water. Selective chemical extraction techniques have been applied to determine the forms of Cd in sediment. However, the extraction procedures rarely are specific for the different forms and can provide only a qualitative picture.

1.1. Water

Metals in solution are classified as free hydrated ions and complexed labile and non-labile metal species. The complexing groups may be both inorganic and organic components that are present in sea water. The predominant inorganic component is chloride. The organic ligands are polyphenols, aminoacids, humates, proteins, and other tissue breakdown products. Synthetic chelating agents like EDTA and NTA may also be present in coastal and estuarine areas. The degree of complex formation and its nature depend upon macrocomponents like alkali earth metals, chelating ligands, and the stability constant ratios for each of the complexes, and is influenced by physicochemical parameters such as salinity and pH. Sillen[130] computed the stable species of several metals in sea water at a representative pH of 8.1 ± 0.2 and concluded that Cd is present primarily as a chloride complex. Other workers[5,42,64,83,150] have also suggested that Cd is present mostly in the 'ionic Cd' form (sum of all free Cd ions and all labile complexes), predominantly in a variety of chloride complexes. Gardner[53] calculated that Cd in anoxic sulphide marine waters would be present as soluble bisulphides, with insignificant complexation with organic compounds.

Considerable differences of opinion exist regarding the presence of Cd bound to organic ligands. Duinker and Kramer[41] suggested that Cd in North Sea water is not bound to organic matter and is completely in the form of chloro complexes. However, Batley and Florence[6] have shown that, in clean surface sea water, 90% of Cd is bound to organic ligands.

1.2 Particulate matter

Cadmium uptake by particulate matter in sea water is negligible. Hydrous manganese oxides show appreciable adsorption of Cd. Most of the Cd in particulates in the marine environment is associated with faecal pellets. Sodium and other ions compete with Cd^{2+} for surface sites of the particulates. The process depends on the pH and ionic strength of the seawater medium.

1.3 Sediment

Sediments may play a key role in determining relative bioaccumulation factors in biota of the marine environment and may determine the metal concentrations in associated waters. The availability of Cd for bioaccumulation by marine organisms is related to the chemical species in the particulate matter and sediment, and also depends upon particle size, organic matter content, and ion-exchange capacity of the sediment[123]. Chemical extraction techniques have been used to determine metals associates with ion-exchangeable, easily reducible, moderately reducible, organic and residual silicate fractions. Metals associated with the organic fraction and deposited as a surface coating may be brought into solution and become bioavailable by physicochemical[87] changes in water.

2. Distribution in biota

A large number of analyses of marine organisms on a world-wide basis has established the presence of Cd in almost all the organisms. The class with the greatest number analyzed is Pisces followed by bivalves and crustaceans. Data on organisms at the lower level in the food chain (phyto- and zooplankton) are scarce. However, the range of marine organisms analyzed for Cd content encompasses bacteria to vertebrates. The observed[113] bioconcentration factors are of the order of 10^4, 10^2–$10^3(10^4)$, 10^3–$10^4(10^5)$, 10^3 and 10^2 for plankton, seaweed, mollusc, crustacean, and fish, respectively. Bryan[25] has collated the concentration of Cd in whole marine organisms (mollusc shell excluded) from uncontaminated areas as follows: phytoplankton – 2, brown seaweed – 1.2, copepod – 4, coelenterate – 0.2, polychaete worm – 0.1, oyster – 15.5, mussel – 5.1, limpet – 1–12, periwinkle – 2, dogwhelk – 23, cephalopod mollusc – 1, tunicate – < 1, and finfish – 0.2 µg Cd/g (dry wt). The mollusc's great ability to concentrate Cd from the environment has led to suggestions[58,59] for their use as sentinel organisms for Cd and several other marine pollutants. The literature on Cd bioaccumulation in marine organisms is replete with unreliable results[66] because of failure to use standard reference materials to ensure quality control of analytical data and because of differences in sampling techniques. The problem is further aggravated by the practice of reporting results on wet- or dry-weight basis, and quite often without reporting the basis. Occasionally, the organims themselves are not identified properly. Most studies have been limited to the Northern Hemisphere and few data are available for organisms from the tropics.

Cadmium accumulation in marine organisms in studies up to about 1977 has been reviewed by Coombes et al.[33] and Phillips[112]. More recent studies are reviewed annually in literature issues of the Journal of Water Pollution Control Federation.

3. *Source of accumulation*

Bioaccumulation of Cd by a number of marine organisms has been studied in the laboratory and in field conditions. The organisms can accumulate Cd directly from water or indirectly from food and detrital particles, and subsequently translocate it throughout the body by active and passive transport mechanisms.

3.1 *Water*

Cadmium bioaccumulation from water can take place either by passive diffusion through body surfaces or from water passing through the gills and subsequently through the body.

Plankton

Little work has been done on Cd bioaccumulation by marine planktonic species. Cossa[34] studied the diatom *Phaeodactylum tricornutum* and observed that Cd uptake varied with the growth phase of the culture and was controlled by adsorption process. Cadmium accumulation by the marine alga *Porphyra umbilicalis*[90] was dependent upon light conditions and had an initial rapid uptake phase followed by a period of slow, sustained uptake.

Molluscs

Brooks and Rumsby[21] found that the concentration factors for Cd in oysters *Ostrea sinuata* decreased steadily with increase in exposure concentration and the concentrations of Cd in the tissues decreased in the order gill > heart > visceral mass > mantle > white muscle > striated muscle. *Crassostrea virginica*[44,129] exposed to Cd rapidly accumulated high levels of the metal in soft tissues. Maximum concentration of Cd accumulated in the animals was independent[129] of exposure concentration.

Crassostrea gigas and *Ostrea edulis* collected from a clean environment and transplanted for a period of 5 months in polluted water containing an average of 5.7 µg Cd/l accumulated[17] 15 and 17 times, respectively, the concentrations of Cd in control samples. *C. virginica*[147,149] exposed to 5, 10, and 15 µg Cd/l sea water for 40 weeks at ambient temperature and salinity had as much as 89, 176, and 292 µg Cd/g (dry wt), respectively, in the whole-body soft tissues without reaching equilibria. The Cd concentration in the soft tissues was linearly related[46,149] to the exposure concentration and had a curvilinear relationship with times of exposure. The accumulation rate also varied with exposure concentration. The gill, mantle, and viscera showed different accumulation rates and uptake patterns, and the tissue concentrations were gill > viscera > mantle while the Cd content was greatest in viscera > gill > mantle.

In *Saccostrea commercialis*[138] exposed to several concentrations of Cd in flowing sea water, the gill tissues always had the highest concentration of Cd at any sampling time, generally in the order gill > viscera ≃ mantle > muscle in all experiments other than in the control group where the viscera consistently had the highest Cd concentration, suggesting that the gill was the critical organ for Cd uptake.

Greig[61] exposed quahaugs, surf clams and oysters to 10 and 20 µg Cd/l sea water for 43 d. At the end of 10 µg Cd/l exposure, the Cd concentrations in quahaugs, clams, and oysters were 2, 2, and 4 times the concentration in control animals. The corresponding values for exposure at 20 µg Cd/l were 2, 4, and 8 times, respectively.

George and Coombs[54] reported an initial lag period in uptake of Cd in mussel *Mytilus edulis*, but a subsequent linear relationship with time and exposure concentration. Similar results were obtained by Jackim et al.[72] for 3 filter-feeding (*M. edulis, Mulinia lateralis,* and *Mya arenaria*), and 1 deposit-feeding bivalve *(Nucula proxima)* and by other workers[62,73,128] for *M. edulis*.

The scallops *Aquipecten irradians* and *Argopecten irradians* were also observed[29,44] to rapidly accumulate Cd when exposed to spiked water.

Crustaceans

Whole-body Cd concentrations in shrimps *Penaeus duorarum*[95], *Palaemonetes vulgaris*[95], and *P. pugio*[47,95], exposed to a range of 75–1000 µg Cd/l were in all cases directly proportional to the concentration in exposure solution and to time of exposure. A similar relationship has also been observed for *Palaemon elegans*[141]. Cadmium bioaccumulation occurred[95] in *P. duorarum* at as low as 2.0 µg Cd/l, and tissue concentration decreased in the order hepatopancreas > gill > exoskeleton > muscle > serum. Selective localization of cadmium in shrimps has also been reported by Ray et al.[120] and by Dethlefsen[39]. Similar results have been reported by Establier[48] with shrimp *Penaeus kerathurus*, by Benayoun et al.[7] with euphausiids *Meganyctiphanes norvegica*, and by Rainbow et al.[116] with barnacle *Semibalanus balnoides*.

O'Hara[101,102] studied Cd bioaccumulation in fiddler crab *Uca pugilator*, and reported gills, hepatopancreas, and green glands to be the major sites for Cd accumulation. Similar distribution patterns for Cd uptake have been reported for several other species of crabs[37,71,74] and lobster *H. americanus*[44,121,134]. The uptake of Cd by the crabs[71,74,101,102] was also reported to be proportional to the level of Cd exposure.

Polychaetes

Several studies have been reported on Cd bioaccumulation by polychaetes. The concentration of Cd was found to increase proportionally with exposure concentration and time in *Nereis diversicolor*[24], *Nereis virens*[119], and *Nereis japonica*[135]. Similar uptake behavior, i.e., dependence on exposure concentration, has been reported in *Ophryotrocha diadema*[77] and *Glycera dibranchiata*[124].

Fish

Cadmium bioaccumulation from solution by plaice *Pleuronectes platessa* and thornback ray *Raja clavata* was found[110] to be extremely low for both species.

3.2 Food

Studies on Cd bioaccumulation from food are scarce and present conflicting results. Benayoun et al.[7] demonstrated that euphausiids can effectively accumulate Cd administered through food and suggested it to be an important route for incorporation of Cd on a long-term basis. However, Jennings and Rainbow[74] observed that only 10% of Cd in food is absorbed by crab, *Carcinus maenas*. Hepatopancreas was of primary importance in uptake of dietary Cd by the crabs, *C. maenas*[74] and *Cancer pagurus*[37], while greater uptake occurs in gill and/or exoskeleton on exposure to Cd in solution. Gutierez-Galindo[63] compared Cd uptake by *C. maenas* and suggested that food is the major source for Cd in the body of the animals.

Nimmo et al.[95] compared Cd accumulation in shrimp *P. pugio* from food and water and reported very low uptake from food by the animals. He calculated that, to produce the same Cd level in shrimp, the level in food had to be 11,750–16,900 times that in water.

Janssen and Scholz[73] studied Cd uptake in solution by mussels *M. edulis* maintained on a diet of the alga *Dunaliella marina*. Some of the Cd from solution was incorporated in food. The Cd concentration in individual organs or whole soft body of fed mussels was 2–3 times more than in unfed animals. The order of distribution of Cd in the tissues remained the same, i.e. midgut $\geqslant$ gland $>$ gills $>$ kidney $>$ mantle $>$ adductor muscle $>$ foot.

The plaice *P. platessa* and thornback ray *R. clavata* retained 5 and 17%, respectively, of the amount of cadmium administered through food[110], mostly in the gut.

4. Excretion

There is very little, if any, excretion of Cd by marine organsims. The biological half-life is usually quite long. Excretion of cadmium by mussels *Mytilus galloprovincialis*[86] and *M. edulis*[54] was reported to be very slow. Biological half-life ($T_{1/2}$) for *M. edulis* has been estimated[128] to be between 14 and 29 days. Greig and Wenzloff[60] and Zaroogian[148] did not observe any lowering in concentration of Cd when contaminated oysters *C. virginica* were held in unpolluted waters for 40 weeks. However, Mowdy[92] showed that oysters exposed in the laboratory eliminated Cd when transferred to uncontaminated estuarine environment and suggested that the elimination rate depended upon salinity and temperature. Similarly, $T_{1/2}$ for oyster *Saccostrea echinata* was found to depend upon temperature and salinity and was estimated[38] to be 30–85 days.

Conflicting results have also been presented for crabs. For example, there was no decrease in Cd content in *Callinectes sapidus*[71] but *C. maenas*[74,143] excreted more than 50% Cd in about 10 days. In the euphausiid *M. norvegica*[7] excretion was rapid and faecal pellets accounted for 84% of the cadmium flux. But $T_{1/2}$ for Cd in shrimp *Lysmata seticaudata* was estimated[49] to be 378 days. Similarly, Ray et al.[120] calculated $T_{1/2}$ for Cd in shrimp *Pandalus montagui* to range from 128 to 5000 days for different tissues. McLeese et al.[91] calculated $T_{1/2}$ for Cd in gill and hepatopancreas of lobster *H. americanus* to be 200 and 500 days, respectively.

Ueda et al.[135] reported 30% Cd loss from polychaete *N. japonica* within 7 days but no subsequent significant loss. In contrast, no Cd excretion from *N. virens* was observed[119] over a period of 75 d, regardless of the initial Cd concentration in the worms.

Eisler[45] reported that in mummichogs *Fundulus heteroclitus* there was a 50% Cd loss within 2–3 days and over 90% by 180 days but $T_{1/2}$ for Cd in plaice *P. platessa* and thornback ray *R. clavata* was 150–200 days and 76–147 days, respectively[110].

5. Factors controlling bioaccumulation

Cd bioaccumulation from sea water by marine organisms in nature may vary for the same species and population even in a clean environment. Several abiotic (viz. chemical form of Cd in solution, metal interaction, salinity, and temperature) and biotic (animal size and characteristics, sex, maturity, etc.) factors affect the bioaccumulation process.

5.1 Chemical form of Cd

It is generally recognized that Cd may be found in a number of different chemical forms in sea water and the relative proportion of each will be an important factor governing bioaccumulation. Several studies have been reported comparing uptake of organically bound Cd in relation to 'ionic' cadmium.

Cossa[34] found that Cd uptake by the diatom *P. tricornutum* in the presence of EDTA is negligible in comparison with exposure to Cd alone. However, increased Cd uptake in the presence of EDTA was reported[114] for the dinoflagellate *Prorocentrum micans*. The phytoplankton *Cricosphaera elongata* exposed to Cd in the presence of natural exudate of phytoplankton accumulated[67] a much lower amount in relation to exposure to Cd without the exudate.

George and Coombs[54] found that prior complexation of ionic Cd with EDTA, humic acid, and alginic acid doubled the final tissue concentration and uptake rate of Cd in *M. edulis* compared with the animals exposed to ionic Cd. But the oyster *C. virginica* accumulated[70] significantly less Cd in the presence of EDTA, NTA, and humic acid.

Ray et al.[118] observed a significant reduction in accumulated Cd in the polychaete *N. virens* and shrimp *P. montagui* exposed to CdEDTA. A similar decrease has been observed[116] in Cd uptake by the barnacle, *S. balanoides* exposed to humate, alginate, and EDTA complexes of Cd, but Cd uptake by the crab *C. maenas* in the presence of sodium salts of EDTA and phosphate was relatively unaffected[63].

5.2 *Metal interaction*

Competition between chemically similar ions for binding sites can significantly affect Cd bioaccumulation by marine organisms. In the bivalves *M. edulis* and *M. lateralis,* the addition of zinc to sea water containing Cd decreased[72] Cd levels in the animals below their values for exposure to Cd alone. But Phillips[111] reported the Cd uptake by *M. edulis* was not affected by the presence of zinc.

Cd uptake by the polychaetes *N. virens*[118] and *N. diversicolor*[24] was significantly reduced by the presence of zinc. Ray et al.[118,120] observed that, when the shrimp *P. montagui* was exposed to Cd in the presence of zinc, the distribution of Cd in the tissues was modified. The uptake by the hepatopancreas was almost doubled while there were no significant differences in other tissues. With an increasing amount of zinc in the solution, the level of Cd in the hepatopancreas decreased. The addition of zinc to Cd containing exposure solution increased[2] the accumulation of Cd by shrimp *Callianassa australiensis,* but the presence of copper in the solution did not affect accumulation of Cd by the shrimp.

Cadmium was found to inhibit[136] uptake of mercury by the crab *U. pugilator,* but mercury had very little effect on Cd uptake. Accumulation of Cd by *C. maenas* was dependent[144] upon calcium metabolism of the animal and the calcium concentration of the exposure medium. Bjerregaard[12] reported increased Cd accumulation in the gills of *C. maenas* in the presence of selenium.

Effects of copper and lead on Cd bioaccumulation by herring eggs have been studied[146]. At 2 exposure concentrations, Cd contents of eggs were enhanced by lead and decreased by copper whereas, at an intermediate concentration, both copper and lead inhibited Cd accumulation.

5.3 *Salinity*

Salinity is an important environmental variable, especially in estuarine and coastal regions. In almost all organisms studied, Cd accumulation increases with decreasing salinity. Jackim et al.[72] reported that decrease in salinity from 30 to 20 ‰ increased Cd uptake by *M. edulis, M. lateralis,* and *N. proxima* by 24 to 400% at 10 and 20 °C. Phillips[111] also reported significantly higher cadmium accumulation by *M. edulis* in dilute sea water. However, Briggs[20] observed that *M. edulis* exposed at 11 ‰ salinity accumulated much less Cd compared with animals exposed to the same concentration at 30 ‰ salinity. Cadmium uptake by the oyster *S. echinata* exposed to the same concentrations of Cd was significantly higher[38] at 20 ‰ than at 30 ‰ salinity.

Wright[143], Hutcheson[71], and O'Hara[101,102] all reported higher Cd accumulation by crab with lower salinity of the exposure solution. Similar effects have been observed in shrimp *P. pugio* by Vernberg et al.[137] and Engel and Fowler[47] but not in hydrozoa *Laomedea loveni*[132].

The common goby *Pomatoschistus minutus* accumulated[9] lower levels of Cd with increased salinity. Similar results have also been observed[145] for Cd bioaccumulation by flounder eggs.

5.4 *Temperature*

Cd bioaccumulation by marine organisms normally increases with temperature because of its effect on the metabolic activity of the animals. Jackim et al.[72] showed that Cd uptake by 3 species of bivalves (*M. arenaria, M. lateralis,* and *N. proxima*) increased with temperature. However, *M. edulis* showed no significant difference in this respect[49,72].

Hung[70] reported that in the oyster *C. virginica* Cd concentrations in whole-body soft tissues as well as in several individual organs increased with temperature during 40-day exposure. The rate of Cd accumulation by the oyster *S. echinata* was signifiantly greater[38] at higher than lower temperature.

The Cd concentration and uptake rate were shown to increase with increasing temperature in crabs *U. pugilator*[101] and *C. sapidus*[7], and in the shrimps *Leander adspersus*[9] and *Lysmata seticaudata*[49].

Accumulation of cadmium by the hydrozoa *L. loveni* gradually increased[132] with increase of temperature over the range 5–20 °C.

5.5 *Seasonal variation*

Seasonal variations in Cd concentrations are likely to occur since the ratio of size and weight of different tissues to the total body weight of organisms varies throughout the year. Boyden[18] observed highest Cd concentration in limpets *Patella vulgata* in January, after spawning, when the body weight was minimum. Phillips[111] attributed observed fluctuations in Cd concentrations in *M. edulis* to change in weight due to the reproductive cycle. The mean concentrations of Cd in mussel *Choromytilus meridionalis* were 3–4 times higher[106] in November than in June. However, Goldberg et al.[59] did not find any seasonal variability in oysters (*C. virginica* and *Ostrea equestirs*) and mussels (*M. edulis* and *M. californianus*).

Zaroogian[147,149] reported that Cd uptake by *C. virginica* followed a seasonal pattern and greatest uptake started in April and continued until August. Weight decreases during spawning resulted in an increase in tissue Cd concentration while the total Cd content remained constant. But Frazier[51,52] reported occurrence of highest Cd concentration in *C. virginica* in April or May followed by a decline through the summer. He suggested the probable reason to be the depletion of glycogen stores in spring followed by growth of the oyster, leading to lowering of Cd concentration. Julshamn[75] observed higher Cd contents in *O. edulis* during maturation and spawning period.

A seasonal pattern of variation of cadmium concentration has also been observed in phytoplankton[78].

5.6 *Effect of body size*

Variations in Cd bioaccumulation due to differing body size (i.e. weight, length, and/or age) have been observed. Nickless et al.[93] and Peden et al.[109] reported increased concentration of Cd in limpets *P. vulgata* with increase in size. Ayling[3], in a study with a single population of oysters *C. gigas* noted that smaller oysters contained higher concentrations of Cd. Boyden[16,18] examined the body size of a variety of mulluscs for their Cd content and reported that in the oyster *C. gigas* and the clam *M. mercenaria*, the smaller animals had higher concentrations of Cd, while in *M. edulis, O. edulis, Chlamys opercularis* and *Littorina littorea*, the concentrations were independent of size. Cadmium concentration in *P. vulgata, P. intermedia*, and the whelk *Buccinum undatum* increased with increasing body weight. However, Cossa et al.[35] and Boalch et al.[14] observed Cd concentrations for *M. edulis* decreased in larger animals while Latouche and Mix[80] reported higher Cd concentration in large *M. edulis*. Bryan and co-workers[26,27] reported higher Cd concentration with increasing weight in the burrowing bivalve *Scrobicularia plana* from 2 contaminated estuaries. The concentration in animals from an uncontaminated estuary was independent of size[27].

Several workers studied size effect on Cd bioaccumulation in oysters and reported that concentrations in *C. gigas[140], O. edulis[140], C. margaritacea[140]*, and *C. commercialis (= Saccostrea cucullata)[84]* decreased with increase in size/age. In contrast, the Cd concentrations in *O. edulis[76]* and *O. lutaria[94]* were independent of size.

Ray et al.[119] observed that small-sized polychaetes *N. virens* accumulated higher concentrations of Cd than larger animals.

Cadmium concentrations in fish[4,36], seal[40,127] *(Phoca vitulina)*, and sea lion[65] *(Eumetopias jubata)* are also related to length, weight and/or age.

5.7 *Characteristic of organisms*

Molluscs are extremely efficient Cd accumulators but specific differences have been observed in bioaccumulation patterns due to species differences. Jackim et al.[72] exposed filter-feeding bivalves *(M. edulis, M. arenaria, M. lateralis)* and a deposit-feeding bivalve *N. proxima* to Cd and found that the filter feeders as a group accumulated greater amounts of Cd than the deposit feeder and exhibited widely varying rates of uptake. *M. edulis* accumulated cadmium nearly 4 times faster than *M. arenaria* with *M. lateralis* at the intermediate level. Furthermore, *M. edulis* did not show any significant difference in Cd uptake when the temperature was increased from 10 to 20 °C, while *M. lateralis* showed a 90% increase in uptake over those animals held at 10 °C.

Brian[23] observed that in the 2 species of scallops *(Pecten maximus* and *Chlamys opercularis)* collected from the same site, there were species variations in Cd accumulation patterns. The Cd concentrations in whole soft tissue (including fluid) of the whole animal, kidney, and digestive gland, of *Pecten* were 32.5, 79, and 321 μg/g(dry), respectively. The corresponding figures for *Chlamys* were 5.5, 41, and 27 μg/g (dry), respectively. The digestive gland and kidney of *Pecten* contributed 89.9 and 1.7% of the total body burden, respectively, while in *Chlamys* they were 41.5 and 7.5%, respectively.

5.8 *Miscellaneous*

Watling and Watling[139] did not find any difference in Cd concentration between male and female mussel *C. meridionalis*. However, the metal uptake rate for the male animals was nearly double of that in the females. Female mussels were reported[51] to have significantly higher Cd concentrations than the males in November.

Coleman[32] compared Cd concentrations between continually immersed and alternately immersed and emersed *M. edulis*. The mussels subjected to emersion accumulated significantly less Cd than those continually immersed but the differences in accumulation were not related to the total time of immersion.

6. *Kinetics of bioaccumulation*

Cadmium bioaccumulation depends upon existence of metal-binding ligands in the proteins, capable of forming highly stable complexes with Cd. If the Cd can move across the permeable membrane and bind to the intracellular ligands, the bioaccumulation will occur until all the binding sites are occupied. Intracellular mobilization can occur simultaneously. This uptake process is by passive diffusion, and the total concentration of Cd taken up is expected to be proportional to the external concentration of Cd in sea water. If saturation equilibrium is observed, then mediated Cd transport is indicated but can be ignored at low Cd concentrations in sea water used in most laboratory experiments or found in nature. Most uptake studies in the laboratory were found to be linear with time of exposure and concentration of Cd, indicating the process is diffusion controlled.

Majori and Petronio[85] used a 1-compartment dynamic equilibrium distribution model to correlate metal accumulation with metal concentration in the solution and the time of exposure. A calculated accumulation factor was used for quantifying affinity of the metals for the biological tissues. Ray et al.[119] determined the total receptor site and binding constant for *N. virens* and noted that the calculated uptake rate for Cd by the worms using the model was in close agreement with the observed values. A similar mechanism based on specific metal binding ligands has been considered[30] for Cd accumulation by gills of *M. edulis*.

However, the Majori and Petronio model[85] is inadequate in dealing with excretion data. The 1-compartment model used by Zitko[151] has been applied[120] to determine uptake and clearance rate constants, accumulation coefficients, and $T_{1/2}$ for Cd bioaccumulation in *P. montagui*.

7. *Storage and metabolism*

Cadmium is not distributed uniformly in the body of marine organisms but is selectively localized in different tissues. The concentrations in individual tissues can vary by several orders of magnitude, but the distribution in animals exposed to Cd in the laboratory normally simulates natural bioaccumulation pattern.

7.1 Metallothioneins

The parallelism between high-storage capacity for Cd in the excretory organs of marine organisms and mammalians has led to the discovery of metalloproteins, called 'metallothioneins' in a variety of aquatic organisms. The metallothioneins are cytoplasmic, low molecular weight (6000–7000 daltons) proteins, having high cysteine content (about 30%), no aromatic acids, and high metal content (6–11% metals).

The presence of definitely characterized metallothioneins and metallothionein-like proteins has been reported to occur in marine biota like mussels[50,55,56,79,128,131], oysters[125], clams[28], crabs[104,105,107,126], limpets[69,97–99], shrimps[104], lobsters[122], chitons[104], whelks[97], barnacles[97], fishes[96,103,108,115,126], sea lions[81,126], seals[103], and whales[126]. Several other low and high molecular weight Cd-binding proteins have also been isolated from marine organisms[28,30,55,56,79,116,122].

The apparent indifference to very high Cd concentrations in several marine organisms is possibly due to induction of metallothionein. The apoprotein, thionein, is an inducible protein and metallothionein synthesis occurs in response to the presence of Cd and several other 'essential' and 'non-essential' trace metals as a 'protective detoxifying mechanism'.

It is assumed that the metals bound to the protein are not available for binding to the enzymes or for damaging intracellular membranes. However, it has been shown with several marine organisms that there may be a threshold limit[71,101,129] of Cd above which the organisms cannot metabolically control the excess.

7.2 Membrane-limited vesicles

Besides storing Cd in the form of metallothionein, many animals may accumulate it in intracellular electron-dense membrane-limited granules or vesicles. The vesicles are formed by effective trapping of Cd and other metals by surrounding membranes and consequently isolating them from cytoplasm and other organelles. These membrane-limited granules have been identified in a number of marine organisms[56,73,88] and were shown to be involved not only in uptake and storage but also in excretion of Cd.

Summary and conclusion

It has been established that, although Cd occurs in the marine environment in only trace concentrations, most marine organisms, especially molluscs and crustaceans, can accumulate it rapidly. Cadmium is not uniformly distributed in the body and selectively accumulates in specific organs like liver, kidney, gills, and exoskeleton. The concentrations in muscle tissues are several orders of magnitude lower. The disposition of Cd in the organisms in the laboratory studies generally parallels those in nature.

A number of biotic factors like body size, maturity, sex, etc. influence bioaccumulation but extensive studies are still lacking.

The chemical form of Cd in the environment is of prime importance in bioaccumulation by marine organisms. Salinity can affect the speciation of Cd, and bioaccumulation is affected by both temperature and salinity. The ultimate level of Cd in the organisms will depend not only on the biotic and abiotic factors but also on metabolism of the metal by the the organisms. A few studies indicate depuration of Cd by some bivalves but other organisms show very effective retention of Cd. Metallothionein formation for detoxification and storage has been observed in a large variety of marine organisms. Recent reports indicate an alternate storage and excretion mechanism in the formation of membrane-limited vesicles or granules. There seems to be a common link between intracellular localisation of Cd in metal-binding proteins and Cd containing vesicles as detoxifying mechanisms in the marine organisms.

Much of what is known about Cd bioaccumulation by marine organisms has come from laboratory studies and there are inherent dangers in trying to extrapolate the results to field situations. In spite of tremendous progress made over the years, the basic understanding of the bioaccumulation process is still very nebulous and will remain so until the uptake, storage, and elimination processes are fully understood.

1 Abdullah, M.I., and Royle, L.G., A study of the dissolved and particulate trace elements in the Bristol Channel. J. mar. Biol. Assoc. U.K. *54* (1974) 581–597.

2 Ahsanullah, M., Negilski, D.S., and Mobley, M.C., Toxicity of zinc, cadmium and copper to the shrimp *Callianasa australiemsis*. III. Accumulation of metals. Mar. Biol. *64* (1981) 311–316.

3 Ayling, G.M., Uptake of cadmium, zinc, copper, lead and chromium in the Pacific oyster, *Crassostrea gigas*, grown in the Tamar River, Tasmania. Water Res. *8* (1974) 729–738.

4 Badsha, K.S., and Sainsbury, M., Uptake of zinc, lead and cadmium by young whiting in the Severn estuary. Mar. Pollut. Bull. *8* (1977) 164–166.

5 Baric, A., and Branica, M., Ionic state polarography of sea water. I. Cadmium and zinc in sea water. J. polarogr. Soc. *13* (1967) 4–8.

6 Batley, G.E., Florence, T.M., Determination of chemical forms of dissolved Cd, Pb, and Cu in sea water. Mar. Chem. *4* (1976) 347–363.

7 Benayoun, G., Fowler, S.W., and Oregioni, B., Flux of cadmium through euphausiids. Mar. Biol. *27* (1974) 205–212.

8 Bender, M.L., and Gagner, C.L., Dissolved copper, nickel, and cadmium in Sargasso Sea. J. mar. Res. *34* (1976) 327–339.

9 Bengtsson, B.-E., Accumulation of cadmium in some aquatic animals from the Baltic Sea. Ambio, special report No. 5 (1977) 69–73.

10 Bewers, J.M., Sundby, B., and Yeats, P.A., The distribution of trace metals in the western North Atlantic off Nova Scotia. Geochim, cosmochim. Acta *40* (1976) 687–697.

11 Bewers, J.M., and Yeats, P.A., Oceanic residence times of trace metals. Nature *268* (1977) 595–598.

12 Bjerregaard, P., Accumulation of cadmium and selenium and their mutual interaction in the shore crab, *Carcinus maenas* (L.). Aquat. Toxic. *2* (1982) 113–125.

13 Bloom, H., and Ayling, G.M., Heavy metals in the Derwent estuary. Envir. Geol. *2* (1977) 3–22.

14 Boalch, R., Chan. S., and Taylor, D., Seasonal variation in the trace metal content of *Mytilus edulis*. Mar. Pollut. Bull. *12* (1981) 276–280.

15 Bowen, H.J.M., Environmental chemistry of the elements. Academic Press, New York 1979.

16 Boyden, C.R., Trace element content and body size in molluscs. Nature *251* (1974) 311–314.

17 Boyden, C.R., Distribution of some trace metals in Poole Harbour, Dorset. Mar. Pollut. Bull. *6* (1975) 180–187.

18 Boyden, C.R., The effect of size upon metal content of shellfish. J. mar. Biol. Assoc. U.K. *57* (1977) 675–714.

19 Boyle, E.A., Sclater, F., and Edmond, J.M., On the marine geochemistry of cadmium. Nature *263* (1976) 42–44.

20 Briggs, L.B.R., Effects of cadmium on the intracellular pool of free amino acids in *Mytilus edulis*. Bull. envir. Contam. Toxic. *22* (1979) 838–845.

21 Brooks, R.R., and Rumsby, M.G., Studies on the uptake of cadmium by the oyster, *Ostrea sinuata* (Lamarck). Aust. J. mar. Freshwater Res. *18* (1967) 53–61.

22 Bruland, K.W., Knauer, G.A., and Martin, J.H., Cadmium in Northeast Pacific waters. Limnol. Oceanogr. *23* (1978) 618–625.

23 Bryan, G.W., The occurrence and seasonal variation of trace metals in the scallops *Pecten maximus* (L.) and *Chlamys opercularis* (L.). J. mar. Biol. Assoc. U.K. *53* (1973) 145–166.

24 Bryan, G.W., and Hummerstone, L.G., Adaptation of the polychaete *Nereis diversicolor* to estuarine sediments containing high concentrations of zinc and cadmium. J. mar. Biol. Assoc. U.K. *53* (1973) 839–857.

25 Bryan, G.W., Heavy metal contamination in the sea, in: Marine pollution, pp.185–302. Ed. R. Johnston. Academic Press, New York 1976.

26 Bryan, G.W., and Uysal, H., Heavy metals in the burrowing bivalve *Scrobicularia plana* from the Tamar estuary in relation to environmental levels. J. mar. Biol. Assoc. U.K. *58* (1978) 89–108.

27 Bryan, G.W., and Hummerstone, L.G., Heavy metals in the burrowing bivalve *Scrobicularia plana* from contaminated and uncontaminated estuaries. J. mar. Biol. Assoc. U.K. *58* (1978) 401–419.

28 Carmichael, N.G., Squibb, K.S., Engel, D.E., and Fowler, B.A., Metals in the molluscan kidney: Uptake and subcellular distribution of ^{109}Cd, ^{54}Mn and ^{65}Zn by the clam *Mercenaria mercenaria*. Comp. Biochem. Physiol. *65A* (1980) 203–206.

29 Carmichael, N.G., and Fowler, B.A., Cadmium accumulation and toxicity in the kidney of the bay scallop *Argopecten irradians*. Mar. Biol. *65* (1981) 35–43.

30 Carpene, E., and George, S.G., Absorption of cadmium by gills of *Mytilus edulis* (L.). Molec. Physiol. *1* (1981) 23–34.

31 Chan, J.P., Cheung, M.T., and Li, F.P., Trace metals in Hong Kong waters. Mar. Pollut. Bull. *5* (1974) 171–174.

32 Coleman, N., The effect of emersion on cadmium accumulation by *Mytilus edulis*. Mar. Pollut. Bull. *11* (1980) 359–362.

33 Coombs, T.L., Cadmium in aquatic organism, in: The chemistry, biochemistry and biology of cadmium, pp.93–139. Ed. M. Webb. Elsevier/North Holland Biomedical Press, Amsterdam 1979.

34 Cossa, D., Sorption du cadmium par une population de la diatomée *Phaeodactylum tricornutum* en culture. Mar. Biol. *34* (1976) 163–167.

35 Cossa, D., Bourget, E., and Piuze, J., Sexual maturation as a source of variation in the relationship between cadmium concentration and body weight of *Mytilus edulis* L. Mar. Pollut. Bull. *10* (1979) 174–176.

36 Cutshall, N.H., Naidu, J.R., and Pearcy, W.G., Zinc and cadmium in the Pacific hake *Merluccius productus* off the western U.S. coast. Mar. Biol. *44* (1977) 195–201.

37 Davies, I.M., Topping, G., Graham, W.C., Falconer, C.R., McIntosh, A.D., and Saward, D., Field and experimental studies on cadmium in the edible crab *Cancer pagurus*. Mar. Biol. *64* (1981) 291–297.

38 Denton, G.R.W., and Burdon-Jones, C., Influence of temperature and salinity on the uptake, distribution and depuration of mercury, cadmium and lead by the black-lip oyster *Saccostrea echinata*. Mar. Biol. *64* (1981) 317–326.

39 Dethlefsen, V., Uptake, retention and loss of cadmium by brown shrimp *(Crangon crangon)*. Meeresforschung 26 (1977/78) 137–152.

40 Drescher, H.E., Harms, U., and Huschenbeth, E., Oranochlorines and heavy metals in the harbour seal *Phoca vitulina* from the German northeast coast. Mar. Biol. *41* (1977) 99–106.

41 Duinker, J.C., and Kramer, C.J.M., An experimental study on the speciation of dissolved zinc, cadmium, lead and copper in River Rhine and North Sea water by differential pulse anodic stripping voltametry. Mar. Chem. *5* (1977) 207–228.

42 Dyrssen, D., and Wedborg, M., Equilibrium calculations of the speciation of elements in sea water, in: The Sea, vol.5: Marine Chemistry, pp.181–195. Ed. E.D. Goldberg. Wiley-Interscience, New York 1974.

43 Eaton, A., Marine geochemistry of cadmium. Mar. Chem. *4* (1976) 141–154.

44 Eisler, R., Zaroogian, G.E., and Hennekey, R.H., Cadmium uptake by marine organisms. J. Fish. Res. Board Can. *29* (1972) 1367–1369.

45 Eisler, R., Radiocadmium exchange with seawater by *Fundulus heteroclitus* (L.) (Pisces: Cyprinodontidae). J. Fish Biol. *6* (1974) 601–612.

46 Engel, D.W., and Fowler, B.A., Copper- and cadmium-induced changes in the metabolism and structure of molluscan gill tissue, in: Marine pollution: functional responses, pp.239–256. Eds W.B. Vernberg, F.P. Thurberg, A. Calabrese and F. Vernberg. Academic Press, New York 1979.

47 Engel, D.W., and Fowler, B.A., Factors influencing cadmium accumulation and its toxicity to marine organisms. Envir. Hlth Prespect. *28* (1979) 81–88.

48 Establier, R., Gutierrez, M., and Rodriguez, A., Accumulatión de cadmio en el músculo y hepatopáncreas del longestino *(Penaeus kerathurus)* y alteraciones histopathologicas pioducidas. Invest. Pesq. *42* (1978) 299–304.

49 Fowler, S.W., and Benayoun, G., Experimental studies on cadmium flux through marine biota, in: Comparative studies of food and environmental contamination IAEA-SM 175/10, pp.159–178. International Atomic Energy Agency, Vienna 1974.

50 Frankenne, F., Noël-Lambot, F., and Disteche, A., Isolation and characterisation of metallothioneins from cadmium-loaded mussel *Mytilus edulis*. Comp. Biochem. Physiol. *66C* (1980) 179–182.

51 Frazier, J.M., The dynamics of metals in the American oyster, *Crassostrea virginica*. I. Seasonal effects. Chesapeake Sci. *16* (1975) 162–171.

52 Frazier, J.M., The dynamics of metals in the American oyster, *Crassostrea virginica*. II. Environmental effects. Chesapeake Sci. *17* (1976) 188–197.

53 Gardner, L.R., Organic versus inorganic trace metal complexes in sulphidic marine waters – some speculative calculations based on availability of stability constants. Geochim. cosmochim. Acta *38* (1974) 1297–1302.

54 George, S.G., and Coombs, T.L., The effects of chelating agents on the uptake and accumulation of cadmium by *Mytilus edulis*. Mar. Biol. *39* (1977) 261–268.

55 George, S.G., Carpene, E., Coombs, T.L., Overnell, J., and Youngson, A., Characterization of cadmium-binding proteins from mussels, *Mytilus edulis* (L.) exposed to cadmium. Biochim. biophys. Acta *580* (1979) 225–233.

56 George, S.G., and Pirie, B.J.S., The occurrence of cadmium in sub-cellular particles in the kidney of the marine mussel, *Mytilus edulis*, exposed to cadmium. Biochim. biophys. Acta *580* (1979) 234–244.

57 Goldberg, E., ed., Baseline studies of pollutants in the marine environment and research recommendations. IDOE Baseline Conf., May 24–26, 1972, New York 1972.

58 Goldberg, E.D., The mussel watch – a first step in global marine monitoring. Mar. Pollut. Bull. *6* (1975) 111.

59 Goldberg, E.D., Bowen, V.T., Farrington, J.W., Harvey, G., Martin, J.H., Parker, P.L., Risebrough, R.W., Robertson, W., Schneider, E., and Gamble, E., The mussel watch. Envir. Conserv. *5* (1978) 101–126.

60 Greig, R.A., and Wenzloff, D.R., Metal accumulation and depuration by the American oyster, *Crassostrea virginica*. Bull. envir. Contam. Toxic. *20* (1978) 499–504.

61 Greig, R.A., Trace metal uptake by three species of mollusks. Bull. envir. Contam. Toxic. *22* (1979) 643–647.

62 Gutierrez-Galindo, E.A., Etude de l'élimination du cadmium par *Mytilus edulis* en présence d'EDTA et de phosphate. Chemosphere (1980) 495–500.

63 Gutierrez-Galindo, E.A., Etude comparée des roles de la nourriture et de l'eau dans l'accumulation du cadmium par le crabe *Carcinus maenas* en présence d'EDTA et de phosphate. Rev. int. Oceanogr. Med. *58* (1980) 69–79.

64 Hahne, H.C.H., and Krontje, W., Significance of pH and chloride concentration on behaviour of heavy metal pollutants: mercury (II), cadmium (II), zinc (II), and lead (II). J. envir. Qual. *2* (1973) 444–450.

65 Hamanaka, T., Itoo, T., and Mishima, S., Age related change and distribution of cadmium and zinc concentrations in the stellar sea lion *(Eumetopias jubata)* from the coast of Hokkaido. Mar. Pollut. Bull. *13* (1982) 57–61.

66 Hamilton, E.I., The analysis of elements - Quality factors. Mar. Pollut. Bull. *12* (1981) 393–394.

67 Härdstedt-Roméo, M., and Gnassia-Barelli, M., Effect of complexation by natural phytoplankton exudates on the accumulation of cadmium and copper by the haptophyceae *Cricosphaera elongata.* Mar. Biol. *59* (1980) 79–84.

68 Holmes, C.W., Slade, E.A., and McLerran, C.J., Migration and redistribution of zinc and cadmium in marine estuarine system. Envir. Sci. Technol. *8* (1974) 255–259.

69 Howard, A.G., and Nickless, G., Heavy metal complexation in polluted molluscs. I. Limpets *(Patella vulgata* and *Patella intermedia).* Chem.-biol. Interactions *16* (1977) 107–114.

70 Hung, Y.-W., Effects of temperature and chelating agents on the cadmium uptake in the American oyster. Bull. envir. Contam. Toxic. *28* (1982) 546–551.

71 Hutcheson, M.S., The effect of temperature and salinity on cadmium uptake by the blue crab, *Callinectes sapidus.* Chesapeake Sci. *15* (1974) 237–241.

72 Jackim, E., Morrison, G., and Steele, R., Effects of environmental factors on radiocadmium uptake by four species of marine bivalves. Mar. Biol. *40* (1977) 303–308.

73 Janssen, H.H., and Scholz, N., Uptake and cellular distribution of cadmium in *Mytilus edulis.* Mar. Biol. *55* (1979) 133–141.

74 Jennings, J.R., and Rainbow, P.S., Studies on the uptake of cadmium by the crab *Carcinus maenas* in the laboratory. I. Accumulation from seawater and a food source. Mar. Biol. *50* (1979) 131–139.

75 Julshamn, K., Studies on major and minor elements in Molluses in western Norway. II. Seasonal variations in the contents of 10 elements in oyster *(Ostrea edulis)* from three oyster farms. Fisk. Dir. Skr., Ser. Ernaering. *1* (1981) 183–197.

76 Julshamn, K., Studies on major and minor elements in molluses in western Norway. III. Effects of size and age on the contents of 10 elements in oyster *(Ostrea edulis)* taken from unpolluted waters. Fisk. Dir. Skr., Ser. Ernaering. *1* (1981) 199–214.

77 Klöckner, K., Uptake and accumulation of cadmium by *Ophryotrocha diadema* (Polychaeta). Mar. Ecol. Prog. Ser. *1* (1979) 71–76.

78 Knauer, G., and Martin, J., Seasonal variation of cadmium, copper, manganese, lead and zinc in water and phytoplankton in Monterey Bay, California. Limnol. Oceanogr. *18* (1973) 597–604.

79 Köhler, K., and Riisgard, H.U., Formation of metallothioneins in relation to accumulation of cadmium in the common mussel *Mytilus edulis.* Mar. Biol. *66* (1982) 53–58.

80 Latouche, Y.D., and Mix, M.C., The effects of depuration, size and sex on trace metal levels in bay mussels. Mar. Pollut. Bull. *13* (1982) 27–29.

81 Lee, S.S., Mate, B.R., Von Der Trenck, K.T., Rimerman, R.A., and Buhler, D.R., Metallothionein and the subcellular localization of mercury and cadmium in the California sea lion. Comp. Biochem. Physiol. *57C* (1977) 45–53.

82 Loring, D.H., Bewers, J.M., Seibert, G., and Kranck, K., A preliminary survey of circulation and heavy metal contamination in Belledune Harbour and adjacent areas, in: Cadmium pollution of Belledune Harbour, New Brunswick, Canada, pp.35–47. Eds J.F. Uthe and V. Zitko. Tech. Rep. Fish. Aquat. Sci. 963 (1980).

83 Lu, J.C.S., and Chen, K.Y., Migration of trace metals in interfaces of seawater and polluted surficial sediments. Envir. Sci. Technol. *11* (1977) 174–182.

84 MacKay, N.J., Williams, R.J., Kacprzac, J.L., Kozacos, M.N., Collins, A.J., and Auty, E.H., Heavy metals in the cultivated oysters *(Crassostrea commercialis=* Saccostrea cuculata) from the esturaries of New South Wales. Aust. J. Mar. Freshwat. Res. *26* (1975) 31–46.

85 Majori, L., and Petronio, F., A simplified model for the dynamic equilibrium distribution of metals between mussels *(Mytilus galloprovincialis* LMK) and seawater. The accumulation factor. Fish. Mar. Serv. Transl. Ser. No.3118. Ig. Mod. *66* (1973) 19–38.

86 Majori, L., and Petronio, F., Accumulation phenomenon which takes place in a mussel *(Mytilus galloprovincialis* LMK) grown in an artificially polluted environment. Verification of a simplified model of the dynamic equilibrium of metal distribution between mussel and sea water. Note II-Pollution from cadmium. Fish. Mar. Serv. Transl. Ser. No.3143 (1974). Ig. Mod. *66* (1973) 39–63.

87 Malo, B.A., Partial extraction of metals from aquatic sediments. Envir. Sci. Technol. *11* (1977) 277–282.

88 Marshall, A.T., and Talbot, V., Accumulation of cadmium and lead in the gills of *Mytilus edulis:* X-ray microanalysis and chemical analysis. Chem.-biol. Interactions *27* (1979) 111–123.

89 Martin, J.H., Bruland, K.W., and Broenkow, W.W., Cadmium transport in California current, in: Marine pollutant transfer, pp.159–184. Eds H.L. Windom and R.A. Duce. D.C. Heath and Co., Lexington, MA, 1976.

90 McLean, M.W., and Williamson, F.B., Cadmium accumulation by marine red alga *Porphyra umbilicalis.* Physiologia Pl. *41* (1977) 268–272.

91 McLeese, D.W., Ray, S., and Burridge, L.E., Lack of excretion of cadmium from lobsters. Chemosphere *10* (1981) 775–778.

92 Mowdy, D.E., Elimination of laboratory-acquired cadmium by the oyster *Crassostrea virginica* in the natural environment. Bull. envir. Contam. Toxic. *26* (1981) 345–351.

93 Nickless, G., Stenner, R., and Terrille, N., Distribution of cadmium, lead and zinc in the Bristol Channel. Mar. Pollut. Bull. *3* (1972) 188–190.

94 Nielsen, S.A., Cadmium in New Zealand dredge oysters: geographic distribution. Int. J. envir. analyt. Chem. *4* (1975) 1–7.

95 Nimmo, D.W.R., Lightner, D.V., and Bahner, L.H., Effects of cadmium on shrimps, *Panaeus duorarum, Palaemonetes pugio* and *Palaemonetes vulgaris,* in: Physiological responses of marine biota to pollutants, pp.131–183. Eds F.J. Vernberg, A. Calabrese, F.P. Thurberg and W.B. Vernberg. Academic Press, New York 1977.

96 Noël-Lambot, F., Gerday, Ch., and Disteche, A., Distribution of Cd, Zn and Cu in liver and gills of the eel *Anguilla anguilla* with special reference to metallothioneins. Comp. Biochem. Physiol. *61C* (1978) 177–187.

97 Noël-Lambot, F., Bouquegneau, J.M., Frankenne, F., and Disteche, A., Le role des métallothionéines dans le stockage des métaux lourds chez les animaux marines. Rev. int. Oceanogr. Med. *49* (1978) 13–20.

98 Noël-Lambot, F., Cadmium accumulation correlated with increase in metallothioneins concentration in the limpet *Patella caerulea,* in: Proc. 1st Congress of the European Society for Comparative Physiology and Biochemistry: animals and environmental fitness, p.83–84. Pergamon Press, Oxford 1979.

99 Noël-Lambot, J.M., Bouquegneau, J.M., Frankenne, F., and Disteche, A., Cadmium, zinc and copper accumulation in limpets *(Patella vulgata)* from the Bristol Channel with special reference to metallothioneins. Mar. Ecol. Progr. Ser. *2* (1980) 81–89.

100 Nriagu, J.O., Human influence on the global cadmium cycle, in: Cadmium in the environment, pp.2–12. Ed. J.O. Nriagu. Wiley-Interscience, New York 1980.

101 O'Hara, J., The influence of temperature and salinity on the toxicity of cadmium to the fiddler crab, *Uca pugilator.* Fish. Bull. *71* (1973) 149–153.

102 O'Hara, J., Cadmium uptake by Fiddler crabs exposed to temperature and salinity stress. J. Fish. Res. Board Can. *30* (1973) 846–848.

103 Olafson, R.W., and Thompson, A.J. Isolation of heavy metal binding proteins from marine invertebrates. Mar. Biol. *28* (1974) 83–86.

104 Olafson, R.W., Sim, R.G., and Boto, K.G., Isolation and chemical characterization of the heavy metal-binding protein metallothionein from marine invertebrates. Comp. Biochem. Physiol. *62B* (1979) 407–416.

105 Olafson, R.W., Kearns, A., and Sim, R.G., Heavy metal induction of metallothionein synthesis in the hepatopancreas of the crab *Scyla serrata*. Comp. Biochem. Physiol. *62B* (1979) 417–424.

106 Orren, M.J., Eagle, G.A., Hennig, H.F-K.O., and Green, A., Variations in trace metal content of the mussel *Choromytilus meridionalis* (Kr) with season and sex. Mar. Pollut. Bull. *11* (1980) 253–257.

107 Overnell, J., and Trewhella, E., Evidence for the natural occurrence of (cadmium, copper)-metallothionein in the crab *Cancer pagurus*. Comp. Biochem. Physiol. *64C* (1979) 69–76.

108 Overnell, J., and Coombs, T.L., Purification and properties of plaice metallothionein, a cadmium-binding protein from the liver of the plaice *(Pleuronectes platessa)*. Biochem. J. *183* (1979) 277–283.

109 Peden, J.D., Crothers, J.H., Waterfall, C.E., and Beasley, J., Heavy metals in Somerset marine organisms. Mar. Pollut. Bull. *4* (1973) 7–9.

110 Pentreath, R.J., The accumulation of cadmium by the plaice, *Pleuronectes platessa* L. and the thornback ray, *Raja clavata* L. J. Exp. mar. Biol. Ecol. *30* (1977) 223–232.

111 Phillips, D.J.H., The common mussel *Mytilus edulis* as an indicator of pollution by zinc, lead and copper. 1. Effect of environmental variables on uptake of metals. Mar. Biol. *38* (1976) 59–69.

112 Phillips, D.J.H., Toxicity and accumulation of cadmium in marine and estuarine biota, in: Cadmium in the environment, part 1, pp.425–569. Ed. J.O. Nriagu. Wiley-Interscience, New York 1980.

113 Preston, A., Cadmium in the marine environment of the United Kingdom. Mar. Pollut. Bull. *4* (1973) 105–107.

114 Prévot, P., and Soyer, M., Action du cadmium sur un Dinoflagellé libre: *Prorocentrum micans* E.: croissance, absorption du cadmium et modifications cellulaires. C.r. Acad. Sci., Paris *D287* (1978) 833–836.

115 Pruell, R.J., and Engelhardt, F.R., Liver cadmium uptake, catalase inhibition and cadmium thionein production in the killifish *(Fundulus heteroclitus)* induced by experimental cadmium exposure. Mar. envir. Res. *3* (1980) 101–111.

116 Rainbow, P.S., Scott, A.G., Wiggins, E.A., and Jackson, R.W., Effect of chelating agents on the accumulation of cadmium by the barnacle *Semibalanus balanoides*, and complexation of soluble Cd, Zn and Cu. Mar. Ecol. Progr. Ser. *2* (1980) 143–152.

117 Raspor, B., Distribution and speciation of cadmium in natural waters, in: Cadmium in the environment, part 1, pp.147–236. Ed. J.O. Nriagu. Wiley-Interscience, New York 1980.

118 Ray, S., McLeese, D.W., and Pezzack, D., Chelation and interelemental effects on the bioaccumulation of heavy metals by marine invertebrates, in: Proc. Int. Conf. Manage. Control, Heavy Metals and the Environment 1979; p.35–38.

119 Ray, S., McLeese, D.W., and Pezzack. D., Accumulation of cadmium by *Nereis virens*. Archs envir. Contam. Toxic. *9* (1980) 1–8.

120 Ray, S., McLeese, D.W., Waiwood, B.A., and Pezzack, D., Disposition of cadmium and zinc in *Pandalus montagui*. Archs envir. Contam. Toxic. *9* (1980) 675–681.

121 Ray, S., McLeese, D.W., and Burridge, L.E., Cadmium in tissues of lobsters captured near a lead smelter. Mar. Pollut Bull. *12* (1981) 383–386.

122 Ray, S., and White, M. Metallothionein-like protein in lobster *(Homarus americanus)*. Chemosphere 10 (1981) 1205–1213.

123 Ray, S., and McLeese, D.W., Factors affecting uptake of cadmium and other trace metals from marine sediments by some bottom-dwelling marine invertebrates, in: Wastes in the Ocean, Vol.II. Dredged material disposal in the ocean, pp.185–197. Eds D. Kester, I.W. Duedall, B.H. Ketchum and P.K. Park. John Wiley and Sons, New York 1983.

124 Rice, M.A., and Chien, P.K., Uptake, binding and clearance of divalent cadmium in *Glycera dibranchiata* (Annelida: polychaeta). Mar. Biol. *53* (1979) 33–39.

125 Ridlington, J.W., and Fowler, B.A., Isolation and partial characterization of a cadmium-binding protein from the American oyster *(Crassostrea virginica)*. Chem.-biol. Interactions *25* (1979) 127–138.

126 Ridlington, J.W., Chapman, D.C., Goeger, D.E., and Whanger, P.D., Metallothionein and Cu-chelatin: characterization of metal-binding proteins from tissues of four marine animals. Comp. Biochem. Physiol. *70B* (1981) 93–104.

127 Roberts, T.M., Heppleston, P.B., and Roberts, R.D., Distribution of heavy metals in tissues of the common seal. Mar. Pollut. Bull. *7* (1976) 194–196.

128 Scholz, N., Accumulation, loss and molecular distribution of cadmium in *Mytilus edulis*. Helgoländer wiss. Meeresunters. *33* (1980) 68–78.

129 Shuster, C.N. Jr, and Pringle, B.H., Trace metal accumulation by the American eastern oyster, *Crassostrea virginica*. Proc. natl Shellfish Assoc. *59* (1969) 91–103.

130a Sillen, L.G., Chemical equilibrium in analytical chemistry. Wiley-Interscience, New York 1962.

130b Sillen, L.G., The physical chemistry of sea water, in: Oceanography, pp.549–582. Ed. M. Sears. Am. Assoc. Adv. Sci. Publ. 67, Washington, D.C. 1961.

131 Talbot, V., and Magee, R.J., Naturally occurring heavy metal binding proteins in invertebrates. Archs envir. Contam. Toxic. *7* (1978) 73–81.

132 Theede, H., Scholz, N., and Fischer, H., Temperature and salinity effects on the acute toxicity of cadmium to *Laomedea loveni* (Hydrozoa). Mar. Ecol. Prog. Ser. *1* (1979) 13–19.

133 Thornton, I., Watling, H., and Darracott, A., Geochemical studies in several rivers and estuaries used for oyster rearing. Sci. total Envir. *4* (1975) 325–345.

134 Thurberg, F.P., Calabrese, A., Gould, E., Greig, R.A., Dawson, M.A., and Tucker, R.K., Response of the lobster, *Homarus americanus*, to sublethal levels of cadmium and mercury, in: Physiological responses of marine biota to pollutants, pp.185–197. Eds F.J. Vernberg, A. Calabrese, F.P. Thurberg and W.B. Vernberg. Academic Press, New York 1977.

135 Ueda, T., Nakamura, R., and Suzuki, Y., Comparison of [115m]Cd accumulation from sediments and sea water by polychaete worms. Bull. Jap. Soc. scient. Fish. *42* (1976) 299–306.

136 Vernberg, W.B., DeCoursey, P.J., and O'Hara, J., Pollution and physiology of marine organisms, pp.381–424. Eds F.J. Vernberg and W.B. Vernberg. Academic Press, New York 1974.

137 Vernberg, W.B., DeCoursey, P.J., Kelly, M., and Johns, D.M. Effects of sublethal concentrations of cadmium on adult *Palaemonetes pugio* under static and flow through conditions. Bull. envir. Contam. Toxic. *17* (1977) 16–24.

138 Ward, T.J., Laboratory study of the accumulation and distribution of cadmium in the Sydney rock oyster *Saccostrea commercialis* (I & R). Aust. J. mar. Freshwater Res. *33* (1982) 33–44.

139 Watling, H.R., and Watling, R.J., Trace metals in *Choromytilus meridionalis*. Mar. Pollut. Bull. *7* (1976) 91–94.

140 Watling, H.R., and Watling, R.J., Trace metals in oysters from Knysna estuary. Mar. Pollut. Bull. *7* (1976) 45–48.

141 White, S.L., and Rainbow, P.S., Regulation and accumulation of copper, zinc and cadmium by the shrimp *Palaemon elegans*. Mar. Ecol. Progr. Ser. *8* (1982) 95–101.

142 Windom, H.L., Gardner, W.S., Dunstan, W.M., and Paffenhofer, G.A., Marine pollutant transfer, pp.135–157. Eds H.L. Windom and R.A., Duce. D.C. Heath and Co., Lexington, MA., 1976.

143 Wright, D.A., The effect of salinity on cadmium uptake by the tissues of the shore crab *Carcinus maenas*. J. exp. Biol. *67* (1977) 137–146.

144 Wright, D.A., The effect of calcium on cadmium uptake by the shore crab *Carcinus maenas*. J. exp. Biol. *77* (1977) 163–173.

145 Westernhagen, H. von, and Dethlefsen, V., Combined effects of cadmium and salinity on development and survival of flounder eggs. J. mar. Biol. Assoc. U.K. *55* (1975) 945–957.

146 Westernhagen, H. von, Dethlefsen, V., and Rosenthal, H., Combined effects of cadmium, copper and lead on develop-

ing herring eggs and larvae. Helgoländer wiss. Meeresunters. *32* (1979) 257–278.
147 Zaroogian, G.E., and Cheer, S., Accumulation of cadmium by the American oyster, *Crassostrea virginica.* Nature *261* (1976) 408–410.
148 Zaroogian, G.E., Studies on the depuration of cadmium and copper by the American oyster *Crassostrea virginica.* Bull. envir. Contam. Toxic. *23* (1979) 117–122.
149 Zaroogian, G.E., *Crassostrea virginica* as an indictor of cadmium pollution. Mar. Biol. *58* (1980) 275–284.
150 Zirino, A., and Yamamoto, S., A pH dependent model for the chemical speciation of copper, zinc, cadmium, and lead in sea water. Limnol. Oceanogr. *17* (1972) 661–671.
151 Zitko, V., Relationships governing the behaviour of pollutants in aquatic ecosystems and their use in risk assessment, in: Proceedings of the 6th annual aquatic toxicity workshop, pp. 243–265. Can. Tech. Rep. Fish. Aquat. Sci. 975 (1980).

Cadmium in freshwater ecosystems

by O. Ravera

Department of Physical and Natural Sciences, Commission of the European Communities Joint Research Centre, Ispra Establishment, I–21020 Ispra (Va, Italy)

Introduction

From an ecotoxicological point of view any substance which substantially modifies population and community characteristics must be considered dangerous. It is evident that a toxic substance may influence the biota if its concentration in the environment is over a certain level. Indeed, some heavy metals, dangerous at high concentrations, are essential to the biota at low concentrations. Unfortunately, it is often difficult to establish this level. Legislation concerning environmental protection is based on these concepts and maximum permissible concentrations are established essentially on the conclusions drawn from short-term experiments carried out under standard conditions. On the other hand, difficulties arising when results from different authors are compared and the great unreliability of extrapolating conclusions from laboratory studies to natural environment are well known.

From these considerations it is clear that the ecotoxicological study of a metal must take into account the fate of the metal from its source to its uptake by organisms and its effects on individuals, populations and communities.

In addition, indirect effects of the metal on the physical environment, such as variations in pH-value and oxygen concentration in water resulting from toxic effects of the metal on photosynthesis must also be considered. The influence of the physical-chemical characteristics of the ecosystem on the metal species (i.e. its physicochemical form) and the influence of biological activity on the metal form and availability, are other important topics of research.

This paper will illustrate these aspects as they are related to the problem of cadmium pollution in freshwater.

Sources and loading

In the lithosphere cadmium is present as the sulphide ore, greenockite (CdS), and it is associated with zinc ores, which are the only sources industrially exploited[42]. The natural cadmium input to the biosphere derives from volcanic activity, exudates from vegetation, forest fires, wind blown dust and leaching of rocks. During recent decades phosphate fertilizers[54], incinerator waste, coal and oil combustion, and most significantly mining and industrial usage of cadmium, have resulted in an important increase of cadmium in the environment. Consequently, cadmium pollution is very hazardous but usually of local importance.

Because in recent years several cases of cadmium poisoning have been reported (for example, Itai-Itai disease)[102] and experimental results confirm its high toxicity, cadmium is considered (together with mercury) by national and international legislations to be the most toxic of metals. Consequently, the maximum permissible concentrations of cadmium in freshwater bodies are always low. For example, the US Department of Health, Education and Welfare established that cadmium concentration in drinking water does not exceed 10 ppb[55]. The Dutch Federation of Water Boards proposed in 1973 that industrial effluents discharged into surface waters should not contain more than 0.1 mg Cd/l[37]. Cadmium concentrations lower than 0.3 mg Cd/l were considered non toxic for waters not used by salmonids and with an alkalinity higher than 80 ppm by New York State legislation[37]. According to the US Envrionmental Protection Agency (EPA), the total amount of cadmium discharged per day through industrial effluents into water courses should not be greater than 5.88 kg into water courses, 4.9 kg into lakes and 39.2 kg into estuaries. In addition, the mean cadmium concentration in an effluent flowing into freshwaters must be lower than 40 ppb, if its flow rate is lower than $\frac{1}{10}$ that of the water body into which it is discharged[37]. In Swiss running waters the legally tolerated cadmium concentrations must not exceed $4.5 \cdot 10^{-8}$ mol/l[34].

Cadmium concentration in the physical environment

In the lithosphere the mean cadmium concentration is 0.2 ppm but with wide variations depending on the rock type. In sea waters cadmium concentrations are commonly low: for example, North et al.[74] reported a mean value of 0.11 ppb, Brooks[16] 0.02 ppb and Taylor[93] 0.1 ppb. Cadmium concentration in freshwaters shows large variations, but the highest values are the effect of industrial wastes. Cadmium concentration in industrial effluents may vary from 0 to 1000 mg/l, whereas that in municipal waste waters is commonly lower than 10 µg/l. The following examples will give an idea of the levels of cadmium found in freshwater. These are normally close to the sensitivity limits of the analytical methods in common use. Taylor[93] measured in the River Tees, one of the major English rivers discharging into the North Sea, a mean cadmium concentration of 1 ppb in the soluble form, corresponding to a total concentration of 1.2 ppb. Cadmium concentration in the water of this river, highly polluted by industrial wastes, was lower than 1.5 ppb in the soluble form[93] measured in the River Rhine.

Cadmium concentration in the water of Lake Balaton (Hungary), slightly contaminated by industrial wastes, in particular by heavy metals, seems in comparison to be too high as Müller[69] reported values ranging from 0.42 to 0.24 ppm for total cadmium and from 0.57 to 0.32 ppm associated with clay fraction < 2 µm. In addition, values measured for the water of its tributary, the River Zala, of 0.37 ppm for total cadmium and 0.73 ppm for cadmium in particulate form, also seem to be high. Gächter[33] reported low cadmium concentrations in the water of two Swiss lakes, Lake Alpnach and Lake Luzern, where values varied from 0.01 to $0.04 \cdot 10^{-8}$ mol/l.

Baudo et al.[7] found a mean annual concentration of less than 1 ppb Cd for the water of Lake Mezzola (Northern Italy); concentrations of the same order of magnitude were measured in the water of the tributaries and the outflow. For the same lake the waters in contact with sediments in the central zone of the basin had a cadmium concentration higher (2.3 ppb) than that in the littoral zone (0.7 ppb)[6]. Cadmium from rainfall amounted to 1.2 ppb in 1976 and 0.3 ppb in 1977; this difference was due to meteorological variations from one year to another. It is notable that in the Lake Mezzola region no significant difference in cadmium concentration has been noted between water bodies lying in non-polluted and those in polluted areas. De Bernardi et al.[27] found cadmium concentrations lower than 1 ppb in the water of 19 small Lombardian lakes (Northern Italy). The distribution of cadmium (and other metals) in water sediments and biota of Lake Maggiore (Northern Italy) has been studied by Muntau[70]; Baudo et al.[8], Locht and Muntau[60] and Gommes and Muntau[38,39]. Cadmium concentration in the tributaries on the east coast of Lake Maggiore varied from 0.32 ppb to 1.32 ppb and in those on the west coast from 0.32 to 2.10 ppb. The mean value in lake water near the east coast was 0.33 ppb and near the west coast 0.30 ppb. The concentration in the water of the lake outflow (Ticino River) reached 0.21 ppb. Seasonal variations of cadmium concentration, from 0.16 ppb to 2.57 ppb, in pelagic lake waters were very large[39]. According to Baudo et al.[8] in the epilimnetic water from different areas of Lake Maggiore, mean cadmium concentrations varied from 0.1 to 1.2 ppb, whereas the seasonal variations for the whole lake ranged from 0.1 to 0.2 ppb. Muntau[70] estimated that the annual load from the tributaries of the southern area of Lake Maggiore was 1.48 t soluble cadmium plus 4.7 t particulate cadmium. The same author using two different models, calculated that the total input of cadmium for this lake ranged from 7.2 t/year to 9.0 t/year. As a consequence, the amount of cadmium accumulated annually resulted in the range from 5.1 to 6.9 t per year. Cadmium uptake from fall-out was not very important, 56 kg/year derived from wet fall-out and only 5 kg from dry fall-out.

An important source of cadmium for Lake Maggiore is the natural load from rocks in its watershed, although the contribution from industry also seems to be of importance (Muntau, unpublished data). For example, cadmium concentrations up to 4.1 ppm measured in the sediments of a small tributary of Lake Maggiore (River Bardello) were due to the industrial wastes discharged into it[60]. Industrial activities probably caused the increase of cadmium concentration in the sediments of the southern area of this lake from 0.63 ppm in 1963 to 3.13 ppm in 1975. In a bay of Lake Maggiore (Pallanza), into which the River Toce flows, cadmium concentration increased from 0.62 ppm in 1963 to 1.96 ppm in 1975[60].

There is evidence that high values are associated with the fine sediments as cadmium concentration in the sediment increased from the mouth of the River Toce towards the Borromeo Isles, where finest mud is located. There are further examples indicating that the highest concentrations were not necessarily measured in the vicinity of the cadmium point source, but in the settling zone of the finest mud and organic material. For example Houba and Remacle[46] found very high concentrations in the sediments of the River Vesdre (Belgium), in an area 2–5 km from a factory that was the source of cadmium pollution. Dall'Aglio[25] analyzed the heavy metal content of water samples from 23 Italian rivers from August to September 1979; cadmium concentrations varied from 0.030 ppb to 0.085 ppb, except for a small river (Entella) which attained 1.8 ppb. In the sediments of Lake Biwa (Japan) Kobayashi et al.[56] found a mean concentration of 0.44 ppm, a value very similar to that of Lake Maggiore measured on sediments collected in 1963[60].

A mean cadmium concentration of 1.1 ppb in the interstitial water from littoral sediments of Lake Mezzola was identical to that from pelagic sediments[7]. In some polluted rivers the cadmium concentration found in the sediments is very high; for example in the Weser estuary a mean concentration of 2.4 mg/kg has been reported[20] and in the River Rhine 9.9 mg/kg dry wt[90]. The highest cadmium concentration in freshwater sediments was measured by Houba and Remacle[46] in the River Vesdre, a small tributary of the Meuse River, where the value reached a maximum of 195 mg Cd/kg sediment and the mean was 86 mg. The point source of this pollution seems to be a smelting works producing metal sheets and located at the confluence of the River Vesdre and the Magne brooks. At the time of the study the factory was no longer active, but in the preceeding years it had discharged daily 100 m^3 of waste water rich in cadmium and zinc into the river.

Concentration in the biota

Cadmium accumulation in aquatic plants and animals has been reported by several authors. A linear correlation between initial cadmium concentration in the growth medium and cadmium concentration in Cd-resistant and Cd-sensitive strains of bacteria has been established. Resistance to high cadmium concentration resulted from the ability of certain strains to immobilize greater amounts of cadmium than their more sensitive counterparts[85].

Gommes and Muntau[39] analyzed the concentration of cadmium in 17 species of macrophytes from Lage Maggiore and found values ranging from 0.20 to 9.60 ppm dry weight with an average of 0.33 for *Nymphaea* (the species with the lowest accumulation capacity) and 3.93 for *Ceratophyllum* (the species with the highest accumulation capacity). Reiniger[84] observed in algae from an experimental field an increase in cadmium concentration from 4.4 µg/g dry wt (if they were living on contaminated sediment with 2 µg Cd/g) to 132 µg/g (if cadmium concentration in the sediment was 24 µg/g). The same author found in three species of aquatic plants *(Elatine hexandra, Althenia filiformis* and *Monita rivularis),* growing in a flooded rice field contaminated with cadmium, far higher concentrations of the metal in the plant tissues than was calculated for leaves and roots of the rice plants.

Gommes and Muntau[38] have measured far higher concentrations of cadmium (15.70 ppm) in soft tissue of *Unio mancus* (bivalve) from Lake Maggiore than in that (0.75 ppm) of *Viviparus ater* (Gastropod) living in the same environment. The high concentration in *Unio* is very similar to that (15.0 ppm) measured in a marine bivalve *(Ostrea)*; this high accumulation is probably due to the filterfeeding activity of these bivalves. Cadmium uptake by *Mytilus edulis* (marine bivalve) was studied by Carpene and George[21] using an isolated gill preparation. From this study the following conclusions may be drawn: a) cadmium uptake increased with time; b) temperature, dissolved oxygen, Zn, Cu, Hg, Pb and Fe seem to have no effect on cadmium uptake; c) cadmium was apparently taken in by diffusion and accumulation and d) cadmium accumulation was facilitated by intracellular binding and sequestration. More difficult to explain are the different concentrations of cadmium found in the shells of *Unio* (1.80 ppm) and *Viviparus* (0.15 ppm) by Gommes and Muntau[38]. It may be that periphyton, which absorbs great amounts of heavy metals, was more abundant on the shell of *Unio* than on that of *Viviparus. Dreissena polymorpha* (bivalve) lives in brackish and freshwater attached to solid surfaces (stones, walls, etc.). Consequently, *Dreissena* may take up cadmium from water and suspended particles but not from sediments, whereas sedimented matter may be an additional source of cadium for other bivalves living in the sediments (e.g. *Unio, Anodonta).*

Marquenie[63] used *Dreissena* for an active biological monitoring programme transferring individuals from unpolluted to polluted areas. From the results obtained it was evident that the cadmium concentration in the soft tissues of *Dreissena* was related to cadmium concentration in the soluble form and not to the particulate form suspended in the water. Consequently, there is evidence that particulate cadmium is less available to the mollusc than the soluble form.

Cadmium accumulation in net plankton is very high, compared with the concentration in lake water. But, because of the comparatively small biomass of plankton, only a small quantity of cadmium present in a unit of water is accumulated by the plankters. Price and Knight[76] reported 15 ppm Cd (dry weight) for the plankton of Lake Washington, and Baccini[4] found a mean value of 5.6 ppm for that of Lake Luzern and Lake Alpnach. Baudo et al.[8] reported values ranging from 6.5 to 39 ppm dry weight for net plankton from Lake Maggiore (mesh size of 126 µm) and values from 6.4 to 16 ppm dry weight for plankton collected with a 76-µm meshed net. Large variations in concentrations have been noted in relation to season. For plankton of Lake Mergozzo, the same authors obtained values of 7.1 ppm for a 126-µm net and 8.3 for a 76-µm net. In the same paper concentration factor values were reported of between 6900 and 12,000 for Lake Maggiore and from 4500 to 5200 for Lake Mergozzo. These values are not 'real concentration factors' but 'observed concentration factors' as they are dependent upon environmental and biological factors existing at the time of plankton sampling.

Lucas et al.[62] calculated that the median concentration in different species of fish from the American Great Lakes was 94 µg Cd/kg wet weight for the total fish and 400 µg for its liver. Uthe and Bligh[95] calculated that the mean cadmium concentration in several species of fish from different Canadian freshwater bodies was 50 µg/kg wet weight. The authors do not agree on the relationship between the level of industrial pollution in water and cadmium concentration in fish. For example, analyses on several species of fish collected from 49 freshwater bodies (New York State) showed that the most part of the samples had a cadmium concentration of about 20 µg/kg wet weight or lower, and that values higher than 100 µg occurred only occasionally in samples collected from areas polluted by ore deposits containing cadmium[61]. Havre et al.[41] reported that cadmium concentration in fish living in a Norwegian fjord, polluted by a zinc factory was higher than in fish from the open sea, but not dramatically so. Cadmium concentration varied with the species and the higher values were associated with benthonic feeders. Jaakkola et al.[50], measuring the cadmium concentration in muscle of *Exos lucius* (pike) from polluted and unpolluted water bodies, found a mean value of 3 µg/kg wet weight for the clean waters and concentrations from 4 to 13 µg/kg wet weight for the polluted environments. Havre et al.[41] found that the cadmium concentrations in fish liver ranged from 80 to 2506 µg/kg wet weight and from 3 to 32 µg/kg wet weight for muscle. The mean ratio between the concentrations in liver and that in muscle was about 42, but no significant correlation between the two concentrations was observed. Other authors (e.g. Mount and Stephan[68] and Calamari and Marchetti[19]) found cadmium in greater concentrations in liver, kidney and gills than in muscle and bones of fish. Using *Lepomis gibbosus* which had been exposed to 40 ppb cadmium for 1 month, Merlini et al.[66] found 7 times more cadmium in the liver and 500 times more in the kidney of those treated than in those of the control experiment. Other authors observed that cadmium was preferentially accumulated in both the liver and kidney of bass and blue gills[22]. Calamari and Marchetti[19] observed that, having attained equilibrium between its tissue and that of its environment, *Salmo gairdneri* lost all its accumulated cadmium after 80 days exposure to unpolluted waters. The same authors found that the biological half-life of cadmium varied for different organs obtaining values, for example, of 70 days for gills and 50 days for liver and kidney. The time needed for attaining the equilibrium between cadmium concentrations in fish and that in water, varied with the species. For example, for *Lepomis macrochirus* the time needed varied from 30 to 60 days[68] and for *Micropterus salmonides* it was about 60 days[22].

Kumada et al.[58] reported that in *Leuciscus leuciscus* cadmium intake was principally via food uptake, but, in general, information on the direct uptake of cadmium from water by fish in comparison to that via food is very scarce. The importance of each pathway is difficult to evaluate as it is dependent on several factors; for example, quality and quantity of food ingested and the physico-chemical form of cadmium in the water.

Effects on the organisms

The high toxicity of cadmium has been established by several authors for different species of aquatic organisms[1, 3, 58, 81, 98]. Informations is rather poor on the effect produced by cadmium in bacteria and algae.

Remacle and Houba[85] compared the effects of cadmium in strains of saprophytic bacteria both resistant and sensitive to cadmium. The most resistant strain belonged to a gram negative genus *(Pseudomonas)* and resistance seemed to be associated with the plasmids. A concentration of 8 ppm cadmium in the medium abolished the growth of the sensitive strain, whereas 300 ppm had no influence on the growth of the resistant strain. Devanos et al.[28] isolated micro-organisms, mainly bacteria, from different polluted environments, and with resistance to cadmium concentrations ranging from 1 to 500 ppm. Of these micro-organisms 94% were resistant to one or more antibiotics and 91% possessed multiple drug resistance. The authors postulated an extrachromosomal linkage of cadmium and antibiotic resistance in the isolated micro-organisms. Zwarum[103] reported that *Escherichia coli* was very tolerant to cadmium, and Bitton and Freihofer[12] studied the protective role of extracellular polysaccharides on the toxicity of copper and cadmium in *Klebsiella aerogenes.*

Population growth of *Euglena gracilis* was reduced in proportion to cadmium concentration in the culture medium; for example, 5 ppm prolonged the duplication time from 16–17 h to 30 h (Albergoni and Piccinni[2]). The same authors observed a significant reduction in motility, but no malformation in *Euglena* cells exposed to cadmium (5 ppm).

Concentration of cadmium from 50 to 500 ppb significantly reduced the growth of *Scenedesmus quadricauda*[17]. A reduction in growth of the same alga was obtained by Klass et al.[55] with 6.1 ppb. Growth inhibition in *Selenastrum capricornutum* was obtained at a cadmium concentration of 50 ppb in a culture with low carbonate content[5]. Cadmium had an inhibitory effect on the growth of a diatom *(Navicula pelliculosa)* at concentrations ranging from 14 to 140 ppm[59] and on that of a green alga (*Scenedesmus* sp.) at 0.1 ppm[14].

In the literature data on the effects produced by cadmium on macrophytes are scarce. Stanley[92] used *Myriophyllum spicatum* to determine the cadmium concentrations which produced a 50% reduction in growth of various organs. The following data were obtained: 7.4 ppm for root weight, 20 ppm for root length, 14.6 ppm for shoot weight and 809 ppm for shoot length. From these values it seems that the growth in length was less sensitive to cadmium than growth in weight. Hutchinson and Czyrska[48] found that the growth of *Lemna valvidiana* was inhibited by 25% and 80% by cadmium concentrations of 10 ppb and 50 ppb, respectively. The same concentrations produced growth inhibition of 50% and 90% respectively in *Salvinia natans*. The same authors[49] observed in both these species that the toxic effects produced by cadmium (10 ppb and 30 ppb) were enhanced if 50 ppb and 80 ppb of zinc were present in the medium. It is noteworthy that these concentrations of zinc alone stimulated the growth of these macrophytes.

More information is available on some species of animals. For example, information on the genus *Daphnia* is relatively abundant. Alabaster and Lloyd[1] stated that some freshwater invertebrates, such as *Daphnia magna*, were very sensitive to cadmium and there is evidence that the young are generally more sensitive than the adult. Baudouin and Scoppa[9] found an LC-50 (48 h)* of 65 ppb cadmium for *Daphnia hyalina* and 3800 ppb for *Cyclops sp.* Cebejszek and Stasiak[23] calculated an LC-50 (48 h) of 68 ppb cadmium for newborn *Daphnia magna;* Anderson[3] observed that the exposure of *Daphnia magna* to 2.6 ppb cadmium produced immobilization after 64 h. The 21-day and 48-h LC-50 values for *Daphnia magna* were calculated as 5 ppb and 65 ppb respectively by Biesinger and Christensen[11]. They found furthermore that cadmium at concentrations as low as 0.17 ppb may influence the reproduction of *Daphnia.* Bellavere and Gorbi[10] measured cadmium toxicity to *Daphnia magna* by the index IC-50, which corresponds to the metal concentration causing the immobility of 50% of the organisms tested in a prefixed time. Using *Daphnia*, kept in water with a hardness of 100 mg/l, they recorded an IC-50 (24 h) of 120 ppb and for water with a hardness of 200 mg/l a value of 160 ppb. When compared with data reported by other authors for the same species, these values seem to be very high, but we must remember that in this experiment the exposure time was particularly short. Increasing the exposure time would probably produce immobility even at concentrations far lower than those used by the authors. However, it is clear that the sensitivity of *Daphnia* decreased with increasing hardness. This relationship is more evident from experiments regarding the influence of cadmium on *Tubifex tubifex* mortality carried out by Brkovic-Popovic and Popovic[15]. LC-50 (48 h) values varied from 2.8 ppb cadmium to 720 ppb cadmium in water with a total hardness (expressed as carbonate) ranging from 0.1 mg/l to 261 mg/l.

The most evident anatomical effects of 3-4 days exposure to cadmium at concentrations lower than 100 ppb on *Biomphalaria glabrata* were the displacement of secreting cells of the epatopancreas and the degeneration of male cells, particularly those at the first stage of maturation[96]. Bellavere and Gorbi[10] determined that *Biomphalaria glabrata* was less resistant to cadmium than *Brachidanio rerio* (cyprinid fish), having an LC-50 of 4.80 ppm cadmium at 24 h and 0.30 ppm at 96 h as compared with LC-50 values for *Brachidanio rerio* of 21.08 ppm and 4.35 ppm, respectively. Decrease in LC value with exposure time demonstrates the limited value of results obtained from experiments of short duration.

Thorp and Lake[94] calculated that the LC-50 (4 days) of the mayfly larva *(Athalophebia australis)* was 0.84 ppm cadmium, and Warnick and Bell[99] found a higher value of 32 ppm cadmium for the stonefly *Acroneura lycorias.*

A wealth of information on the effects of cadmium on fish has been collected by Alabaster and Lloyd[1]. According to these authors many sublethal effects of cadmium in fish are caused by the modification of ionic balance induced by this metal. Examples are degeneration of muscle, lesions in the spinal cord, convulsion, tetanic and neuromuscular disturbances.

The range of lethal concentrations for fish is very large: from 10 to 100,000 ppb cadmium. These authors[1] attributed this to inconsistency in length of exposure to cadmium, the species, the behavioral aspects of the fish and the slope of the concentration response curve. Salmonids seem to be the least resistant fish to cadmium, whereas cyprinids are the least sensitive. For example, 50% of the alevins of rainbow trout died after an exposure of 2 days at 150 ppb cadmium and 33% at 25 ppb[19] and an LC-50 (48 h) of 3000 ppb was calculated for trout by Jung[51], whereas for *Cyprinus carpio* (common carp) the LC-50 (4 days) was 240 ppb[83]. Kumada et al.[57] observed neither higher mortality nor growth reduction in rainbow trout exposed to 5 ppb cadmium for more than 30 weeks.

Certain developmental stages of *Xenopus laevis* (Amphibia) cannot tolerate cadmium concentrations higher than 2 ppm, but may survive concentrations lower than 1.5 ppm. The most evident embryonic anatomical alterations occurred at the epidermis, encephalon, eyes, pronephros and somites[96].

An experiment carried out by Merlini[67] showed that pumpkinseed sunfish *(Lepomis gibbosus)* exposed for 2 weeks at 40 ppb cadmium (as sulphate) contained less vitamin B_{12} in the liver than the control fish. These, however, had lower concentrations of this vitamin B_{12} in the gall-bladder, intestine and gills than the treated fish. Since vitamin B_{12} is eliminated through the intestine, it is probable that cadmium accelerated this process by stimulating the biliary excretion of vitamin into the intestine and its elimination through gill epithelium. Merlini et al.[66] drew the following conclusions from the experiments carried out on the effects produced by cadmium on the Zn-65 uptake by *Lepomis gibbosus*: a) Zn-65 concentration in fish exposed for 31 days to 40 ppb cadmium was $\frac{1}{4}$ that of the control fish; b) the percentage loss was similar in the control and treated fish; c) the control fish had higher cadmium concentrations in interrenals (2.85 µg/g wet weight), bone (1.48 µg/g wet weight) and brain (1.06 µg/g wet weight); d) contaminated fish had small cadmium concentrations in eye, brain, fin, scale and skin, one tenth that in the gonads, gills and bone and one hundredth of that in the heart, spleen and kidney. From these results it is evident that cadmium influenced zinc uptake, but not its elimination. Some organs (i.e. heart, spleen and kidney) concentrated a large amount of cadmium in treated fish but not in the control fish. As a consequence, it was clear that the distribution pattern of this element in fish exposed to cadmium was different from that in fish living in unpolluted waters.

The protective effect of selenium against cadmium in rats and mice is well known and there is information on the protective effect of cadmium against selenium in chickens[43,44]. We have found only one paper (by Van Puymbroeck et al.[97]) on the antagonism between selenium and cadmium in the freshwater snail *Lymnaea stagnalis*. These authors found that the effects of cadmium were reduced to about 50% in the presence of sublethal amounts of selenium. In addition, sublethal concentrations of cadmium protected *Lymnaea*

---

* LC-50 (lethal concentration 50%) is the concentration of any toxic substance reducing by mortality the number of tested individuals to 50% in a prefixed time.

against high concentrations of selenite (3 ppm Se) and selenate (15 ppm Se). For aquatic invertebrates, the biochemical mechanisms involved in this antagonism are unknown, but for mammals it seems that proteins (thiol groups) play a fundamental role in the Cd/Se interaction.

Effects at population and community level

Most of the information on the effects of cadmium in aquatic biota concerns short term experiments carried out in the laboratory. This information is insufficient to determine the real damage to individuals and populations and for the assessment of an acceptable unharmful metal concentration in the environment (e.g. Sprague[91], Nobbs and Pearce[71], Ravera[80], Hoppenheit[45]). On the other hand, there is much uncertainty about the real effects of a pollutant when it is observed in the field because toxic substances are generally present as mixtures in water bodies[78]. Long-term studies on laboratory populations concerning the life span of the tested species produce more useful and interesting results than experiments on acute toxicity.

Experiments carried out by Hoppenheit[45] on the effects of cadmium at population level in 20 successive generations of *Tisbe holothuriae* (coastal and brackish water copepod) give a clear example of the value of long-term population studies in ecotoxicology. The most important results produced from the research were: a) concentrations of 148 ppb and 222 ppb delayed by 8 weeks the decrease of the population density found in the control but, at the end of the experiment, the mean population density of the contaminated populations was similar to that of the control ones. According to the author this was probably due to the adaptation of the contaminated population within 20 generations; b) significant differences in density were observed between populations exposed to 148 and to 222 ppb; c) variability within treated populations was larger than that of the control populations and d) reduction in density of populations exposed to cadmium did not result in an increase in the number of nauplii, as, generally, occurred in control populations. In conclusion, *Tisbe holothuriae* tolerated relatively high concentrations of cadmium, compared with other freshwater and marine crustaceans (e.g. Andersen[3], Eisler[29]).

Marshall[64] carried out experiments on populations of *Daphnia galeata mendotae,* kept for 22 weeks with contaminating concentrations of 2 ppb cadmium and higher. He observed a significant reduction in population density and biomass and an increase in average brood size, and the ratios between number of ovigerous females and total number of females, and between the number of eggs and total number of females. The same author noted that 1 ppb cadmium dramatically increased embryo mortality.

A long-term experiment on the effects of cadmium on laboratory populations of a tropical freshwater snail *(Biomphalaria glabrata)* was carried out by Ravera et al.[81]. All adults were killed after 6 and 3 h by 2 and 4 ppm cadmium, respectively. Survival time for 50% (ST-50) of the tested snails was 15 h at 1 ppm concentration, 31 h at 0.5 ppm and 63 h at 0.1 ppm. Eggs were deposited by adults kept at 0.1 and 0.5 ppm, but their number was about $\frac{1}{4}$ that of the control. Although the eggs produced by contaminated adults were transferred to clean water, all the embryos died at 'morula' stage. No embryo completed its development when eggs from untreated adults were exposed to 3 cadmium concentrations: 0.1, 0.5 and 1.0 ppm. In this experiment the viability of the control was about 90%. A comparison of the sensitivity of the embryo and adult on the basis of ST-50, showed the embryo to be more resistant to cadmium, but this difference decreased with increasing cadmium concentration and at 1.0 ppm both ST-50 values were similar.

Ricci and Pozzoli[86] studied the chronic effects of cadmium ($CdCl_2$) and zinc ($ZnCl_2$) on the life span and reproduction of a Bdelloid rotifer *(Phylodina roseola).* The rotifers were exposed to a series of cadmium concentrations ranging from 10 ppb to 10,000 ppb. This species seems to be exceptionally resistant to cadmium as, even at the highest concentrations, egg hatchability and life span were not significantly different from those of the control. In addition, neither cadmium nor zinc seemed to influence the age at which the rotifer attained reproductive maturity.

Premazzi et al.[75] studied the effects of cadmium and 5 other metals (Cu, Ni, Hg, Zn and Pb) on *Selenastrum minutum.* The culture medium was a very dilute solution of macro- and microelements with no chelating substances. From the results obtained for cadmium the following conclusions may be drawn: a) cadmium concentrations lower than 10 ppb had no effect at population level, concentrations between 60 and 200 ppb abolished algal reproduction and concentrations higher than 200 ppb killed all cells; b) concentrations between 15 ppb and 20 ppb caused a reproductive depression during the first 4–6 days, after which a population increase was evident. This increase probably followed a decrease of cadmium concentration in the solution, due to adsorption onto free surfaces (e.g. cells, wall of the container); Cd, Hg and Cu were the most toxic of the 6 metals; d) NTA addition to the medium had no effect on the toxicity of cadmium, whereas EDTA and humic acids strongly reduced the effects of this metal; e) at 4–5 ppb cadmium cell-size began to increase; this increase always occurred during the period of reproductive depression (at 15–20 ppb). It is probable that cadmium reduced or abolished reproduction by autospores indirectly causing an increase in cell-size; f) the hypothesis postulated in e) is supported by the cell-size decrease observed when cells were transferred from a solution with an algostatic concentration of the metal to a medium without cadmium; g) it is interesting to note that an increase in cell-size in association with a depression of reproduction was also observed in cultures contaminated by the other metals, and h) the high sensitivity of *Selenastrum* to cadmium in these experiments if compared with those of other authors, may be due to the low nutrient concentration of our medium and to the absence of chelating substances (e.g. EDTA).

Davies[26] noted that the presence of mercury in cultures caused cell division to be uncoupled from growth so that abnormally large cells were produced and he proposed that the action of mercury was to inhibiit intracellular production of methionine and amino acid associated with normal cell division.

An interesting hypothesis has been proposed by Foster and Morel[32] in order to understand the mechanism of cadmium toxicity on a marine diatom. They measured a 3-day delay between inoculation and the development of toxic effects in *Thalassiosira weissflogii.* Therefore it seems that toxic effects occurred only after cadmium had reached a certain concentration in the cell. This hypothesis was also supported by the shape of the uptake curve showing an accelerating accumulation of cadmium with time. A similar curve has been reported by Cain et al.[18] for *Scenedesmus obliquus,* but uptake curves from Break and Jenesen[13] for marine diatoms and Sakaguchi et al.[88] for freshwater green algae showed a rapid initial cadmium uptake, suggesting injury of the cell membrane. According to Foster and Morel[32] the rapid recovery of *Thalassiosira,* following the lowering of cadmium concentration in the medium, is explicable by the hypothesis that population growth cannot be inhibited if the metal concentration within the cell has not reached a given value. This suggests that one or more functions of the algal cell were blocked and the absence of these functions was not immediately lethal. As a consequence, the algae were able to recover soon after their cadmium load was reduced. In addition, EDTA did not produce a massive cadmium release from the cell, but presented further uptake. At sublethal cadmium concentration in the medium, the cells were balanced between inhibition and growth, a state that can be maintained for a long time without permanent damage.

Interesting results on the direct and indirect effects produced by pollutants at community level have been obtained by means of 'enclosure' techniques ('micro-ecosystems'). This method has 3 main advantages: a) the community is sampled over a relatively long period of time; b) several populations belonging to at least 2 trophic levels are available in naturally occurring proportions and c) replicated enclosed populations can be tested in semi-natural conditions. By this method the effects produced by one or more pollutants on a natural community may be predicted with greater ease than from field observations and more realistically than from laboratory experiments. On the other hand, the disadvantages of this technique derive from the 'enclosure' itself, which prevents water renewal and horizontal water transport in the enclosed column. In addition, the inner wall of the 'enclosure' offers a relatively large free surface for the attachment of periphyton and the adsorption of various substances (for example metals).

In spite of this the method represents a useful compromise between laboratory experiments and field observation and a series of experiments using 'enclosures' has been carried out to evaluate the effects of cadmium in planktonic community and water quality[52]. To this end 8 'enclosures' were anchored in the shallow and eutrophic Lake Comabbio (Lombardy, Northern Italy) at 3 m depth. The water columns in 4 enclosures were maintained in contact with lake sediments (open series), while those in the remaining 4 were isolated from the sediment by PVC bases (closed series). Two enclosures were kept as controls. At the beginning of the experiment $CdCl_2$ was added to 3 enclosures of each series to produce initial concentrations of 10, 50 and 100 ppb cadmium. The duration of the experiment was 11 days (September 30–October 11, 1979) and the mean sampling interval was 2 days.

The results obtained were as follows. Dissolved oxygen (DO) and pH decreased progressively in all contaminated 'enclosures', but this decrease was more evident in those with higher initial concentrations of cadmium. In the 'open' contaminated enclosures the DO decrease was greater than in the corresponding 'closed' enclosures. As the experiment progressed cadmium concentration in the water of the contaminated enclosures decreased until it reached values similar to those of the controls (less than 5 ppb) at the end of the experiment, at which time the following cadmium concentrations were measured in the sediments: 0.80 µg/g dry weight in the control; 1.00 µg in the 10 ppb enclosure; 1.18 µg in the 50 ppb enclosure and 3.50 µg in the 100 ppb enclosure. At the beginning of the experiment an increase in phytoplankton biomass in both controls and 10 ppb enclosures was observed, but was absent from the enclosures with higher initial cadmium concentrations in which phytoplankton biomasses decreased to 20% of the initial values. Chlorophyll concentration followed a similar pattern to phytoplankton. In the contaminated 'enclosures' phytoplankton biomass decreased while orthophosphate and ammonia concentrations incrased. Diatoms were more resistant to cadmium than Cyanophyta and Pyrrophyta. During the experiment Cladocerans increased in both controls. This effect was noticeable but less marked in the 'enclosures' contaminated by 10 ppb, whilst in those with 50 ppb the population declined dramatically. Concentrations of 10 and 50 ppb had no apparent effect on Copepods, whereas 100 ppb reduced Copepod population density and eliminated the Cladocerans. Rotifers numbers also dropped in contaminated 'enclosures'.

To evaluate the influence of cadmium on various important demographic parameters (i.e. mortality, fecundity, fertility and embryo viability) of *Physa acuta* (Gastropoda, Pulmonata) population, a series of experiments was carried out partly in semi-natural conditions (enclosure) and partly in the laboratory (Ravera, Giannoni and Agostini; in press). The 'enclosure' were established in Lake Comabbio from which the material to be tested was collected. Initial cadmium ($CdCl_2$) concentrations ranged from 125 ppb to 1000 ppb and experiments were carried out in different periods of the year from June to October. All the concentrations tested produced toxic effects and a significant increase of mortality was measured even at the lowest concentrations. For example, at the end of the August experiment (20th day), 125 ppb caused a mortality 55% higher than that of the control. From the results obtained it is evident that adult mortality, calculated for each experiment, increased with cadmium concentration, but a given cadmium concentration produced different mortalities in different seasons. For example, the ST-50 value for the lot exposed to 250 ppb was 23.2 days in October, 11.2 in June and only 5.8 days in August. As variations in temperature could explain these differences, a series of laboratory experiments was carried out at 3 temperatures: 15 °C, 20 °C and 25 °C. Cadmium concentrations used were 250 ppb, 500 ppb, 750 ppb and 1000 ppb. From the results obtained it was evident that at constant temperature mortality increased with cadmium concentration, but the influence of temperature upon mortality increased with decreasing cadmium concentration. Although most of the higher ST-50 values came from experiments at 15 °C and few from the 25 °C experiments, for the same cadmium concentrations the relationship between ST-50 values and temperature was not clear. It seems that temperature may influence cadmium toxicity towards *Physa* but that other factors may also be involved. As the physiological condition of adult *Physa* varies with the time of year, it is likely that its resistance to cadmium (and probably to other pollutants) also varies with the season[77].

All eggs produced during the experiment were collected and counted. Samples of egg-capsules were chosen at random from each experiment and kept in the laboratory until the end of their development under conditions of temperature and cadmium concentration identical to those of the 'enclosures' from which they were collected, enabling fertility and embryo viability to be measured in addition to the mortality and fecundity data already measured in semi-natural conditions. Fecundity was strongly reduced even by the lowest cadmium concentration (125 ppb) and this reduction increased with concentration and also with time, increasing from the beginning to the end of each experiment. Fertility and embryo viability were completely abolished in all the contaminated 'enclosures', with one exception, the 125 ppb enclosure, where there was a viability value of 2.85%. Fertility of another freshwater snail *(Biomphalaria glabrata)* was abolished by 100 ppb of cadmium[79]. From these results there is evidence that *Physa* as well as *Biomphalaria* populations, living in an environment contaminated by a concentration even as low as about 100 ppb cadmium, are unlikely to survive.

Cadmium speciation

The chloride, sulphate and nitrate of cadmium are soluble compounds, whereas carbonate and hydroxide are not. In soft waters with low pH, cadmium in ionic form seems to be more abundant than in complexed forms[36] but the ionic form is readily adsorbed onto free surfaces[31]. Since such adsorption decreases with increasing salinity[89], the removal of cadmium by this process would be 'ceteri paribus'

more important in freshwaters than in estuaries and marine waters. In addition, adsorption obviously increases with concentration of suspended matter, and Williams et al.[101] found that a large percentage of the cadmium present in river waters was adsorbed onto particles. According to some authors the uptake by lake phytoplankton is the major sink for dissolved heavy metal salts (e.g. Gächter[35]).

Iron and manganese hydroxides and floccules of organic matter provide additional surface for adsorbing cadmium. In surface waters, rich in oxygen, dissolved iron and manganese are oxidized and coprecipitated with cadmium.

Chelated toxic metals become less toxic than the corresponding free metal ions[40] whereas chelated trace metal nutrients are more available to algae[100]. Cadmium may be chelated by several organic compounds, of which the humic substances seem to be the most widespread and quantitatively important[30]. Several compounds released by planktonic algae chelate metals; for example amino acids, polypeptides, proteins, porphyrins, purines (for example, Khailov[53]). The metal chelating effects of extracellular products of larger aquatic plants have been investigated to a lesser extent (for example, McKnight and Morel[65]).

From these considerations it is evident that dissolved cadmium concentration and its consequent effects on the community depend not only on the external loading of the metal, but also on the characteristics of the water body into which the cadmium is discharged. These characteristics are often interconnected; for example, high phytoplankton concentration and the resultant photosynthetic activity increases dissolved oxygen concentration and pH, as well as sestonic particles and extra-cellular products. Consequently, adsorption and chelation of cadmium would be more active in eutrophic than in oligotrophic water bodies. Attention must be paid also to the cadmium adsorbed by suspended particles before and after their sedimentation. In fact, if dissolved cadmium damages an organism by entering through its membranes (for example, Alabaster and Lloyd[1]), cadmium adsorbed onto particles may be taken up by filter feeders and benthonic organisms and penetrate their tissues through the intestine wall. Under changing redox conditions sedimented material can release important quantities of metals to interstitial water, and then in the soluble form to water in contact with the sediments.

A detailed study of heavy metal speciation in river sediments from different parts of the world was carried out by Salomons and Förstner[90]. From the analysis of cadmium forms in sediments from ten European rivers considerable variations in the proportions of each form in the various rivers was observed.

Although a metal may be accumulated in an organism in the same form in which it was taken up, it is more commonly transformed to another form, bound, for example, to proteins or lipids, etc. Howard and Nickless[47] found cadmium and zinc bound to proteins in aquatic molluscs and other invertebrates, whereas Noël-Lambot[73] found the same metals bound to corpuscles in the intestinal lumen of some species of fish. Metallothioneins are low molecular weight proteins characterized by a high cysteine and metal content (Zn, Cd, Hg, Cu). These proteins have been found at low concentrations in the organs of untreated animals, but in animals exposed to high concentration of heavy metals metallothionein content increased very rapidly (for example, Chen et al.[24]). Metallothioneins were found both in warm-blooded (mammals and birds) and cold-blooded animals (fish and invertebrates). Noël-Lambot et al.[72] studied the Cd, Zn and Cu distribution in the soluble fraction of liver and gills of eel *(Anguilla anguilla)* contaminated by cadmium. In chronic contamination experiments, most of the cadmium was concentrated in the liver and gills bound to metallothioneins, but in short-term experiments cadmium bound to metallothioneins was accumulated only in the liver. Small quantities of metallothioneins were found also in the liver of untreated eels.

Albergoni and Piccinni[2] studied detoxification mechanisms in *Euglena gracilis* exposed to cadmium and copper solutions. Cadmium was accumulated in larger amounts than copper and was the metal to which *Euglena* showed less resistance. This occurred because cadmium present in the *Euglena* cell was bound to organic molecules with high molecular weight and, consequently, its elimination from the cell was prevented. Conversely, copper was complexed by low molecular weight molecules and, subsequently, easily eliminated. This difference produced a greater cellular accumulation of cadmium than copper and, consequently, *Euglena* showed a higher resistance to copper than to cadmium.

The transformation of one metal form to another, which occurs at tissue level, commonly modifies the toxicity of a metal. The result of this process, if it reduces the original toxicity of the metal, is called 'detoxification'. Some authors (for example, Rugstad and Norseth[87]) have implicated metallothioneins in cadmium and mercury detoxification processes.

Discussion and conclusions

There is a considerable gap between present knowledge and the information necessary to understand the influence of cadmium on freshwater ecosystems and thus to obtain reliable basis for legislation to protect the environment against cadmium pollution.

In figure 1 the fundamental topics of the ecotoxicological research on cadmium are schematized. It is obvious that this scheme may be adopted also for any other non-volatile metal.

From a scientific point of view all aspects of the scheme are important but, for practical purposes, (protection of the environment against potential damage by cadmium pollution) research must be focused on a few fundamental points. For example, the establishment of a quantitative relationship between cadmium concentration in water (or the cadmium loading of a water body) and the consequent effects at population and community level is essential to environmental protection. According to all the legislation, protection of the ecosystem means maintaining a reasonable likelihood that the indigenous populations will survive and preventing any significant modification of the structure and functioning of the community. Therefore, the death of individuals does not concern environmental protection, as long as the death rate is not high enough (if compared with birth rate) to eliminate the population as a whole. Consequently, certain levels of pollution are acceptable, but must not produce dramatic consequences on the ecosystem.

Information concerning the relationship between amount of cadmium released from a source (or sources) and the cadmium loading a waterbody is very scant. Unfortunately, there are also limited data on the relationship between cadmium loading and its concentration in the ecosystem.

Information on cadmium concentration in the various compartments of the ecosystem (for instance, water, sediments, aquatic plants and animals) is more abundant. For example, a scheme reported in figure 2 shows the relative capacity of certain important compartments of a freshwater ecosystem (Lago Maggiore) to concentrate cadmium. From a scientific and practical point of view the values reported in this figure must be considered in association with biomass and

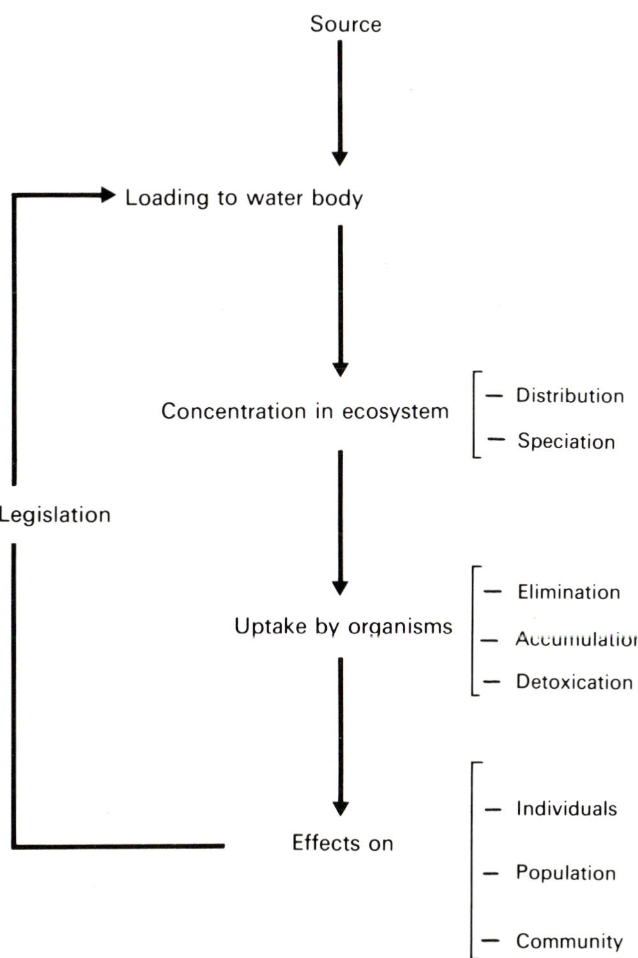

Figure 1. Fundamental research topics on the ecotoxicological effects of cadmium.

production values for each important compartment and to the volume of the water and sediments of the ecosystem. In addition, to estimate cadmium mass balance, information on the input and output of cadmium from the water body is needed (e.g. Muntau[70]).

Since the physico-chemical form (species) of cadmium influences its availability and toxicity to the organism, the determination of the metal species present in an environment is desirable. On the other hand, metal form separation is a difficult problem and many analytical aspects are presently unknown. The difficulties increase if these techniques are applied routinely to ecological research. In addition, the interrelationships between physical environment, biota and metal species make the research on metal speciation more difficult (fig. 3). As a result, information on this important ecotoxicological aspect is lacking and the difficulties involved in establishing a relationship between cadmium concentration in water and the concentration inside of the organism, are obvious.

Cadmium is taken up by an aquatic organism via two pathways: food and/or body surfaces (cell membrane, epithelia and skin). The relative importance of each of these pathways has attracted little research and the data available mostly concern fish. In both cases metal must pass through the intestinal wall or external membranes.

In spite of the fundamental role played by the membranes, cadmium transfer mechanisms through artificial and natural membranes has been the object of few studies. Cadmium, having been taken up by an organism, may be totally or partially excreted or may accumulate in the body. The cadmium content of an organism is the product of the equilibrium between the metal uptake and loss. Because some animal and plant species are able to accumulate relatively large amounts of cadmium, if compared with the concentration in the water, they may be used as 'cadmium accumulating organisms', for example freshwater plankton and bivalves. Detoxification mechanisms are an important subject of research, as the results indicate fairly reliably why one species is more resistant to cadmium than another, even if both live in the same environment. Results obtained from the limited studies carried out on this subject are encouraging, although they cover only a very small number of species. From an ecotoxicological point of view the effects produced by cadmium must be measured at individual level, but overall effects are seen at population and community level. There is a great deal of data on the cadmium toxicity at individual level but data are scarce for populations and almost nonexistent for natural communities[80]. Most of the information at individual level concerns a small number of species (e.g. fish, *Daphnia)* and comes from short-term experiments carried out under laboratory conditions. More information on the relationship between the cadmium content in an organism and the effects is needed. Almost all population studies have been carried out on unicellular algae and a few species of Metazoa. Promising results on the cadmium toxicity at community level have been obtained by the 'micro-ecosystem' method. The advantages and disadvantages of this technique have been discussed in the preceeding pages. Effects at population level must be considered in relation to demographic characteristics, especially birth rate, as sublethal concentrations may eliminate a population if its fertility is heavily reduced by cadmium pollution. Cadmium (or any other pollutant) affects community structure both directly and indirectly and indirect effects are more difficult to identify. For example, a population may increase its density, after the contamination of the environment, if its sensitivity to the pollutant is lower than that of its predators. Damage produced by pollution on a certain trophic level may be expected to influence the

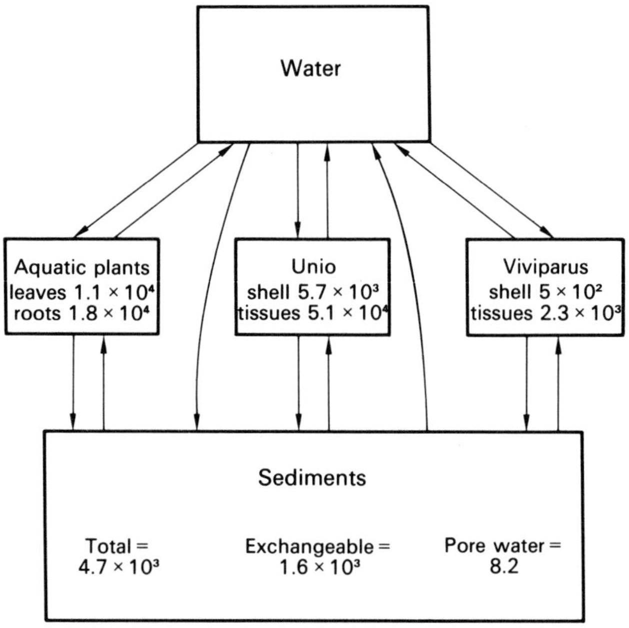

Figure 2. Relative capacity of certain compartments of a freshwater ecosystem (Lago Maggiore) to concentrate cadmium. (from: Ravera et al.[81], modified)

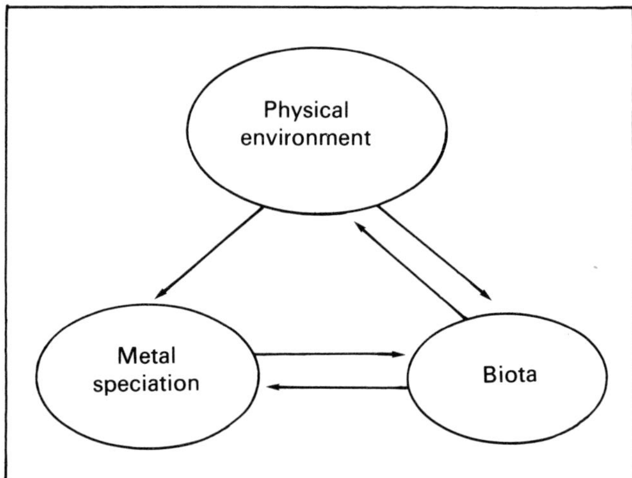

Figure 3. Interrelationships between physical environment, biota and metal species.

lower and higher levels. In addition, species resistant to cadmium may show no evident damage from exposure to the metal, but may transfer cadmium to their predators. For this reason studies on the cadmium transfer along the food chains are important and, particularly, when man is at the top.

Since control of cadmium discharge in effluents reduces but does not abolish cadmium loading to aquatic ecosystems, metal concentration will increase with time. Consequently, monitoring of cadmium-polluted environments and studies on biological effects must be continued. In particular, information on the effects produced by this metal at population and community level (fig. 1) are urgently needed.

1 Alabaster, J.S., and Lloyd, R., Cadmium, in: Water quality criteria for freshwater fish, pp.221-251. Butterworths, London 1980.

2 Albergoni, V., and Piccinni, E., Regolazione e tolleranza per i metalli in *Euglena gracilis*. Lincei-Rend. Sc. fis. mat. e nat., *66* (1979) 209-213.

3 Anderson, B.G., The apparent thresholds of toxicity to *Daphnia magna* for chlorides of various metals when added to Lake Erie water. Trans. Am. Fish. Soc. *78* (1950) 96-113.

4 Baccini, P., Untersuchungen über den Schwermetallhaushalt in Seen. Schweiz. Z. Hydrol. *38* (1976) 121-158.

5 Bartlett, L., Rabe, F.W., and Funch, W.H., Effects of copper, zinc and cadmium on *Selenastrum capricornutum*. Wat. Res. *8* (1974) 179-185.

6 Baudo, R., Galanti, G., Guilizzoni, P., and Varini, P.G., Relationships between heavy metals and aquatic organisms in Lake Mezzola hydrographic system (Northern Italy). 2. Heavy metals in rainfall. Mem. Ist. ital. Idrobiol. *37* (1979) 185-195.

7 Baudo, R., Galanti, G., Guilizzoni, P., and Varini, P.G., Relationships between heavy metals and aquatic organisms in Lake Mezzola hydrographic system (Northern Italy). 3. Metals in sediments and exchange with overlying water. Mem. Ist. ital. Idrobiol. *39* (1981) 177-201.

8 Baudo, R., de Bernardi, R., Soldarini, E., Locht, B., and Muntau, H., Spatial and temporal variations of metal concentrations in plankton of Lago Maggiore and Lago Mergozzo. Mem. Ist. ital. Idrobiol. *38* (1981) 79-100.

9 Baudouin, M.F., and Scoppa, F., Acute toxicity of various metals to freshwater zooplankton. Bull. envir. Contamin. Toxic. *12* (1974) 745-751.

10 Bellavere, C., and Gorbi, J., A comparative analysis of acute toxicity of chromium, copper and cadmium to *Daphnia magna*, *Biomphalaria glabrata* and *Brachydanio rerio*. Envir. Technol. Lett. *2* (1981) 119-128.

11 Biesinger, K.E., and Christensen, G.M., Effects of various metals on survival, growth, reproduction and metabolism of *Daphnia magna*. J. Fish. Res. Bd Can. *29* (1972) 1691-1700.

12 Bitton, G., and Freihofer, V., Influence of extracellular polysaccharides on the toxicity of copper and cadmium toward *Klebsiella aerogenes*. Microbiol. Ecol. *4* (1978) 119-125.

13 Braek, G.S., and Jenesen, A., Heavy metal tolerance of marine phytoplankton. 4. Combined effect of zinc and cadmium on growth and uptake in some marine diatoms. J. exp. mar. Biol. Ecol. *42* (1980) 39-54.

14 Bringmann, G., and Kühn, R., Vergleichende wassertoxikologische Untersuchungen an Bakterien, Algen und Kleinkrebsen. Gesundheitsingenieur *80* (1959) 115-120.

15 Brkovic-Popovic, I., and Popovic, M., Effects of heavy metals on survival and respiration rate of tubificid worms. 1. The effects of survival. Envir. Pollut. *13* (1977) 65-72.

16 Brooks, R.R., The use of ion exchange enrichment in the determination of trace elements in sea water. Analyst *85* (1960) 745-748.

17 Bumbu, Y.V., and Mokryak, A.S., Effects of some trace elements on the development of the alga *Scenedesmus quadricauda*. Biol. Khim. Nauk. *1* (1973) 82-83.

18 Cain, J.R., Paschal, D.C., and Hayden, C.M., Toxicity and bioaccumulation of cadmium in the colonial green alga *Scenedesmus obliquus*. Arch. envir. Contam. Toxic. *9* (1980) 9-16.

19 Calamari, D., and Marchetti, R., Accumulo di cadmio in *Salmo gairdneri* Rich. N. Ann. Ig. *28* (1977) 425-436.

20 Calmano, W., Wellershaus, S., and Förstner, V., Dredging of contaminated sediments in the Weser estuary: chemical forms of some heavy metals. Environ. Technol. Lett. *3* (1982) 199-208.

21 Carpene, E., and George, S.G., Adsorption of cadmium by gills of *Mytilus edulis*. Molec. Physiol. *1* (1981) 23.

22 Cearley, J.E., and Coleman, R.L., Cadmium toxicity and bioconcentration in largemouth bass and bluegill. Bull. environ. Contam. Toxic. *11* (1974) 146-151.

23 Cebejszek, I., and Stasiak, M., Studies on the effect of metals on water biocenosis using the *Daphnia magna* index. Part 2. Rocz. Panstw. Zakl. Hyg. *11* (1960) 533-540.

24 Chen, R.W., Whanger, P.D., and Weswig, P.H., Biological function of metallothionein. I Synthesis and degradation of rat liver metallothionein. Biochem. Med. *12* (1975) 95.

25 Dall'Aglio, M., Distribution of trace elements in major Italian rivers. Water Qual. Bull. *7* (1982) 16.

26 Davies, A.G., An assessment of the basis of mercury tolerance in *Dunaliella tertiolecta*. J. mar. biol. Ass. U.K. *56* (1976) 39.

27 de Bernardi, R., Giussani, G., Guilizzoni, P., and Mosello, R., Indagine conoscitiva per una caratterizzazione limnologica dei 'piccoli laghi lombardi'. Regione Lombardia, in press, 1983.

28 Devanos, M.A., Litchfield, C.D., Mc Lean, C., and Gianni, J., Coincidence of cadmium and antibiotic resistance in New York Bight (USA) apex benthic micro-organisms. Mar. Poll. Bull. *11* (1980) 264.

29 Eisler, R.J., Cadmium poisoning in *Fundulus heteroclitus* (Pisces: Cyprinodontidae) and other marine organisms. J. Fish. Res. Bd Can. *28* (1971) 1225-1234.

30 Elder, J.F., Complexation side reactions involving trace metals in natural water systems. Limnol. Oceanogr. *20* (1975) 96-102.

31 Farrah, H., and Pickering, W.F., Influence of clay-solute interactions on aqueous heavy metal ion levels. Wat. Air Soil Pollut. *8* (1977) 189-197.

32 Foster, P.L., and Morel, F.M.M., Reversal of cadmium toxicity in a diatom: an interaction between cadmium activity and iron. Limnol. Oceanogr. *27* (1982) 745-752.

33 Gächter, R., Untersuchungen über die Beeinflussung der planktischen Photosynthese durch anorganische Metallsalze im eutrophen Alpnachersee und der mesotrophen Horwer Bucht. Schweiz. Z. Hydrol. *38* (1976) 97-119.

34 Gachter, R., MELIMEX, an experimental heavy metal pollution study: goals, experimental design and major findings. Schweiz. Z. Hydrol. *41* (1979) 169-176.

35 Gächter, R., MELIMEX, an experimental metal pollution study. EAWAG News *11* (1980) 1.

36 Gardiner, J., The chemistry of cadmium in natural water. 1. A study of cadmium complex formation using the cadmium specific-ion electrode. Wat. Res. *8* (1974) 23-30.

37 Gatta, M., and Calvi, R., Legislazione internazionale per la tutela delle acque. Aspetti esecutivi di interesse per le attività industriali. MONTEDISON, Milano 1977.

38 Gommes, R., and Muntau, H., Cadmium levels in biota and abiota from Lake Maggiore. EUR 5411 (1976).

39 Gommes, R., et Muntau, H., La composition chimique des limnophytes du Lac Majeur. Mem. Ist. ital. Idrobiol. *38* (1981) 237-307.

40 Hart, B.T., Trace metal complexing capacity of natural waters: a review. Envir. Technol. Lett. *2* (1981) 95-110.

41 Havre, G.N., Underdal, B., and Christiansen, C., The content of lead and some other heavy elements in different fish species from a fjord in Western Norway, in: Int. Symp. Environmental Health Aspects of Lead, Amsterdam, 1972. CEC Luxemburg EUR 5004 d-e-f (1973) 99.

42 Hawkes, H.E., and Webb, J.S., eds Geochemistry in mineral exploration. Hawker and Row, New York 1962.

43 Hill, C.H., Reversal of selenium toxicity in chicks by mercury, copper and cadmium. J. Nutr. *104* (1974) 593.

44 Hill, C.H., Interrelationships of selenium with other trace elements. Fedn Proc. *34* (1975) 2096.

45 Hoppenheit, M., One the dynamics of exploited populations of *Tisbe holothuriae* (Copepoda, Harpacticoida). Helgoländer wiss. Meeresunters. *29* (1977) 503-523.

46 Houba, C., and Remacle, J., Distribution of heavy metals in

sediments of the river Vesdre (Belgium). Envir. Technol. Lett. *3* (1982) 237–240.

47 Howard, A.G., and Nickless, C., Heavy metal complexation in polluted molluscs. 3. Periwinkles (*Littorina littorea*, cokles *(Cardium edule)* and scallops *(Chalmys opercularis).* Chem. Biol. Interactions *23* (1978) 227.

48 Hutchinson, T.C., and Czyrska, H., Cadmium and zinc toxicity and synergism to floating aquatic plants. Water pollution research in Canada 1972, in: Proc. 7th Can. Symp. Water Pollution Res., pp. 59–65.

49 Hutchinson, T.C., and Czyrska, H., Cadmium and zinc toxicity and synergism to floating aquatic plants. In: Water Pollution Research in Canada, vol. 7, 1978.

50 Jaakkola, T., Takahash, H., Soininen, R., Rissanen, K., and Miettinen, J.K., Cadmium content of sea water, bottom sediment and fish and its elimination rate in fish. IAEA Panel Proc. Ser. (PL-469/7 Sta-Pub/332), (1972) 69–75.

51 Jung, K.D., Extrem fischtoxische Substanzen und ihre Bedeutung für ein Fischtest-Warnsystem. GWF Wasser/Abwasser *114* (1973) 232.

52 Kerrison, P.H., Sprocati, A.R., Ravera, O., and Amantini, L., Effects of cadmium on an aquatic community using artificial enclosures. Envir. Technol. Lett. *1* (1980) 169–176.

53 Khailov, K.M., The formation of organometalic complexes with the participation of extracellular metabolites of marine algae. Dokl. Akad. SSSR., Biol. Sci. *155* (1964) 237.

54 Kjellstrom, T., Lind, B., Linnaman, L., and Elinder, C.G., Variation of cadmium concentration in Swedish wheat and barley. Archs envir. Hlth *30* (1975) 321.

55 Klass, E., Rowe, D.W., and Massaro, E.J., The effect of cadmium on population growth of the green alga *Scenedesmus quadricauda.* Bull. envir. Contam. Toxic. *12* (1974) 442.

56 Kobayashi, J., Moru, F., Muramoto, S., Nakashima, S., Teraoka, H., and Horie, S., Distribution of arsenic, cadmium, lead, zinc, copper and manganese contained in the bottom sediment of Lake Biwa. Jap. J. Limnol. *36* (1975) 6–15.

57 Kumada, H., Kimura, S., Yokata, M., and Matida, Y., Acute and chronic toxicity, uptake and retention of cadmium in freshwater organisms. Bull. Freshwat. Fish. Res. Lab., Tokyo *22* (1972) 157–165.

58 Kumada, H., Kimuru, S., Yokote, M., and Matida, Y., Acute and chronic toxicity, uptake and retention of cadmium in freshwater organisms. Bull. Freshwat. Fish. Res. Lab., Tokyo *22* (1973) 157.

59 Lewin, J.C., Silicon metabolism in diatoms. 1. Evidence for the role of reduced sulphur compounds in silicon utilization. J. gen. Physiol. *37* (1954) 589–599.

60 Locht, B., and Muntau, H., La composition chimique des sédiments profonds du Lac Majeur. Mem. Ist. ital. Idrobiol. *38* (1981) 101–186.

61 Lovett, R.J., Gutenmann, W.G., Pakkala, J.S., Youngs, W.D., Lisk, D.J., and Burdick, G.E., A survey of the total cadmium content of 406 fish from 49 New York State fresh waters. J. Fish. Res. Bd Can. *29* (1972) 1283–1290.

62 Lucas, H.F., Edgington, D.N., and Colby, P.J., Concentrations of trace elements in Great Lake fish. J. Fish. Res. Bd Can. *27* (1970) 677–684.

63 Marquenie, J.M., The freshwater mollusc *Dreissena polymorpha* as a potential tool for assessing bio-availability of heavy metals in aquatic systems. in: Proc. Int. Conf. Heavy Metals in the Envir., Amsterdam 1981; p. 409.

64 Marshall, J.S., Population dynamics of *Daphnia galeata mendotae* as modified by chronic cadmium stress. J. Fish. Res. Bd Can. *35* (1978) 461–469.

65 McKnight, D.M., and Morel, F.M., Release of weak and strong coppercomplexing agents by algae. Limnol. Oceanogr. *25* (1979) 823–837.

66 Merlini, M., Argentesi, F., Brazzelli, A., Oregioni, B., and Pozzi, G., The effects of sublethal amounts of cadmium and mercury on the metabolism of Zn-65 by freshwater fish, in: Internat. Symp. Radioecology applied to the Protection of Man and his Environment, 1972, pp. 1327–1344.

67 Merlini, M., Hepatic storage alteration of vitamin B$_{12}$ by cadmium in a freshwater fish. Bull. envir. Contam. Toxic. *19* (1978) 767–771.

68 Mount, D.I., and Stephan, C.E., A method for detecting cadmium poisoning in fish. J. Wildl. Mgmt *31* (1967) 168–172.

69 Müller, G., Heavy metals and nutrients in sediments of Lake Balaton, Hungary. Envir. Technol. Lett. *2* (1981) 39–48.

70 Muntau, H., Heavy metal distribution in the aquatic ecosystem Southern Lake Maggiore. II. Evaluation and trend analysis. Mem. 1st. ital. Idrobiol. *38* (1981) 505–529.

71 Nobbs, C.L., and Pearce, D.W., The economics of stock pollutants: the example of cadmium. Int. J. envir. Stud. *8* (1976) 245–255.

72 Noël-Lambot, F., Gerday, Ch., and Disteche, A., Distribution of Cd, Zn and Cu in liver and gills of the eel *Anguilla anguilla* with special reference to metallothioneins. Biochem. Physiol. *61 C* (1978) 177–187.

73 Noël-Lambot, F., Presence in the intestinal lumen of marine fish of corpuscles with a high cadmium, zinc, and copper-binding activity: a possible mechanism of heavy metal tolerance. Mar. Ecol. Prog. Ser. *4* (1981) 175–181.

74 North, W.J., Stephens, G.C., and North, B.B., Marine algae and their relations to pollution problems. FAO Technical Conference on Marine Pollution and its Effects on Living Resources and Fishing, Rome 1970.

75 Premazzi, G., Bertone, R., Freddi, A., and Ravera, O., Combined effects of heav metals and chelating substances on *Selenastrum* cultures, in: Proc. Seminar 'Ecological tests relevant to the implementation of proposed regulations concerning environmental chemicals: evaluation on research needs, Berlin, December 1977, pp. 169–187.

76 Price, R.E., and Knight, L.A. Jr, Mercury, cadmium, lead and arsenic in sediments, plankton and clams from Lake Washington and Sarids Reservoir, Mississippi, Oct. 1975–May 1976. Pest. Mon. Jour. *11* (1978) 182.

77 Ravera, O., The effect of X-rays on the demographic characteristics of *Physa acuta* (Gastropoda: Basommatophora). Malacologia *5* (1967) 95–109.

78 Ravera, O., Critique of concepts and techniques regarding biological indicators. Boll. Zool. *42* (1975) 111–121.

79 Ravera, O., Effects of heavy metals (cadmium, copper, chromium and lead) on a freshwater snail: *Biomphalaria glabrata,* say (Gastropoda, Pulmonata). Malacologia *16* (1977) 231–263.

80 Ravera, O., Consideration on the effects of pollution at community and population level. Experientia *35* (1979) 710–713.

81 Ravera, O., Gommes, R., and Muntau, H., Cadmium distribution in aquatic environment and its effects on aquatic organisms, in: European Colloquium Problems of the Contamination of Man and his Environment by Mercury and Cadmium, Luxemburg, 3–5 July 1973. CEC EUR-5075 (1974) 317–330.

82 Ravera, O., Giannoni, L., and Agostini, T., Effects of cadmium on demographical characteristics of a freshwater snail *Physa acuta,* Drap. (in press).

83 Rehwoldt, R., Menapace, L.W., Nerbie, B., and Alessandrello, D., The effect of increased temperature upon the acute toxicity of some heavy metal ions. Bull. envir. Toxic. *8* (1972) 91–96.

84 Reiniger, P., Concentration of cadmium in aquatic plants and algal mass in flooded rice culture. Envir. Pollut. *14* (1977) 297–301.

85 Remacle, J., and Houba, C., The influence of cadmium upon freshwater saprophytic bacteria. Envir. Technol. Lett. *1* (1980) 193–200.

86 Ricci, C., and Pozzoli, E., Life tables of *Philodina roseola* (Rotifera) under conditions of chronic cadmium and zinc stress. Boll. Zool. *46* (1979) 209–216.

87 Rugstad, H.E., and Norseth, T., Cadmium resistance and content of Cd-binding protein in cultured human cells. Nature, Lond. *257* (1975) 136.

88 Sakaguchi, T., Tsuji, T., Nakajima, A., and Horikoshi, T., Accumulation of cadmium by green microalgae. Eur. J. appl. Microbiol. Biotechnol. *8* (1979) 209–215.

89 Salomons, W., Adsorption processes and hydrodynamic conditions in estuaries. Envir. Technol. Lett. *1* (1980) 356–365.

90 Salomons, W., and Förstner, U., Trace metal analysis on polluted sediments. Part II: Evaluation of environmental impact. Envir. Technol. Lett. *1* (1980) 506–516.

91 Sprague, J.B., Measurement of pollutant toxicity to fish. III. Sublethal effects and 'safe' concentrations. Wat. Res. *5* (1971) 245–266.

92 Stanley, R.A., Toxicity of heavy metals and salts to Eurasian watermilfoil (*Myriophyllum spicatum* L.). Arch. envir. Contam. Toxic. *2* (1974) 331–341.

93 Taylor, D., Distribution of heavy metals in the water of a major industrialised estuary. Envir. Technol. Lett. *3* (1982) 137–144.

94 Thorp, V.J., and Lake, P.S., Toxicity bio-assays of cadmium on selected freshwater invertebrates and the interaction of cadmium and zinc on the freshwater shrimp (*Paratya tasmanensis* Ric.). Aust. J. mar. Freshwat. Res. *25* (1974) 97–104.

95 Uthe, J.F., and Bligh, E.G., Preliminary survey of heavy metal contamination of Canadian freshwater fish. J. Fish. Res. Bd Can. *28* (1971) 786–788.

96 Vailati, G., Teratogenic effects of differently polluted waters. Riv. Biol. *72* (1979) 229–256.

97 Van Puymbroeck, S.L.C., Stips, W.J.J., and Vanderborght, O.L.J., The antagonism between selenium and cadmium in a freshwater mollusc. Arch. envir. Contam. Toxic. *11* (1982) 103.

98 Vernberg, W.B., de Coursey, P.J., and O'Hara, J., Multiple environmental factor effects on physiology and behaviour of the fiddler crab, *Uca pugilator*, in: Pollution and physiology of marine organisms, pp.381–425. Eds F.J. Vernberg and W.B. Vernberg. Academic Press, New York 1974.

99 Warnick, S.L., and Bell, H.L., The acute toxicity of some heavy metals to different species of aquatic insects. J. Wat. Pollut. Control Fed. *41* (1969) 280–284.

100 Wetzel, R.G., Limnology, W.B. Saunder Co., Philadelphia 1975.

101 Williams, L.G., Joyce, J., and Monk, T.J., Stream-velocity effects on the heavy metal concentrations. J. Air Wat. Wks Assoc. *65* (1973) 275–279.

102 Yamagata, N., and Shigematsu, I., Cadmium pollution in perspective. Bull. Inst. publ. Hlth, Tokyo *19* (1970) 1–27.

103 Zwarum, A.A., Tolerance of *Escherichia coli* to cadmium. J. Envir. Qual. *2* (1973) 353.

Cadmium contamination in agriculture and zootechnology

by R. Van Bruwaene, R. Kirchmann and R. Impens

Department of Radiobiology, CEN/SCK, B–2400 Mol (Belgium), and Department of Plant Biology, Faculté Sciences Agronomiques de l'Etat, B–5800 Gembloux (Belgium)

Introduction

Fortunately, acute Cd toxicity caused by food consumption is rare, but chronic exposure to significant Cd levels in food (and specially in plant materials) may be more frequent. It could significantly increase the accumulation of this heavy metal in certain body organs of man or animals (kidney, liver, etc.).

For populations not exposed to Cd from fall out or to professional contamination, the main source of Cd body burden is also from food. The Food and Drug Administration reported an average daily ingestion of 39 µg/day for 15–20-year old males in the USA[44].

Drinking water and ambient air contribute relatively little to the daily intake. Cigarette smoking is another risk for Cd intake by inhalation. Occurrence of Cd in the food chain and in tobacco is certainly the main source for human or animal contamination.

The concentration of Cd in foods is controlled by its level in the soil, its availability for plants, and by the physical and chemical properties of the growing substrate.

Some agricultural practices, as phosphatic fertilizers, sewage sludge disposal, town-refuse composts applications etc., may increase Cd accumulation in soil and lead to heavy metal transfer to crops and to the food chain.

1. Cd additions to the soil through environmental pollution and agricultural practices

The concentration of Cd in most soils is in the range of 0.5–1.0 mg/kg, although concentrations above 20 mg/kg occur naturally in some mining areas. Following Davies and Roberts[11] we may consider that concentrations of Cd, in agricultural soil, in excess of 2.4 mg/kg are anomalously high. Raised concentrations of Cd in soil may be found naturally or as a result of pollution from mining or smelting (e.g. in the Meuse Valley[14,15,45]) or from moving sources (principally automobiles).

The dispersal of metal rich waste around mine and smelting plants has led to high concentrations of Cd in top soils at various localities (e.g. La Calamine, Plombières, Engis, in Belgium). Atmospheric fall-out raise levels of Cd in industrial and urban areas, even in rural areas. Secondary metal refining activities, waste incineration, tyre and oils residues from vehicles, also dissipate Cd into the environment[12].

Agricultural soils are mainly contamined by phosphatic fertilizers and sludge disposal. The Cd content of rock phosphate is variable and depends on its geographical origin[70]. In a Belgian survey of 31 common phosphatic fertilizers Beaufays and Nangniot[6] found Cd concentrations ranging from 0.1 to 80.8 mg/kg. In glasshouses, where intensive cultures of vegetables are performed, an excess of phosphatic fertilizers presents an important hazard of Cd pollution, especially when leafy vegetables (lettuce, spinach etc.) are grown. Andersson[3] calculated that if the Cd concentration in phosphatic fertilizers exceeds 8 mg/kg, Cd levels in top soil may be increased.

The use of Cd-containing agricultural sprays (plant protection chemicals containing Cd are rare) or soil amendments is limited to phosphatic fertilizers or micro nutrients solutions.

In flooded rice culture, as practiced in the Po Valley, the irrigation water is the most important vector of the heavy metals Cd, Cu and Cr. The maximum yearly

Table 1. Total trace elements contents of composted town refuse (ppm dry matter). After Gray and Biddlestone[21] and Delcarte et al.[13]

Sample	Pb	Cu	Cd	Ni	Zn	Fe	Mn	B	Cr
(Gray-Biddlestone)[21]	1630	610	7.5	–	1350	–	–	174	–
Roma, Italy[13]	57	130	0.8	6	365	2430	29	–	19
Switzerland[13]	396	483	7.1	41	1875	3500	100	–	65
Belgian (common)[13]	595	251	5.2	87	1363	24548	856	–	69
Belgian (arrosol)[13]	360	318	5.1	25	846	1615	39	–	8

increase in the Cd concentration of the top 20 cm of soil, as a percentage of the actual concentration, would be 6%[7].

Another important source for Cd contamination of soils and crops are the sewage sludges used as fertilizers (cf. chapter written by Davies on cadmium in sludges used as fertilizers).

The composting of town refuse, in order to produce a material which would be of value as a soil conditioner or manure, becomes a new recycling process for organic waste. The nutrient content of such composts is low, but they contain considerable amounts of trace elements as shown in table 1.

Farmyard manures from animals receiving dietary supplements is not known as a possible source for Cd transfer to soils.

Prediction of increases in Cd concentrations in agricultural soils due to high inflows by atmospheric fallout, intensive culture practices, combined with low outflows by leaching and crop removal is important in assessing future Cd load and availability to the animals and to the human populations.

Some authors have attempted to simulate the progressive enrichment of Cd in soils, with a predicted average retention time for Cd in the ploughing layer (25 cm depth) in the range 5–50 years[57,64,68]. The cadmium balance on normal agricultural soil indicates a much longer retention, and actually inflows are much greater than outflows (see fig. 1)[28].

A regular survey of soil pollution by Cd must be initiated where important fall out occurs and when high doses of phosphatic fertilizers or sludge applications are repeated. This survey will include chemical soil analysis, permitting to characterize the enrichment, but a real diagnosis should be based on determination of the mobile fraction of Cd and on plant analysis. Indeed, only the mobile part of Cd is available to plants, and this fraction depends on pH, organic matter content, redox-potential, precipitation with phosphates, carbonates, competition with other heavy metals, etc.

2. Cadmium in plants

2.1 Distribution within plants

Cadmium is a non-essential element for both plants and animals, so there are no lower critical concentrations below which deficiency of the element would occur. Upper critical concentrations mark the beginning of phytotoxicity.

All plants contain detectable concentrations of Cd. In general, the background concentration analyzed in crops grown on soils not subjected to any known outside source of Cd pollution, seems to be in the range 0.1–1.0 mg Cd/kg dry matter. In experimental results, Cd concentrations found in plant tissue are expressed on a dry-weight basis. However, for convenience, wet weight is frequently used as the basis for Cd concentrations in dietary studies. Page et al.[55] have reviewed and summarized the concentrations of Cd in various leafy, legume and root vegetables, fruits and various grains grown on non-polluted soils. Generally, the variation of Cd concentrations among the various plant parts analyzed is decreasing in the order roots > leaves > fruiting parts > seeds = storage organs. Differences in Cd concentration between these organs become accentuated as exposure to Cd increases.

The concentration of Cd in foods or fodder grown in soils are related to the properties of the soil on which they are grown, to the Cd concentrations in the soil, to different absorption characteristics of species and cultivars, the nature of the edible part, the age of the plant, and some environmental factors.

2.2 Uptake of Cd

There are 2 possible pathways for disposed Cd to enter into the plant tissues: the intake of soil's Cd by the roots and the foliar absorption of Cd present in deposited dust.

In the Meuse Valley investigations were carried out in order to identify and measure separately the transfer of Cd from soils to the roots and from air to the leaves (aerial deposition).

A crossed culture pattern was designed to distinguish the two routes of plant contamination; lettuce and italian rye grass were used as indicators of the intake (table 2).

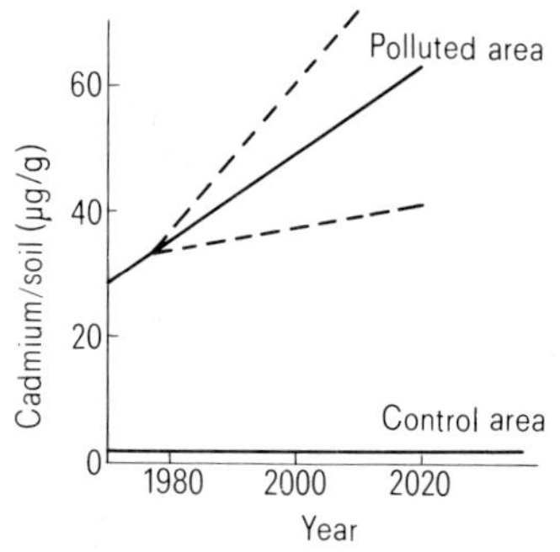

Figure 1. Evolution of Cd accumulation in soil of 2 different areas: polluted or not. The dotted lines are representative of the dispersion due to standard deviation[28].

Table 2. Experimental scheme of crossed cultures pattern[28]

Crops grown on Cd polluted site Bulk contamination of crops (pattern 1)	
Without atmospheric Cd source Crops grown on polluted soil in clean air area (pattern 2)	*Without soil Cd source* Crops grown on non-Cd polluted soil, exposed in an industrial area (pattern 3)
Physiological contamination by soil-plant transfer	Contact contamination by aerial deposition on leaves
Crops grown as reference on soil in a non-polluted area (pattern 4)	

In the 3 exposed patterns (1, 2 and 3), there is a regular enrichment in Cd, during the growth period. Contamination kinetics of plants growing in polluted soil is characteristic of root absorption with a peak during the first weeks of exposure and a decrease later (pattern 2). This decrease does not appear when plants are cultivated on polluted soil in presence of aerial fall out (pattern 1).

Crops grown according to the 3rd pattern allow the detection of the actual air pollution and to follow quantitatively the progressive Cd contamination of plant tissues, as shown in figure 2. In this figure it can be seen how crops growing slowly and submitted to aerial deposition (measured by Owen gauges and analyzed for their Cd content) might be contaminated during the vegetation period[28].

Comparison between the two pathways were also made using a radioactive tracer (^{109}Cd). Foliar uptake and translocation of ^{109}Cd is limited to a few percent of the disposed quantity[23,26]. Root absorption remains the major pathway for Cd enrichment in plant tissues as shown by pot trials, greenhouse's experiments or open field observations.

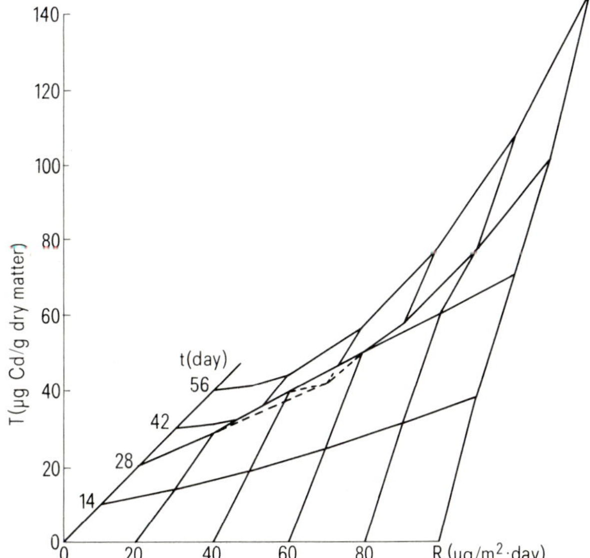

Figure 2. Relation between Cd content in plant tissues (T) and atmospheric fall out (R) (µg Cd/m²/day) and term of exposure (t)[28].

The soluble inorganic salts are much more readily available to plant roots than Cd originating from organic material (sludge, town refuse compost etc.). Extreme caution is essential in extrapolating from results with inorganic salts under controlled conditions to what might happen in living soils, with an important organic matter content, and a wild microflora, under natural environmental conditions. Physiological experiments are performed with inorganic and usually soluble salts of Cd (e.g. $CdCl_2$, $CdSO_4$). The trials use various substrata (water, sand, natural or artificial soils) and are justified for examining the physiological effects of Cd within plant tissue (phytotoxicity) and how the Cd enters and accumulates in the tissue.

2.3 *Plant uptake of Cd from soil*

Chemically, Cd may be dissolved in the soil solution, adsorbed into organic or inorganic colloidal surfaces, occluded into soil minerals, precipitated with other compounds in soils, and incorporated into biological structures. As long as Cd remains tightly bound to solid soil constituent and that the available contents (i.e. soluble forms) remain low, there will be little effects on the environment. When soil conditions change in such a way that Cd goes into solution, the raised Cd content of soils imposes a direct environmental hazard. The increasing availability (or mobility) of Cd in soils may have adverse effects on plant growth. In addition, there may be a leaching down to groundwater or to surface water, and hence reach man and animals through the drinking water.

The mobility of Cd in a soil, and the hazards associated to its availability for plants growing on that soil depends on:
- chemical form of the Cd in that soil, plants may take up some Cd by direct contact between roots and solid soil constituents (in order for root uptake to occur, soluble forms must exist adjacent to the root membrane for some finite periods),
- pH of the soil,
- occurrence of other elements, notably competing heavy metals (Zn) and of complexing ligands, and adsorption sites associated with the solid phase,
- amount of available Cd in the soil,
- environmental conditions (soil temperature and moisture contents, and other factors which affect microbial activity),
- plant species[23].

Field and greenhouse experiments have shown that Cd concentrations and pH of the soil are the two main factors influencing uptake of Cd by food crops. Although oxidation reduction potential is important on Cd absorption by roots, it is a relatively unimportant factor in food production as most food crops, except for rice, will not grow under reducing soil conditions. Numerous authors have described the extent to which Cd is accumulated by plants in relation to soil pH (a review is given by Page et al.[55]). If other soil conditions remain unchanged the plant tissue Cd concentration would decrease as the pH of the soil increases. One of the most effective means of minimizing the absorption of Cd by plants grown on acid soils is to increase their pH by liming. Bingham et al.[9] reported that liming soil from pH 5.2 to pH 6.7 reduced the Cd content of wheat grain by about 50% and also reduced Cd concentrations in lettuce and swiss chard[8].

Organic matter has been thought to play a role in the binding of Cd in soil through both chelation and adsorption mechanisms. It seems that chelation is much less important than adsorption, so that the ability of organic matter to immobilize Cd is largely due to its CEC (cation-exchange capacity)[33]. As result, the complexing of Cd with organic matter affects its uptake by plants[40] which decreases mainly because of the CEC of the organic matter.

The synergistic and antagonistic effects of other trace metals in the soil substrate on the absorption of Cd by plants has been examined by a number of investigators. For example, abundant available zinc in soil might depress plant uptake of Cd[1]. However, the effect of Zn depended upon the Cd concentration in the soil. At low Cd levels (1 ppm), increasing levels of Zn reduced the concentration of Cd in lettuce leaves, but at higher Cd levels the added Zn either shows no effect or increases the Cd concentration.

The interactive effects of Cd, Zn, Cu and Ni were studied by Mitchell et al.[50]. Increasing concentrations of both Cu and Ni in soil consistently reduced concentrations of Cd in the leaves of lettuce plants. Plant macronutrients, e.g. phosphorus, have also been found to affect Cd uptake and concentrations in crops, but the relationship is complex.

2.4 Factors affecting Cd uptake

a) Effect of substrate Cd concentration

Under similar soil conditions (pH, CEC, etc.), amounts of Cd absorbed by plants tend to increase as the concentration of Cd in the soil increases. There may or may not be a linear relation between the increases of Cd concentration in plant tissues and in soil. This relation is influenced by biological and environmental factors.

The greater part of information collected about Cd accumulation in plants are known with soil that has been fertilized with municipal sewage sludges containing Cd. Reviews of Cd absorption by crops grown in sludge-amended soils are numerous and recent[12, 42]. They will not be discussed here.

Results obtained in experiments conducted in greenhouses, to evaluate the effect of Cd level in soil on its concentration in plants have to be improved in field trials. Generally the plants grown on Cd enriched soils or in flowing culture solution, containing soluble Cd salts, absorb more Cd than the same plants grown on the same soil amended with identical amounts of Cd in the field.

Root development is different: the roots of the container-grown crops are exposed to contaminated soil or solution exclusively and uniformly, whereas in the field the roots may extend to depths below the Cd contamined layer or to soil volumes not uniformly polluted.

Not only the amount of Cd added, but also its chemical form will influence the amount absorbed by roots. Easily extractable Cd content of a soil is important to predict uptake of the metal by plants.

A growth experiment using corn (maize), rye grass, red clover, turnip and dandelion on a sandy loam soil was carried out in greenhouse by Gupta and Haeni[22]. The Cd doses ranged between 3.6 and 58.1 µg Cd/g soil. There exists a statistically significant relationship between 0.1 M NaNO$_3$ extractable soil Cd, dry matter yield and Cd uptake for all 5 plants.

Pot trials were made on rye grass to see both influences of Cd addition to soil and duration of exposure to soil contamination (fig. 3). There is a progressive enrichment of Cd content in plant tissues. When successive harvests of rye grass are made, the highest tissue concentrations are found at the second harvest[29].

A number of studies suggest that the length of time of soils incubations with Cd also influences the availability of Cd to crops. Bates et al.[5] have followed the Cd concentrations from successive planting of crops grown on soils which were treated with the same amount of Cd prior to each planting; the concentrations of Cd in the first crop of rye grass were approximatively the same as their concentrations in the following 3 crops.

b) Biological factors affecting Cd uptake

Plants differ in their tolerance to heavy metals. On old mine spoil tips and on heavy polluted soils, one may find a special calamine-tolerant flora (e.g. *Viola calaminaria* (D.C.) Lej.). Metal tolerance in plants is variable but genetically controlled. Simon[60] has shown that a Cd tolerance occurs within populations of *Agrostis tenuis* Sibth and *Festuca ovina* L. This tolerance is due to the ability of the plant to accumulate the metal in an inert form associated with the cell walls of its roots. When the capacity of absorbing and accumulating sites in the cell walls are exceeded, translocation to stems, leaves and fruits becomes possible with an important hazard of phytotoxicity when Cd concentrations are high.

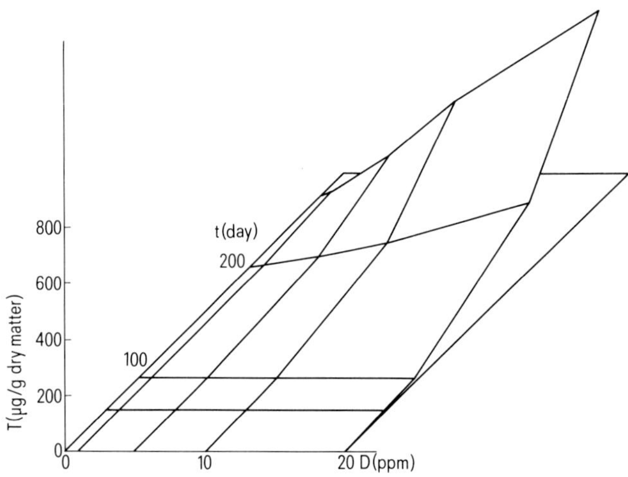

Figure 3. Cd content of rye grass (T) in relation with the concentration of metal in the soil (D), and the time of culture (t)[28].

Many workers have reported that concentrations of Cd in plant parts vary between plant species growing in the same substratum under the same environmental conditions. This reflects the different genetically fixed abilities of the plant roots to restrict the transfer of Cd from roots to epigeal organs[31].

A classical experiment comparing these abilities was realized by Jarvis et al.[30]: 23 different species of crops were grown in water culture solution for between 45 and 62 days, during which they were exposed for 3 days to 0.01 mg Cd/l (as $CdCl_2$) in the flowing culture solution. At harvest, Cd concentrations in the shoots ranged from 21.1 mg/kg (fodder beet) to 1.8 mg/kg (radish).

Pot trials using naturally or artificially contamined soils were performed by many authors: e.g. John[31] observed concentrations of Cd in the leaves of plants grown on the soil supplemented with 40 mg Cd/kg ranged from 264.7 mg/kg (radish) to 18.5 mg/kg (cauliflower).

At the Water Research Center (Stevenage, U.K.) 39 crop plants, including representatives of all the major botanical groups, were grown to maturity in a pot trial using contaminated soil. Leaves and edible parts were analyzed. An ornamental cultivar of tobacco showed the highest concentration of Cd in its leaves. Leaves of Chenopodiaceae, and lettuce, tomato, potato and celery also showed comparatively high foliar concentrations of Cd.

Concentrations of metal in the edible roots tubers and grains were usually lower than in the leaves of the same plant and this effect was often very acute, e.g. potato. Concentrations in edible parts ranged from 0.1 mg/Cd in pea seeds to 9.0 mg Cd/kg in lettuce[12].

Some experiments performed in our laboratory have given similar results (fig.4). There is an important variation amongst cultivars of a single species grown on the same substratum as shown by John and Van Laerhoven[33]. The authors grew 9 cultivars of lettuce in a sand culture experiment, with a range of Cd concentrations from 0.1 to 50 mg Cd/l. After 3 weeks of culture, tissue concentrations in the leaves of the

various cultivars receiving 0.1 mg Cd/l in their nutrient solution ranged from 0.4 to 26.6 mg/kg, this kind of variation appears at all treatments.

As leafy vegetables (lettuce, spinach) accumulate the highest concentrations of Cd in their tissues, they may be used, in polluted areas, as indicators to prevent hazards of foodchain contamination[29].

Although much remains to be learned concerning the uptake of Cd by crops, and the accumulation of Cd in edible parts of crops, it should be reasonable to build a model involving the interaction between soil physico-chemical properties, which determine how much Cd is potentially available for root absorption, and biological characteristics (plant genotype) which determine how much Cd is really taken up.

3. Cadmium in farm animals

3.1 Uptake and absorption

In the environment, farm animals and wild animals can be exposed to cadmium pollution by 2 main routes: inhalation of polluted air and ingestion of polluted food.

Respiratory deposition clearance and respiratory absorption were studied principally in laboratory animals[20]. From these researches it may be concluded that cadmium is absorbed and retained to a considerable degree in the body after inhalation. The respiratory absorption is primarily from the lungs; an absorption between 10 and 40% of inhaled cadmium can be expected as a considerable difference might well exist for different cadmium compounds. Cadmium metabolism has been studied in a variety of mammalian species. Intestinal absorption is low: 0.3% in goat[46,47], 0.035–0.2% in the lactating dairy cow[51,66], 5% in swine[10]. Cadmium absorption is influenced by different dietary factors: calcium, protein and vitamin D deficient diet increase the absorption of cadmium[56,62,71].

Suckling and young animals have higher absorption than adults; milk diet could be an important factor in the increased cadmium absorption in sucklings and youngs since some authors found also a higher cadmium absorption in older animals fed a milk diet[35].

3.2 Metabolism

a) *Faecal and urinary excretion*

Faecal excretion is the major excretion pathways for cadmium. After a single oral dose in goats[46,47] and after a single or repeated dose in cows[48,51,66] about 80–90% of the total ingested cadmium is excreted in the faeces within 14 days, after the end of the application.

Faecal excretion of cadmium after an i.v. injection in goats[47] and in cow[66] indicate that metabolized cadmium is principally excreted by the gastro-intestinal tract: after a single injection of $^{109}CdCl_2$ in goats, 5.6% of the cadmium was excreted via feces within 5 days and, in cows, about 5.5% was found in feces within 10 days.

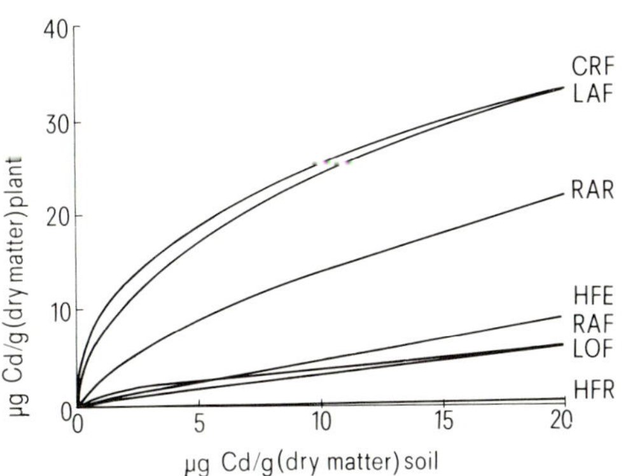

Figure 4. Caption: Evolution of Cd content in plants in relation with Cd addition to soils. CRF, *Lepidium sativum* L. leaves; LAF, *Lactuca sativa* L. leaves; RAR, *Raphanus sativus* L. roots; HFC, *Phaseolus vulgaris* L. leaves; RAF, *Raphanus sativus* L. leaves; LOF, *Lolium multiflorum* L. leaves; HFR, *Phaseolus vulgaris* L. fruits[28].

Studies on cadmium metabolism in laboratory animals and ruminants indicated that daily excretion of cadmium in the urine is very low prior to tubular dysfunction. Total urinary excretion observed in goats was 0.031% of the administered dose after oral ingestion and i.v. injection. Authors reported that a contamination with faecal excretion, in the case of oral dosing was not excluded.

In cows a total urinary excretion of 0.05% of dose[51] and as low as 2.5×10^{-3}% of dose was observed after an oral dose of $^{109}CdCl_2$[66]. Experiments on laboratory animals show that once tubular dysfunction has occurred the urinary excretion of cadmium associated to urinary proteins (probably metallothionein) increases markedly[4].

b) *Secretion and distribution in milk*

Cadmium secretion and distribution in cow milk was studied by various authors[48,51,66].

Miller et al.[48] found that after a daily administration of 3 g cadmium as $CdCl_2$, for 2 weeks, the concentration of cadmium in milk samples stayed < 0.1 mg/l of milk and the excretion/day was less than 2.2×10^{-2}% of the daily administered dose. Effectively, cadmium concentration in milk after various single oral doses of ^{109}Cd in lactating cows suggests much lower secretion. Total cadmium secretion in milk either as a single oral dose of cadmium as $^{109}CdCl_2$ in gelatine capsule or as a single oral dose of *Zea mays* leaves, contaminated by $^{109}CdCl_2$ drops, was respectively 1.8×10^{-3}% and $7.0\ 10^{-4}$% of the administered dose[66]. Neathery et al.[51] on the other hand found that the level was below the detection limit: 8×10^{-5}% of dose/day.

The cadmium distribution in milk after "in vitro" incubation of milk with cadmium[48] or after a single oral administration of ^{109}Cd to lactating cows[66] is essentially the same and indicate that cadmium is preferentially bound to the protein fractions with about equal concentrations in casein and albumin. The concentration in lactose is about one half and in mineral-water fraction about one order of magnitude lower than the concentration in proteins, the milkfat does not seem to have any affinity for cadmium (table 3).

c) *Retention and distribution of cadmium in organs of farm animals*

Cadmium retention and distribution in farm animals were reported by several authors: Miller et al.[46,47] presented data obtained from young goats 14 days after an oral or intravenous administration, Neathery et al.[51] from pregnant lactating cows 14 days after an oral application and Van Bruwaene et al.[66] from lactating cows 131 days after an oral application.

Total retention observed in goats was about 0.3–0.4% of administered dose. In cows total retention was estimated to be about 0.75% of the dose, 14 days after an oral dosing; 131 days after dosing, 0.13% of dose was retained after an oral administration of ^{109}Cd solution in gelatine capsules and 0.025% of dose was retained after administration of contaminated *Zea mays* leaves contaminated by externally applied $^{109}CdCl_2$ drops.

At sacrifice of goats and of cows, the highest cadmium concentrations were found in kidneys followed by liver, pancreas and small intestine. In goats a high concentration of cadmium was also found in abomasum. Distribution of cadmium within organs was similar in young goats and in cows sacrificed respectively 14 days and 131 days after dosing; about 50% of the total retained dose was present in the liver and almost 25% in the kidneys. In goats a large part of the remainder was found in the gastrointestinal tract and its contents. In cows, 131 days after dosing, still 4–5% of body burden were found in the following tissues: abomasum, duodenum and jejunum. No differences in distribution was observed after administration of either a $^{109}CdCl_2$ solution or contaminated *Zea mays* leaves. The distribution of cadmium in cows 14 days after dosing was slightly different: 32% was found in liver, 18% in GI contents, 16% in GI tissues and 10% in kidneys.

Cadmium distribution in goats after intravenous and oral dosing[47] indicated that the pathway of dosing has affected the metabolism. Accordingly it seems reasonable to suspect that cadmium absorbed from the GI tract is probably in association with some types of protein compounds.

Feeding a level of cadmium which approaches the upper limits which could be fed without developing noticeable toxicity symptoms does not greatly influence the percentage of radiocadmium absorbed or the rate at which it is lost[46]. Thus it seems evident that goats do not have homeostatic control mechanism which would be dependent on dietary or tissue cadmium levels for changing amounts of cadmium absorbed and reexcreted.

In conclusion, the liver and the kidneys appear to be the 2 organs of greatest interest with regard to cadmium storage, however cadmium is found in the most compartments of the body. Since relatively little cadmium is transferred to muscle or to milk it is apparent that the main food products (other than liver and kidneys) from ruminants are quite well protected from cadmium accumulation.

3.3 *Transfer in animal products*

a) *Transfer coefficients*

Transfer coefficients are expressed as the fraction of daily nuclide intake present in 1 l of milk or 1 kg of organ.

Table 3. Distribution of cadmium (in percent of total activity in milk) in the principal organic constituents of milk collected 2 days after a single oral dose of $^{109}CdCl_2$ (Van Bruwaene et al.[66])

Administration mode of Cd	Casein	Albumin	Lactose	Fat
In gelatine capsules	28.8 ± 0.08	5.5 ± 0.04	3.0 ± 0.34	Not detectable
In maize leaves	17.7 ± 0.40	4.0 ± 0.13	3.3 ± 0.26	Not detectable

Table 4. Cadmium concentrations in beef muscle, liver and kidneys (expressed as mg/kg fresh weight) observed in various countries

Muscle	n*	Liver	n*	Kidneys	n*	Countries	Years	References
0.047–0.183	4	0.108–1.16	4	0.220–39.88	4	FRG	1968	39
0.011 ± 0.079 (0.041 ± 0.022)	30	–	–	–	–	FRG	1972	24
< 0.005	141	0.005–0.3 (0.06 ± 0.05)	141	0.04–1.66 (0.3 ± 0.3)	141	FRG	1972/74	38
< 0.005	280	0.005–1.658 (0.05)	280	0.005–1.95 (0.23)	283	FRG	1972/75	37
< 0.04	8	0.1–0.38 (0.23)	8	0.4–2.0 (1.1)	8	Belgium	1980	65
0.01–1.00	1947	** 0.01–3.17	2316	** 0.01–7.82	2.553	USA	1971/74	53
0.035 ± 0.034	71	(0.183 ± 0.228)	71			USA	1974	53
0.001–0.04	13	0.071–0.3	13	0.12–1.2	13	The Netherlands	1977	67
				0.18–5.69 (1.11)	150	Denmark	1975	32

* n = number of animals; ** with measurable concentration.

In the case dealing with the transfer of cadmium in milk, a broad range of values of transfer coefficient are proposed. A transfer coefficient value of about 1.10^{-3} day/l is reported by Ng et al.[52], based on the concentrations of cadmium in forage and milk. A survey on transfer of cadmium in cow milk conducted in Belgium[65] suggests a value of $< 1.10^{-6}$ day/l.

From some experimental works, daily administration for 14 days of 3 g of cadmium chloride in a cow[48] and of 9 mg $CdCl_2$/kg body weight[43] values of transfer coefficients of respectively $< 3.3 \ 10^{-5}$ day/l and 5.9×10^{-8} day/l are suggested. After a single oral dose the transfer coefficient can be calculated indirectly taking into account that the activity time integral/kg in milk or organ equals the concentration at equilibrium reached under continuous dosage. For example, after a single oral dose of ^{109}Cd chloride in cow, transfer coefficients of 1.10^{-4} day/l[15] and $< 2.7 \ 10^{-6}$ day/l[51] were obtained.

The transfer coefficients estimated by Van Bruwaene et al.[66] for liver and meat, were 0.02–0.2 day/kg (liver) and 1×10^{-4} to 1×10^{-3} day/kg (muscle), the values of the transfer coefficient for meat being in the same order of magnitude as the value reported in the NRC[54] (5.3×10^{-4} day/kg).

b) Cadmium concentrations observed in farm animals

It is generally recognized that ruminant and especially cattle are more exposed to local pollution situation than pigs or other intensive breeded domestic animals for human consumption. Ruminants are depending for 90% and even more on their daily ration of local produced forage and feeds.

A review of cadmium concentrations observed in beef muscle, liver and kidneys in different countries is presented in table 4.

Although less interest has been devoted to the measurements of the cadmium concentration in pig tissues, some data are available in the literature: Table 5 reports cadmium concentration in pig muscle, liver and kidneys observed in Federal Republic of Germany. The data presented in tables 4 and 5 demonstrate that meat contained the lowest cadmium level and kidneys the highest, the cadmium concentration observed in kidneys being 2–5 times more important

Table 5. Cadmium concentrations in pork muscle, kidneys and liver observed in Federal Republic of Germany[24]

	Number of samples	X̄	S	Range of concentration
Muscle (in µg/kg fresh wt)	82	12.3	11.6	1.46–72.1
Kidneys (in mg/kg fresh wt)	72	0.54	0.277	0.1–1.67
Liver (in mg/kg fresh wt)	82	0.19	0.061	0.086–0.4

than the cadmium concentration in liver. A large variation around the mean values is observed.

Some authors[37] observed in cattle a relationship between the age of the animals and the cadmium content in kidneys; the cadmium content in kidneys and in livers are also closely correlated. These authors observed no indication that breed or sex had any effect on Cd content of kidney and liver.

3.4 Cadmium toxicity in domestic animals

Cadmium toxicity has been demonstrated experimentally in numerous animal species[16, 17]. Manifestations of toxicity include loss of weight, reduced food intake, anemia, hypertension, proteinuria, poor bone mineralization, testicular necrosis, aborted fetuses, neonatal death and youngs with birth defects.

a) Cadmium toxicity in ruminant (cattle and sheep)

A plant fungicide (cadmiate) was given in doses of 50, 100, 200, 300 and 500 mg/kg food to adult cows for maximum 49 weeks of less and to mature ewes for 41 weeks[72]. A dose of 50 mg Cd/kg food caused only a slight reduced food intake and wight loss but above a dose of 200 mg Cd/kg food, anemia was observed. In sheep, levels of 100 mg Cd/kg food decreased RBC (erythrocyte), PCV (packed cell volume) and Hb (hemoglobin) values. A dietary concentration greater than 200 mg Cd/kg food for both cattle and sheep had BUN (blood urea nitrogen) levels that increased later on during the experimental period. The BUN levels increased gradually in the blood of cattle but increased sharply in blood of sheep.

In pregnant cows and sheep for all doses (50–500 mg Cd/kg food) aborted fetuses, neonatal death and youngs with birth defects were observed. In sheep infertility was also observed. Placenta which acts as a barrier to low doses of cadmium can be overcome at high doses; residues of cadmium appeared in the tissues of fetuses.

Lactating cows, given 3 g of cadmium daily, lost considerable weight and milk production declined sharply for several days and then increased appreciably but stayed substantially lower than for control cows. When cadmium treatment was stopped milk production increased within 10 days to normal production[48]. Powell et al.[59] studied cadmium toxicity in calves. Calves were given: 40, 160, 640 and 2560 mg Cd/kg food with and without supplementation of 100 mg Zn/kg food. It seems that calves can tolerate considerable Cd concentrations in food. There was immediate reduction in feed consumption with weight loss for all groups fed as much as 160 mg Cd/kg food. For the lowest dose (40 mg Cd/kg) these effects were annulated by the addition of 109 mg Zn/kg food. Calves fed 640 or 2560 mg Cd/kg food exhibited unthrifty appearance, rough hair coat, severe body dehydration, dry and scally skin, loss of hair from legs, thighs, chest floor and brisket mouth lesions, oedematous shrunken and scaly scrotum, sore and enlarged points, impaired sight, extreme emaciation.

With 160 mg Cd/kg food, the testicle growth was decreased; whereas at 640 mg Cd/kg good there was very little testicle growth. The results of this study indicate that a very severe and extended Cd toxicity may cause serious damage to the testicle development in the bovine.

After stopping the Cd treatment, improvement in appearance feed consumption, growth were quite rapid, also sperm-producing tubules were not completely destroyed as suggested by other studies.

Sheep appear to be very sensitive to cadmium, even concentrations of 5 mg Cd/kg food ($CdCl_2$) during 163 days result in a slight reduced body weight[19]. Interactions of cadmium with copper, iron, zinc and manganese were studied in sheep by different authors[18,49].

Significant numbers of lambs fed low levels of dietary Cd had markedly depressed blood and liver Cu concentrations and ceruloplasmin levels indicating a significant derangement of Cu metabolism; the data also indicate significant disruption of Fe, Zn and Mn metabolism in various parts of the body. Low Cu and Fe levels in liver are associated with anemia, whereas decreased Mn concentrations in body tissues are associated with skeletal defects[41].

b) *Cadmium toxicity in swine*

Cadmium toxicity was studied in 8-week-old swine at levels of 0, 50, 150, 450 and 1350 mg Cd/kg feed during a 6-week comparison period[10]. There was no mortality during the comparison period. Animals receiving 450 and 1350 ppm cadmium exhibited signs of toxicity. The skin covering the inner portion of the hindlegs and the ears were red and scaly and in these areas small leasions similar to those found in the early stages of parakeratosis were observed. Growth rate was decreased as a function of Cd level and was inhibited in the 1350 ppm group. Hematocrit values were the most sensitive criteria of toxicity, they decreased in all Cd fed animals. Bone ash content was decreased as a function of Cd intake. The addition of Zn (52 mg Zn/kg) had little influence on the toxicity symptoms of Cd in young pigs[58].

c) *Cadmium toxicity in poultry*

Effects of cadmium administration were studied in chickens by Krampitz[36]. LD_{50} in chikens is observed at levels of 165–188 mg Cd/kg b.wt and the lethal dose is about 216 mg Cd/kg b.wt. Toxicity was already observed at Cd concentrations of 60–90 mg Cd/kg feed, but toxicity can be reversed by addition of zinc, manganese, copper and cobalt to the ration[63]. Broiler and laying rations containing 20 and 40 mg Cd/kg reduced the body weight respectively by 8 and 24% and reduced the feed conversion efficiency by 3 and 11%[69]. By addition of 20–200 mg of zinc, no difference with control animals was observed[61].

Bone decalcification was observed in consumption chickens fed rations with 5 mg Cd/kg[69]. Enteritis and nephritis in chickens was observed after feeding ration with 3 mg Cd/kg[36].

Contraceptive effect was observed in laying hens, when fed 7–10 days a ration containing 50 mg Cd/kg a total stop of laying was observed[27].

1   Allaway, W. H., Agronomic controls over the environmental cycling of trace elements. Adv. Agronomy *20* (1968) 235–274.

2   Andersen, A., and Engberg, A., Cadmium, bly, kobber og zink i oksenyrer fra Jylland. Sl. statens levnedmiddel insitut, Afdelingen for pesticide og forureninger, 1977.

3   Andersson, A., On the influence of manure and fertilizers on the distribution and amounts of plant available Cd in soils. Swed. J. agric. Res. *6* (1976) 27–36.

4   Axelson, and Piscator, M., Renal damage after prolonged exposure to cadmium. Archs envir. Hlth *12* (1966) 360–373.

5   Bates, T. E., Hag, A., Soon, Y. K., and Royer, J. A., Uptake of metals from sludge amended soils. Proc. Int. Conf. Heavy metals in the Environment, Toronto, Canada 1975, pp. 403–416.

6   Beaufays, J. M., and Nangniot, P., Etude comparative du dosage du Cd dans les eaux, les engrais et les plantes par polarographie impulsionnelle différentielle et spectrométrie et absorption atomique. Analysis *4* (1976) 193–199.

7   Bigliocca, C., Girardi, F., Reininger, P., and Rossi, G., Transfer of heavy metals in irrigated cultures. Results of 1975–1976 studies. CEE-Ispra 1977, working paper, 17 pp.

8   Bingham, F. T., Bioavailability of Cd to food crops in relation to heavy metal content of sludge amended soil. Envir. Hlth Perspect. *28* (1979) 39–43.

9   Bingham, F. T., Page, A. L., Mitchell, G. A., and Strong, J. E., Effects of liming an acid soil amended with sewage sludge enriched with Cd, Cu, Ni and Zn on yield and Cd content of wheat grain. J. envir. Qual. *8* (1979) 202–207.

10 Cousins, R. J., Barber, A. K., and Trout, J. R., Cadmium toxicity in growing swine. J. Nutr. *103* (1973) 964–972.

11 Davies, B. E., and Roberts, L. J., Heavy metals in soils and radish in a mineralised limestone area of Wales, Great Britain. Sci. total Envir. *4* (1975) 249–261.

12 Davis, R. D., and Coker, E. G., Cadmium in agriculture, with special reference to the utilisation of sewage sludge on land. Water Research Center, technical report TR 139. June 1980. Stevenage UK.

13 Delcarte, E., Anid, P., and Impens, R., Présence de métaux lourds dans les composts. Anns Gembloux *88* (1982) 133–144.

14 Delcarte, E., Impens, R., Kirchmann, R., Fagniart, E., and Nangniot, P., Etude de la contamination de fourrages et légumes par le plomb et les métaux lourds. C.r. Symp. Intern. Assessment of Health Effects of Environmental Pollution, Paris, 24–28 June 1974; pp. 1675–1683.

15 Delcarte, E., Tilman, J., Clarembeaux, A., Impens, R., and Nangniot, P. (1976). Recherche du cadmium, plomb, zinc et fluor chez les végétaux provenant de la région d'Engis. C.r. AFAS, Bruxelles, July 1976, Communication No 1712.

16 Decleire, M., and De Cat, W., Aperçu bibliographique de la sensibilité des animaux domestiques aux aliments contaminés par le cuivre, le cadmium et le plomb. Revue Agric. Brux. *32* (1979) 1209–1216.

17 Doyle, J.J., Effects of low levels of dietary cadmium in animals. A review. J. envir. Qual. *6* (1977) 111–116.

18 Doyle, J.J., and Pfander, W.H., Interactions of cadmium with copper, zinc, iron and manganese in ovine tissues. J. Nutr. *105* (1975) 599–606.

19 Doyle, J.J., Pfander, W.H., Grebing, S.E., and Pierce, J.O., Effect of dietary cadmium on growth cadmium absorption and cadmium tissue levels in growing lambs. J. Nutr. *104* (1974) 160–166.

20 Friberg, L., Piscator, M., Nordberg, G., and Kjellstroem, T., Cadmium in the Environment. Chapter 4: Metabolism, pp. 23–27. CRC Press, Cleveland 1974.

21 Gray, K.R., and Biddlestone, A.J., Agricultural use of composted town refuse, in: Inorganic Pollution and Agriculture, MAFF Ref. Book 326 MMSO, pp. 279–305. London 1980.

22 Gupta, S., and Haeni, H., Easily extractable Cd-content of a soil, its extraction, its relationship with the growth and root characteristics of test plants, and its effects on some of the soil microbiological parameters, in: Characterization Treatment and Use of Sewage sludge, p. 665–676. Proc. Soc. Eur. Symp., CEE Vienna Oct. 21–23, 1980. D. Reidel, Amsterdam 1981.

23 Harmsen, K., Behavior of heavy metals in soils. Agricultural Research Reports 866. Centre for Agricultural Publishing and Documentation. Wageningen 1977, 171 pp.

24 Hecht, H., Untersuchungen über Spurenelemente in Fleisch. Arch. Lebensmittel Hyg. *24* (1973) 255.

25 Hecht, H., Der Gehalt des Fleisches an toxischen Elementen. Ber. Landw. *55* (1977) 828–834.

26 Hemphill, D.D., and Rule, J., Foliar uptake and translocation of ^{210}Pb and ^{109}Cd. Intern. Conf. Heavy Metals in the Environment. Toronto, Ontario 27–31-X-1975. Abstracts Nos C 239–240.

27 Hennig, A., Hartmann, G., Gruhn, K., and Anke M., Contraceptive effect of cadmium in laying hens. Naturwissenschaften *55* (1968) 551.

28 Impens, R.A., and Piret, T., Transfert des métaux lourds dans le sol et les plantes. Programme National R.D. Environnement Air. CIPS. 8, rue de la Science, Bruxelles, Belgium 1981; 106 pp.

29 Impens, R.A., Piret, T., Kooken, G., and Benko, A., Monitoring of the air quality by analysis of biological indicators and accumulators. Atmospheric Pollution: Proc. 14th Intern. Coll. Paris, France, May 5–8, 1980. Stud. envir. Sci. pp. 417–424. Elsevier, Amsterdam 1980.

30 Jarvis, S.C., Jones, L.H., and Hopper, M.J., Cadmium uptake from solution and its transport from roots to shoots. Plant Soil *44* (1976) 179–191.

31 Jastrow, J.D., and Koeppe, D.E., Uptake and effects of Cd in higher plants in: Cadmium in the Environment, part 1, Ed. J.O. Nriagu. Wiley Interscience Publ., New York 1980, pp. 607–638.

32 John, M.K., Cadmium uptake by eight food crops as influenced by various soil levels of cadmium. Envir. Pollut. *4* (1973) 7–15.

33 John, M.K., and Van Laerhoven, C.J., Differential effects of Cd on lettuce varieties, Envir. Pollut. *10* (1976) 163–173.

34 John, M.K., Van Laerhoven, C.J., and Chuah, H.H., Factors affecting plant uptake and phytotoxicity of Cd added to soils. Envir. Sci. Technol. *6* (1972) 1005–1009.

35 Kostial, K., Kello, D., Jugo, S., Rabar, I., and Maljkovic, T., Influence of age on metal metabolism and toxicity. Envir. Hlth Perspect. *25* (1978) 81–86.

36 Krampitz, G., Suelz, M., and Hardebeck, H., Effect of cadmium doses in chickens. Toxicity in chickens. Arch. Geflügelk. *38* (1974) 86–90.

37 Kreuzer, W., Kracke, W., Sansoni, B., and Wissmath, P., Untersuchungen über den Blei(Pb)- und Cadmium(Cd)-Gehalt in Fleisch und Organen von Schlachtrindern. 1) Rinder aus einem wenig umweltbelasteten Gebiet. Fleischwirtschaft *58* (1978) 1022–1030.

38 Kreuzer, W., Sansoni, B., Kracke, W., and Wissmath, P., Cadmium in Fleisch und Organen von Schlachttieren. Fleischwirtschaft *55* (1975) 387.

39 Kropf, R., and Geldmacher- von Mallinckrodt, M., Der Cadmiumgehalt von Nahrungsmitteln und die tägliche Cd-Aufnahme. Arch. Hyg. *152* (1968) 218.

40 Lagerwerff, J.V., Biersdorf, G.T., Milberg, R.P., and Brower, D.L., Effects of incubation and liming on yield and heavy metal uptake by rye from sewage sludged soils. J. envir. Qual. *6* (1977) 427–431.

41 Leach, R.M., and Muenster, A.M., Studies on the role of manganese in bone formation. I. Effects upon the mucopolysaccharide content in chick bone. J. Nutr. *78* (1962) 51–56.

42 Lhermite, P., and Ott, H., Characterization Treatment and Use of Sewage Sludge. Proc. Second European Symposium, 803 pp. C.E.E., Vienna, Oct. 21–23, 1980. D. Reidel, Amsterdam 1981.

43 Lynch, G.P., Excretion of cadmium and lead into milk; trace element metabolism in animals. 2. Proceedings of 2nd intern. Symposium on trace element metabolism in animals, Madison, Wisconsin, 18–22 June 1973 p. 470–472. Eds Hoestra et al. Baltimore University Park Press, Baltimore 1974.

44 Mahaffey, K.R., Corneliussen, P.E., Jelinek, C.F., and Fiorino, J.A., Heavy metal exposure from foods. Envir. Hlth Perspect. *12* (1975) 63–69.

45 Mathy, P., Piret, T., Kooken, G., and Impens, R., Heavy metal contaminations in the Meuse Valley (Belgium). C.r. Intern. Conf. Management and Control of Heavy Metals in the Environment, London, 18–21 September 1979, pp. 549–552.

46 Miller, W.J., Blackmon, D.M., Gentry, R.P., and Pate, F.M., Effect of dietary cadmium on tissue distribution of ^{109}Cadmium following a single oral dose in young goats. J. Dairy Sci. *52* (1969) 2029–2035.

47 Miller, W.J., Blackmon, D.M., and Martin. J.G., ^{109}Cadmium absorption, excretion and tissue distribution following single tracer oral and intravenous doses in young goats. J. Dairy Sci. *51* (1968) 1836–1839.

48 Miller, W.J., Lampp, B., Fowell, G.W., Salotti, C.A., and Blackmore, D.M., Influence of a high level of dietary cadmium on cadmium content in milk excretion and cow performance. J. Dairy Sci. *50* (1967) 1404–1408.

49 Mills, C.F., and Dalgarno, A.C., Copper and zinc status of ewes and lambs receiving increased dietary concentrations of cadmium. Nature *239* (1972) 171–173.

50 Mitchell, G.A., Bingham, F.T., Page, A.L., and Nash, P., Interactive effects of Cd, Cu, Ni and Zn on phytotoxicity to and metal intake by lettuce. J. envir. Qual, in: Page A.L., et al.[55].

51 Neathery, M.W., Cadmium 109 and methyl mercury-203 metabolism, tissue distribution and secretion into milk of cows. J. Dairy Sci. *57* (1974) 1177–1183.

52 Ng, Y.C., Colsher, C.S., Quinn, D.J., and Thompson, S.E., Transfer coefficients for the prediction of the dose to man via the forage – cow – milk pathway from radionuclides released to the biosphere UCRL II – 51939, 1977.

53 N.N. Compliance program evaluation. Bureau of foods, Washington D.C., 1975.

54 N.R.C. Nuclear Regulatory Commission, March 1976. Regulatory Guide 1–109.

55 Page, A.L., Bingham, F.I., and Chang. A.C., Cadmium, in: Effect of heavy metal pollution on plants – vol. 1. Effects of trace metals on plant function, pp. 77–109. Ed. N.W. Lepp. Applied Science Publ., London 1981.

56 Piscator, M., and Larsson, S.E., Effects of long term cadmium exposure in calcium deficient rats. In: Proc. 2nd Int. Symp. on Trace Elements Metabolism in animals, Madison, Wisconsin, 1973, p. 687. Eds Hoestra. Baltimore University Park Press, Baltimore 1974

57 Poelstra, P , Frissel, M.J., and El-Bassam, Transport and accumulation of Cd ions in soils and plants. Z. Pfl. Ernähr. Düng. Boden. *142* (1979) 848–864.

58 Pond, W.G., Chapman, P., and Walker, E., Influence of dietary zinc corn oil and cadmium on certain blood components: weight gain and parakeratosis in young pigs. J. Anim. Sci. *25* (1966) 122–127.

59 Powell, W.G., Miller, W.J., Morton, J.D., Clifton, C.M., Influence of dietary cadmium level and supplemental zinc and cadmium toxicity in the bovine. J. Nutr. *84* (1964) 205–213.

60 Simon, E., Cadmium tolerance in populations of *Agrostis tenuis* and *Festuca ovina*. Nature *265* (1977) 328–330.

61 Supplee, W.C., Antagonistic relationship between dietary cadmium and zinc. Science 139 (1963).

62 Suzuki, S., Taguchi, T., and Yokohashi, G., Dietary factors influencing upon the retention rate of orally administered $^{115m}CdCl_2$ in mice with special reference to calcium and protein concentrations in diet. Ind. Hlth 7 (1969) 155.

63 Taucins, E., Swilane, A., and Valdmanis, A., Relation between zinc and cadmium in chick nutrition. Chem. Abstr. 74 (1971) 29457.

64 Tjell, J.C., Hansen, J.Aa., Christensen, T.H., and Hovmand, M.F., Prediction of Cd concentrations in danish soils. Proc. Soc. European Symposium Characterization, treatment and use of sewage sludge. Vienna, October 21–23, 1980; pp.652–664.

65 Van Bruwaene, R., Gerber, G.B., Kirchmann, R., and Colard, J., Coefficient de transfert pour divers métaux "in situ". Rapport final du programme national "Environment-Air", pp.81–83. CIPS, 8 rue de la Science, Bruxelles, Belgique 1982.

66 Van Bruwaene, R., Gerber, G.B., Kirchmann, R., and Colard, J., Transfer and distribution of radioactive cadmium in dairy cows. Int. J. envir. Stud. 9 (1982) 47–51.

67 Van De Ven, W., Gerbens, J., Van Driel, W., De Goeij, J., Tjioe, P., Holzhauer, C., and Verweij, J., Spoorelementgehalten in koeien uit gebieden langs Rijn en Ijssel. Landbouwk. Tijdschrift 89 (1977) 262–269.

68 Van Enk, R.H., The pathway of Cd in the European Community. EUR 6626 EN Ispra, 1979.

69 Vogt, H., Nezel, K., and Matthes, S., Effects of various lead and cadmium levels on broiler and laying rations on the performance on the birds and the residues in tissues and eggs. Nutr. Metabol. 21 (1977) suppl. 1, 203–204.

70 Williams, C.H., and Davids D.J., The effect of superphosphate on the Cd content of soils and plant. Aust. J. Soil Sci. 11 (1973) 43–56.

71 Worker, N.A., and Migicovsky, B.B., Effect of vitamin D on the utilisation of Zinc, Cadmium and Mercury in the chick. J. Nutr. 75 (1961) 222.

72 Wright, F.C., Palmer, J.S., Riner, J.C., Haufler, M., Miller, A., and McBeth, C.A., Effects of dietary feeding of organocadmium to cattle and sheep. J. agr. Fd Chem. 25 (1977) 293–297.

Pathways and distribution of cadmium in grasslands

by F. F. Munshower

Reclamation Research Unit, College of Agriculture, Montana State University, Bozeman (Montana 59717, USA)

Introduction

Rangelands or grasslands comprise over 45% of the earth's land surface. In general these areas are restricted to grazing because of physical limitations which prevent their cultivation[20, 64]. These grasslands, shrub-grasslands, or open forest grasslands are not normally fertilized or subjected to sewage sludge application. They are, however, subjected to airborne pollutants including Cd (cadmium) in increasing amounts from smelters, refineries, power stations, or other industrial developments as world population and consumer demands continue to grow. It is in part because of these trace metal inputs to grasslands and their subsequent uptake by grazers that attention has been directed to the transport of Cd through grassland food chains.

All ecosystems contain some quantity of Cd. Although plant root zone concentrations are usually low (less than 1.0 ppm), they contribute to measurable levels in plants and subsequently in herbivores. This movement of Cd through various pathways to man is of concern because of the chronic toxicity, organ specific accumulation and long halftime of this particular metal in higher animals. Numerous anthropogenic sources of this element have recently increased its potential contamination of ecosystems and subsequently of human foodstuffs. These characteristics place Cd high on any list of potentially hazardous inorganic substances.

Cadmium in the soil

The total concentration of Cd in most soils varies from approximately 0.1 to less than 2.0 ppm[45, 60, 70]. Soils in metalliferous zones, those surrounding industrial areas, and fields receiving sewage sludge or phosphate fertilizers will commonly reveal several ppm Cd[15, 16, 38]. While the primary determinate of soil Cd is the parent material[29], several distinct reservoirs of this metal exist. The soil Cd pool consists principally of a solid inorganic phase derived from the parent material, an exchangeable phase adsorbed by clay particles or organic matter, mobile Cd in the soil solution readily available to plant roots and an organic phase complexed in humus[28, 69]. The constant weathering of the inorganic reservoir or aerial fallout replenishes any mineral losses from the soil Cd pools.

Pollution by sewage sludge contributes Cd to the soil primarily in the form of the organically complexed or adsorbed metal. Over time this organic fraction decomposes releasing active Cd to the other soil reservoirs. A rise in the Cd concentration of the soil solution is followed by an increase in plant Cd levels. Fortunately, some of this soil Cd may slowly revert to less soluble or more immobile forms[6, 31]. Aerial fallout and phosphate fertilizers also contribute to Cd pollution of grasslands. These Cd sources are primarily inorganic and the severity of pollution is determined by the Cd concentrations of the particulate fallout or fertilizer and characteristics of the receiving soil.

Total or strong acid extractable soil metal levels are usually quoted in the soil Cd literature. These techniques show little relationship to plant uptake of the metal[8, 27, 41] because of the number of factors which influence soil Cd activity. Plant available Cd is controlled by pH, organic matter, phosphorus, cation exchange capacity, precipitation of various Cd compounds, and other factors[5, 34, 40]. Soil pH is among the more important factors influencing Cd activity. As the soil becomes more acid most metals including Cd become increasingly mobile. On the other hand, Cd activity is reduced by increasing the soil organic fraction or cation exchange capacity. These parameters are measures of the ability of the soil to complex Cd and remove it from the soil solution[37, 65]. Carbonate precipitates also appear to regulate Cd solubility in calcareous grassland soils[57].

Cadmium in plants

Discussions of trophic level transfer of any metal must be cognizant of total plant elemental levels. That is, the amount of an element available for transfer between trophic levels is the sum of the elemental concentration within plant tissues and dust or particulates containing the element and adhering to leaves and stems. With respect to Cd, root uptake is a function of the available or soluble Cd pool in the root zone and the plant species. The surface particulate load is a function of leaf size, shape and surface structure, and the Cd concentration of resuspended soil particles or airborne particulate pollutants. In nonpolluted grasslands both plant Cd sources (root zone Cd and dust particles) have such low levels as to pose no significant problems to grazing animals or higher consumers. Grasslands receiving inputs from any source of Cd (smelter, fertilizer, etc.) rarely contain phytotoxic elemental levels, but the possibility of increased plant uptake, plant surface contamination, and accumulation at higher trophic levels are realities that must be considered in any study of food chain Cd transfer.

Plant Cd concentrations in non-mineralized and nonpolluted grasslands rarely exceed 1.0 ppm and are usually less than 0.2 ppm[24, 47]. There are plant species differences, in uptake of this element[3, 4] however, and these differences are accentuated when root zone Cd concentrations are elevated[54]. For example, various species leaf Cd levels varied from 0.1 to 3.9 µg/g in controls in Bingham et al.'s[3] study but at 5.0 ppm Cd in the root zone leaf Cd concentrations ranged from 0.1 to 23 µg/g. There are also variations in Cd concentrations of different plant organs. They generally reveal Cd concentrations in the sequence roots > leaves > fruiting parts > seeds[67].

Cadmium in grazing animals

Most Cd consumed by animals is excreted with the feces within a few days. Of the total quantity of Cd in or on ingested forages, only 2–8 % is retained for a longer interval[17, 44, 46]. Most of this absorbed Cd is stored in the kidneys or liver[43, 52]. The halftime for this element once it has been absorbed and chemically fixed in higher organisms is quite long[7, 9], usually on the order of 15–30 years in man[18].

Acute Cd toxicity to grazing animals from the forage they consume has not been recorded in natural environments and chronic toxicity has only been produced in experimental animals. The transfer of this element through food chains has been documented, however, in polluted and unpolluted ecosystems. Since grazing animals (e.g., cattle, sheep, goats) are not normally maintained for entire life cycles, the accumulation of several ppm Cd in their internal organs does not pose a serious threat to the husbandry of these animals, but the transfer of this metal to consumers (including man) may pose serious long-term problems.

Researchers have estimated that 50–70 % of the body burden of Cd is located in the liver and kidneys of animals[18, 42]. The accumulation of Cd with age in these two primary receptor organs has also been noted[55]. Even with low metal concentrations in the diet, receptor organ Cd levels continue to increase throughout the life of the organism[50], but with indications of decreasing concentrations in humans past the age of 50[58]. Kidney and occasionally liver Cd levels of older horses and cattle from areas not subjected to known pollutant sources may exceed several ppm Cd[12, 66], but concentrations in younger animals are much lower. The concentrations of Cd in muscle, liver, and kidneys of several animals from nonpolluted grasslands are shown in table 1. Tissue levels in mule deer *(Odocoileus hemionus)* and antelope *(Antilocapra americana)*, big game animals found on grasslands of the Great Plains of North America, are very similar to the levels in these same tissues of cattle from similar nonpolluted areas.

Although muscle and fat of grazers consistently reveal very low Cd concentrations, (less than 0.1 ppm), there are elemental increases in these tissue levels when cattle or sheep are fed high Cd diets[11]. Small increments in herbivore muscle Cd concentrations are of much greater significance to humans than larger increments in kidney or liver tissues. These organs may be discarded but herbivore muscle is a major protein source for part of the world's population. The 'average' diet of most humans in areas for which data are available is near 50 µg Cd/person/day[5, 18]. The maximum dietary levels suggested by F.A.O./W.H.O. is approximately 70 µg/person/day[14]. Any increase in Cd concentrations of animal tissues commonly used as human food must be avoided. Increases in herbivore muscle Cd levels although very small must be viewed with caution because of the amounts consumed by humans.

Grassland cadmium pathways and compartments

Natural, unpolluted soils and vegetation usually reveal very low Cd levels along all segments of the food chain; of concern, therefore, are grasslands receiving additional inputs of Cd from anthropogenic sources. In both examples, polluted and unpolluted grasslands, pathways and compartments are similar; they differ only in amounts transferred between trophic levels and concentrations in the various compartments.

In arid or semiarid grasslands with soil pH values near or

Table 1. Cadmium concentrations in animals from nonpolluted grasslands (ppm wet wt)

| | Mean Cd concentration | | | |
	Muscle	Liver	Kidney	Reference
Cattle		0.06 (13)	0.22 (13)	47
	0.005 (141)	0.17 (141)	0.42 (141)	32
	0.08 (1765)	0.21 (2122)	0.55 (2157)	66
Mule deer		0.15 (30)	0.42 (24)	50
Antelope		0.06 (20)	0.27 (21)	50

( ), number of animals.

above seven, movement of Cd through the soil is restricted to the upper few centimeters of the profile[30, 48]. In more mesic regions, it is usually carried deeper into the soil profile. This increased mobility of Cd is attributed to lower soil pH values and the infiltration of precipitation. It has also been shown that earthworms encourage the mixing of Cd in the root zone of more mesic ecosystems[1]. The leaching of this metal from grassland soils does not appear to occur[62, 73], therefore in any soil receiving Cd inputs from industrial activities or agricultural practices there will be an accumulation of Cd near or at the soil surface. At this soil-atmosphere interface Cd has some inhibitory influence on seed germination[39, 56]; however, the major impact of pollutant Cd is to be found in its influence on plant Cd concentrations.

Vegetation Cd concentrations are a function of the activity of Cd in the root zone and the plant species. Increased soil Cd levels are accompanied by increases in plant tissue levels unless other soil factors vary the availability of the metal. Plant species and even varieties show remarkable differences in their absorption and transport of Cd[25, 26, 35]. Some grasses appear to have low Cd uptake and translocation rates[10]. Grasses, especially those from undisturbed rangelands, reveal lower Cd concentrations than other plants from the same soils[51, 59], but this may be a function of surface contamination of the broader leaves of shrubs and forbs.

Mention was made of the ability of earthworms to aid in the distribution of Cd throughout the root zone. While ingesting soil and organic material, earthworms were found to accumulate Cd over 10 times soil levels[19, 21]. As a lower link in the grassland food chain, earthworms are a readily available source of Cd to small predatory vertebrates.

Grassland insects do not show the strong accumulation of Cd that has been reported in earthworms. Van Hoook and Yates[68] found no whole body accumulation of Cd in crickets or spiders above the levels in their food sources. However, slight biomagnifications of Cd (concentration factor: Cd in organism/Cd in food source = 1.3) were found in grasshoppers near a zinc smelter[48]. Another investigation of pristine rangelands in southeastern Montana also revealed very slight biomagnification of Cd in grasshoppers[49]. Apparently Cd is accumulated to concentrations above background food levels in only some insects. In general arthropods simply serve as a food chain link or transfer organism for this element in the grassland ecosystem.

Small mammals such as meadow mice (*Apodemus* sp.) and voles (*Microtus* sp.) are among the major food sources of predatory birds and larger carnivorous mammals of grasslands. Whole body Cd determinations revealed no biomagnification of this element in the rodent segment of grassland food chains[2] despite very slow turnover of this metal in small mammals[7, 13]. If analyses of Cd concentration increases from one trophic level to the next are based upon receptor organ Cd levels, however, biomagnification of this element is observed. Two studies have reported relatively high Cd concentrations in liver and kidneys of these organisms[23, 72]. Because of the presence of Cd in these herbivores, and its increasing accumulation with age the potential for subsequent transfer of this element to carnivores must be assumed.

Of greater concern in terms of food chain Cd transfer in animals of this size are the very high Cd concentrations reported in insectivorous shrews (*Sorex* sp.)[2]. Whole body Cd levels in these organisms revealed concentration factors greater than two in Cd-polluted and non-polluted environments. Because of the storage of over 50% of the whole body Cd burden of mammals in the liver and kidneys, the concentration of this element in these organs must be very high. Indeed, Hunter and Johnson[23] found 190 and 280 µg Cd/g dry wt in kidneys and liver of shrews they collected from polluted areas.

Cottontail rabbits (*Sylvilagus* sp.) from pristine grasslands revealed a slight biomagnification of this element in their liver and kidneys[51]. The livers of grouse and pheasants from the same area showed a similar Cd accumulation pattern but as in rabbits, no whole body biomagnification of this element. The trace element concentration in species of grassland birds in table 2 reveal liver Cd concentrations reflective of their diets. The pheasants (*Phasianus colchicus*) are primarily seed eaters. Plant seeds usually reveal lower Cd levels than leafy plant tissues and pheasant liver Cd concentrations were the lowest recorded in birds in the rangeland study. Sharptailed grouse (*Pedioecetes phasianellus*) feed on a mixed diet consisting of seeds, berries, leafy plant tissues, and insects. Their liver Cd levels were slightly higher than those found in pheasants. Sage grouse (*Centrocercus urophasianus*) feed extensively on leafy tissues of various species of *Artemisia*. These plants revealed the highest vegetative Cd concentrations in the study area. Liver samples from this species also showed the highest avian Cd levels.

Other studies of avian species[1, 36, 71] revealed similar Cd accumulation patterns. The Oak Ridge study found no biomagnification of Cd in chipping sparrows (*Spizella passerina*). The nationwide starling (*Sturnus vulgaris*) monitoring program has also failed to reveal any biomagnification of Cd in this species. Elemental levels in these birds were always less than levels anticipated in their food sources. More noteworthy was a relationship found between thin eggshells and elevated Cd in accipiter eggs[63]. Eggshell thinning has also been reported in laying hens but only at very high dietary Cd levels (48 ppm)[33].

Cattle liver and kidney Cd levels reflect the elemental content of the diet as well as the age of the animal[32, 61], but few range cattle are kept on pasture longer than 6–8 years. Slaughter steers and heifers are removed at 1.5–2 years and cows die or are removed for slaughter usually before they reach 6 years of age. Thus, elemental levels in these grazers are probably increasing throughout their entire life span. Table 3 illustrates liver and kidney Cd levels in animals from a known polluted area. Comparison with those cattle organ Cd concentrations in table 1

Table 2. Small animal Cd concentrations* (ppm wet wt)

	Mean Cd concentration	
	Liver	Kidney
Cottontail rabbit	0.17 (60)	0.72 (58)
Sharptail grouse	0.34 (15)	
Sage grouse	1.26 (4)	
Ringnecked pheasant	0.22 (19)	

( ), number of animals.
* Munshower et al.[49].

reveals the increase in liver and kidney levels to be expected in large herbivores grazing several km from a known pollutant source.

Accumulation of Cd by sheep or horses is less well documented than in cattle. Elemental biomagnification in liver and kidney samples of sheep has been shown to occur, however, as a function of age and diet[11]. Elinder and his associates[12] reported increased Cd concentrations with the age of horses and very high Cd concentrations in kidneys. As in smaller mammals, whole body biomagnification of Cd was not apparent in these animals but tissue specific biomagnification was shown in older animals.

Table 3. Liver and kidney cadmium concentrations in cattle from a rangeland polluted by emissions from a smelting complex (ppm wet wt)

Age (years)	Mean Cd concentration		
	Liver	Kidney	Reference
1.5–2	0.34 (9)	1.67 (9)	47
9.0	1.40 (1)	5.70 (1)	48

( ), number of animals.

Literature citations quoting Cd concentrations in internal organs of grassland predators are rare. Three carnivores from pastures polluted by a smelter complex are shown in table 4. The fox *(Vulpes vulpes)* and weasel *(Mustela* sp.) revealed kidney Cd concentrations approximately equal to grass Cd levels in the area in which these animals were collected. The badger *(Taxidea taxus)* kidney Cd concentration was slightly higher. Biomagnification of Cd at this trophic level is not apparent. As with herbivores, older animals may reveal kidney biomagnification.

Table 4. Carnivore cadmium concentrations* (ppm wet wt)

Animal	Kidney Cd concentration	Approx. age (months)
Red fox	1.35	24
Badger	2.04	48
Weasel	1.87	12

* Munshower[48].

Cadmium contents in various compartments of a grassland ecosystem

The contents of Cd in various compartments of a hypothetical grassland may be estimated. This distribution is schematically represented in the figure. The plant and animal concentrations in this ecosystem are based upon studies conducted by the author in two areas of the Northern Great Plains of North America. One site was 24 km from a smelter, the other 150 km from any industrial development. The polluted site was used as the source of Cd concentrations in soils, plants and animals. Due to accompanying human disturbance, however, the more remote site was selected for the estimation of plant and animal composition of the hypothetical grassland.

The soil Cd reservoir of the figure is based upon a 15-cm furrow slice and a Cd concentration of 2 mg/kg. Within these restrictions a soil Cd pool in the plow layer of 4.48 kg/ha may be calculated.

The soil Cd level indicates only minor pollution or a slightly metalliferous parent material, therefore, Cd concentrations in plant growth of current years were relatively low. Forage Cd concentrations were also variable depending upon the species involved. Grass levels were 1.7 mg/kg, forb Cd levels 2.5 mg/kg, and shrubs 3.4 mg/kg oven dry weight. Typical dry weight production levels on these semiarid rangelands were: grasses, 750 kg/ha, forb, 170 kg/ha, and 200 kg/ha of shrubs. Within 1 ha the plant Cd pools are: grasses, 1275 mg; forbs, 425 mg; and shrubs, 680 mg; or a total plant Cd pool of 2.4 g/ha.

The herbivore compartment of this grassland was composed of insects, primarily grasshoppers *(Acrididae),* small mammals (rodents and lagomorphs), birds, cattle, and large wildlife such as deer and antelope. Small mammal and resident bird populations are usually low and Cd does not accumulate in these animals above vegetation levels. Their Cd contribution to grassland food chains is significant but total Cd in this segment of the model is very small. It may be estimated on the order of 0.1 mg/ha. Grasshoppers show only slight biomagnification of Cd but they are so numerous they are the major herbivore Cd compartment of grassland food chains. Basic parameters controlling this arthropod contribution to the Cd food chain are: the number of insects – 5/m² [22], Cd concentration slightly higher than grasses or estimated at 2 mg/kg, and the biomass of insects. Average dry weight of these grasshoppers is 82 mg/insect[53]. Total grasshopper Cd in this hypothetical grassland is 8.2 mg/ha.

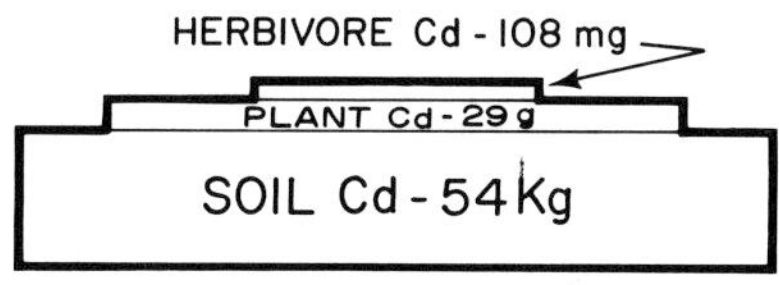

Cadmium contents of each compartment of a 12 ha semiarid grassland ecosystem.

Cattle Cd contents are stored primarily in the liver and kidneys. The weight of Cd in this segment of the herbivore Cd compartment may be estimated from known tissue levels if it is assumed that 50 percent of the body burden is stored in these organs. A kidney Cd concentration of 1.67 mg/kg and organ weight of 810 g will yield 1.4 mg of Cd. A liver Cd level of 0.34 mg/kg and weight of 5 kg will yield 1.7 mg of Cd. Total Cd body burden of a range cow, is, therefore, approximately 6.2 mg.

Other large range herbivores (deer, antelope, etc.) accumulate Cd at a rate and to levels comparable to cattle[50]. Since these animals are about one fifth the size of range cattle, it can be assumed that the Cd contribution to the herbivore compartment is 1.2 mg from each animal.

The area required to graze one cow for 1 year in this ecosystem is approximately 12 ha. A comparable range may be estimated for the wild ungulates. Therefore, within a 12 ha grassland containing on cow, deer and antelope, the large mammal Cd burden is 8.6 mg. The Cd content of each trophic level of this 12 ha grassland is shown in the figure. Total Cd contents are: soils, 54 kg; vegetation, 29 g; animals, 108 mg.

Warm-blooded herbivores in the grassland ecosystem are

protected from increasing Cd body burdens by two effective Cd filtration or isolation mechanisms. The first protective barrier is the isolation of soil Cd in a non-plant-available form. Most grasslands are semiarid or arid and many are calcareous. The high pH and carbonates of these soils provide an insoluble sink for pollutant derived metals. The second filter is the poor absorption, low retention and organ specific isolation of Cd in mammals. Formation of renal or hepatic metalloproteins effectively removes the toxic metal from sites of action in higher animals. Of that fraction of ingested Cd that is absorbed by grazers the vast majority is isolated in the liver and kidneys, not muscle or skeletal tissues. This selective storage of Cd in internal organs does pose a problem for long-lived animals. Tissue levels may exceed damage thresholds but this only occurs after prolonged ingestion of elevated dietary levels. Fortunately, small mammals have very rapid turnover rates in natural ecosystems and large herbivores are harvested for human food. Thus, the effects of long-term Cd accumulation are not felt at the herbivore trophic level unless pollution is extremely severe. Human contamination is averted by the absence of Cd accumulation in muscle tissue or animal fat. Further protection to humans would be afforded by exclusion of the liver and kidneys from human diets.

1 Anderson, S. H., Andren, A. W., Baes, C. F. III, Dodson, G. J., Harris, W. F., Henderson, G. S., Reichle, D. E., Story, J. D., Van Hook, R. I., Van Winkle, W., and Yates, A. J., Environmental monitoring of toxic materials in ecosystems, in: Ecology and Analysis of Trace Contaminants, pp. 95–139. Oak Ridge National Laboratory. ORNL-NSF-EATC-6. Oak Ridge, TN. 1974.

2 Andrews, S. N., Johnson, N. S., and Cooke, J. A., Cadmium in small mammals from grassland established on metalliferous mine wastes. Envir. Pollut. (Series A) *33* (1984) 153–162.

3 Bingham, F. T., Page, A. L., Mahler, R. J., and Ganje, T. J., Growth and cadmium accumulation of plants grown on soil treated with a cadmium-enriched sewage sludge. J. envir. Qual. *4* (1975) 207–211.

4 Bingham, F. T., Page, A. L., Mahler, R. J., and Ganje, T. J., Yield and cadmium accumulation of forage species in relation to cadmium content of sludge-amended soil. J. envir. Qual. *5* (1976) 57–60.

5 Bureau of Foods, Compliance Program Evaluation, Total Diet Studies: 1973, U.S.D.A. Washington, D.C. 1975. Chaney, R. L., and Hornick, S. B., Accumulation and effects of cadmium on crops, in: Proceedings First International Cadmium Conference, pp. 125–140. Metal Bulletin Limited. London, England, 1978.

6 Chang, A. C., Page, A. L., and Bingham, F. T., Heavy metal adsorption by winter wheat following termination of cropland sludge application. J. envir. Qual. *11* (1982) 705–708.

7 Cotzias, G. C., Borg, D. C., and Selleck, B., Virtual absence of turnover in cadmium metabolism: Cd109 studies in the mouse. Am. J. Physiol. *201* (1961) 927–930.

8 Davis, R. D., Cadmium – a complex environmental problem Part II, cadmium in sludges used as fertilizer. Experientia *40* (1984) 117–126.

9 Decker, C. F., Byerrum, R. U., and C. A. Hoppert, A study of the distribution and retention of cadmium 115 in the albino rat. Archs Biochem. Biophys. *66* (1957) 140–145.

10 Dijkshoorn, W., vanBroekhoven, L. W., and Lampe, J. E., Phytotoxicity of zinc, nickel, cadmium, lead, copper, and chromium in three pasture plant species supplied with graduated amounts from the soil. Neth. J. agric. Sci. *27* (1979) 241–253.

11 Doyle, J. J., Pfander, W. H., Gebring, S. E., and Pierce, J. G. III, Effect of dietary cadmium on growth, cadmium absorption, and cadmium tissue levels in growing lambs. J. Nutr. *104* (1974) 160–166.

12 Elinder, C. G., Jonsson, L., Piscator, N., and Rehnster, B., Histopathological changes in relation to cadmium concentration in horse kidneys. Envir. Res. *26* (1981) 1–21.

13 Exon, J. H., Lamberton, J. S., and Koller, L. D., Effect of chronic oral cadmium exposure and withdrawal on cadmium residues in organs of mice, Bull. envir. Contam. Toxic. *18* (1977) 74–76.

14 FAO/WHO Expert Committee on Food Additives, Evaluation of certain food additives and the contaminants mercury, lead, and cadmium. WHO Tech. Rep. Ser. No. 505, 1972.

15 Fleischer, M., Sarofin, A. F., Fassett, D. W., Hammond, P., Shacklette, H. T., Nisbet, I. C., and Epstein, S., Environmental impact of cadmium. Envir. Health Perspect. *7* (1974) 253–323.

16 Frank, R., Ishida, K., and Suda, P., Metals in agricultural soils of Ontario. Can. J. Soil Sci. *56* (1976) 181–196.

17 Friberg, L., Piscator, N., and Nordberg, G., Cadmium in the Environment. CRC Press. Cleveland, OH 1971.

18 Friberg, L., Piscator, N., Nordberg, G. F., and Kjellstrom, T., Cadmium in the Environment. 2nd edn. CRC Press Inc., Cleveland, OH 1974.

19 Gish, C. D., and Christensen, R. E., Cadmium, nickel, lead, and zinc in earthworms from roadside soil. Envir. Sci. Technol. *7* (1973) 1060–1062.

20 Heady, H. F., Rangeland Management. McGraw-Hill Book Co. N.Y. 1975.

21 Helmke, P. A., Robarge, W. P., Korotev, R. L., and Schomberg, P. J., Effects of soil-applied sewage sludge on concentration of elements in earthworms. J. envir. Qual. *8* (1979) 322–327.

22 Hewitt, G. B., and Onsager, J. A., Control of grasshop pers on rangeland in the United States – A perspective. J. Range Manag. *36* (1983) 202–207.

23 Hunter, B. A., and Johnson, N. S., Food chain relationships of copper and cadmium in contaminated grassland ecosystems. Oikos *38* (1982) 108–117.

24 Jenkins, D. W., Biological Monitoring of Toxic Trace Metals. Vol. 2; Toxic Trace Metals in Plants and Animals of the World. Part I. U.S.E.P.A. Environmental Monitoring Systems Laboratories, EPA-600/3-80-090 Las Vegas, NV. 1980b.

25 John, M. K., Cadmium uptake by eight food crops as influenced by various soil levels of cadmium. Envir. Pollut. *4* (1973) 7–15.

26 John, M. K., and VanLaerhoven, C. J., Differential effects of cadmium on lettuce varieties. Envir. Pollut. *10* (1976) 163–173.

27 John, M. K., VanLaerhoven, C. J., and Chuah, H. H., Factors affecting plant uptake and phytotoxicity of cadmium added to soils. J. envir. Sci. Technol. *6* (1972) 1005–1009.

28 Jurinak, J. J., and Santillan-Medrano, J., The Chemistry and Transport of Lead and Cadmium in Soils. Utah Agric. Exp. Sta. Resch. Rpt. 18, Utah State Univ, Logan 1974.

29 Kabata-Pendias, A., and Pendias, H., Trace Elements in Soils and Plants. CRC Press, Inc., Boca Raton, FL 1984.

30 Kobayashi, J., Air and water pollution by cadmium, lead, and zinc attributed to the largest refinery in Japan, in: Trace Substances in Environmental Health, Vol. V, pp. 117–128. Ed. D. D. Hemphill. Univ. of Missouri 1972.

31 Korcak, R. F., Gowin, F. R., and Fanning, D. S., Metal content of plants and soils in a tree nursery treated with composted sludge. J. envir. Qual. *8* (1979) 63–68.

32 Kreuzer, W., Samsoni, B., Kracke, W., and Wissmath, P., Cadmium in Fleisch und Organen von Schlachttieren. Die Fleischwirtschaft *3* (1975) 387–396.

33 Leach, R. N. Jr, Wei-Li Wang, K., and Baker, D. E., Cadmium and the food chain: The effect of dietary cadmium on tissue composition in chicks and laying hens. J. Nutr. *109* (1979) 437–443.

34 Levi-Minzi, R., Soldatini, G. F., and Riffaldi, R., Cadmium adsorption by soils. J. Soil Sci. *27* (1976) 10–15.

35 Maclean, A. J., Cadmium in different plant species and its availability in soils as influenced by organic matter and additions of lime, P, Ca, and Zn. Can. J. Soil Sci. *56* (1976) 129–138.

36 Martin, W. E., and Nickerson, P. R., Mercury, lead, cadmium, and arsenic residues in starlings-1971. Pestic. Monit. J. *7* (1973) 67–72.

37 McBride, M. B., Tyler, L. D., and Hovde, D. A., Cadmium absorption by soils and uptake by plants as affected by soil chemical properties. Soil Sci. Soc. Am. J. *45* (1981) 739–744.

38 Miesch, A. T., and Huffman, C. Jr, Abundance and distribution of lead, zinc, cadmium, and arsenic in soils, in: Helena Valley, Montana, Area Environmental Pollution Study, pp. 65–80. U.S.E.P.A. Research Triangle Park, N. C. 1972.

39 Miles, L. J., and Parker, G. R., Effect of soil Cd addition on germination of native plant species. Plant Soil *54* (1980) 243–247.

40 Miller, J. E., Hassett, J. J., and Koeppe, D. E., Uptake of cadmium of soybeans as influenced by soil cation exchange capacity, pH, and available phosphorous. J. envir. Qual. *5* (1976) 157–160.

41 Miller, R. J., and Koeppe, D. E., Trace elements and plant growth, in: Chemistry and Biology of Trace Metals in the Environment, pp. 175–188. Univ. of Illinois, Urbana. NTIS PB-291368, 1971.

42 Miller, W. J., Cadmium absorption, tissue and product distribution, toxicity effects and influence on metabolism of certain essential ele-

ments, in: Proceedings of the Georgia Nutrition Conference, pp. 58–69. Univ. of Georgia, Atlanta 1971.

43 Miller, W. J., Blackmon, D. N., Gentry, R. P., and Pate, F. N., Effect of dietary cadmium on tissue distribution of 109 cadmium following a single dose in young goats. J. Dairy Sci. *52* (1969) 2029–2035.

44 Miller, W. J., Blackmon, D. N., and Martin, Y. G., Cadmium absorption, excretion, and tissue distribution following single tracer oral and intravenous doses in young goats. J. Dairy Sci. *51* (1968) 1836–1839.

45 Mills, J. G., and Zwarich, M. A., Heavy metal content of agricultural soils in Manitoba. Can. J. Soil Sci. *55* (1975) 295–300.

46 Moore, W., Jr, Stara, J. F., Crocker, W. C., Malanchuk, N., and Iltis, R., Comparison of 115 m cadmium retention in rats following different routes of administration. Envir. Res. *6* (1973) 473–478.

47 Munshower, F. F., Cadmium accumulation in plants and animals of polluted and nonpolluted grasslands. J. envir. Qual. *6* (1977) 411–413.

48 Munshower, F. F., Cadmium compartmentation and cycling in a grassland ecosystem in the Deer Lodge Valley, Montana. Ph.D. thesis, Univ. of Mont., Missoula 1972.

49 Munshower, F. F., DePuit, E. J., and Neuman, D. R., Effects of Stack Emissions on the Range Resource in the Vicinity of Colstrip, Montana: Pre-operational Description of the Range Ecosystem. Mont. Agric. Exp. Sta. Res. Rpt. 126, Mont. State Univ., Bozeman 1978.

50 Munshower, F. F., and Neuman, D. R., Metals in soft tissues of mule deer and antelope. Bull. envir. Contam. Toxic. *22* (1979) 827–832.

51 Munshower, F. F., and Neuman, D. R., The Effects of Stack Emissions on the Range Resource in the Vicinity of Colstrip, Montana After Five Years of Operation of Units 1 and 2. Mont. Agric. Exp. Sta. Special Rpt. No. 1, Mont. State Univ., Bozeman 1983.

52 Neathery, N. W., Cadmium-109 and methyl mercury-204 metabolism, tissue distribution and secretion into milk of cows. J. Dairy Sci. *57* (1974) 1177–1183.

53 Onsager, J. A., A method for estimating economic injury levels of or control of rangeland grasshoppers with malathion and carbaryl. J. Range Manag. *37* (1984) 200–203.

54 Page, A. L., Bingham, F. T., and Nelson, C., Cadmium absorption and growth of various plant species as influenced by solution cadmium concentration. J. envir. Qual. *1* (1972) 288–291.

55 Perry, H. M. Jr, Tipton, J. H., Schroeder, H. A., Steiner, R. L., and Cook, M. J., Variation in the concentration of cadmium in human kidney as a function of age and geographic origin. J. chron. Dis. *14* (1961) 259–271.

56 Rothenberger, S. A., Effects of cadmium, zinc, lead, and molybdenum on germination and growth of selected grasses. Ph.D. thesis, North Dakota State Univ., Fargo 1978.

57 Santillan-Medrano, J., and Jurinak, J. J., The chemistry of lead and cadmium in soil: Solid phase formation. Soil Sci. Soc. Am. Proc. *39* (1975) 851–856.

58 Schroeder, H. A., and Balassa, J. J., Abnormal trace metals in man: Cadmium, J. chron. Dis. *14* (1961) 236–258.

59 Severson, R. C., Gough, L. P., and McNeal, J. M., Availability of elements in soils to native plants, Northern Great Plains, in: Geochemical Survey of the Western Energy Regions, pp. 98–143. U.S.G.S. Open-file Report 77-872. Denver, CO, 1977.

60 Shacklette, H. T., Cadmium in Plants. U.S. Geological Survey Bulletin 1314-G, Government Printing Office, Washington, D.C. 1972.

61 Sharma, R. P., Street, J., C., Verma, M. P., and Shupe, J. L., Cadmium uptake from feed and its distribution to food products of livestock. Envir. Health Perspect. *28* (1979) 59–66.

62 Sidle, R. C., and Sopper, W. E., Cadmium distribution in forest ecosystems irrigated with treated municipal waste-water and sludge. J. envir. Qual. *5* (1976) 419–422.

63 Snyder, N. F., Snyder, H. A., Lincer, J. L., and Reynolds, R. T., Organochlorides, heavy metals, and the biology of North American accipiters. Bioscience *23* (1973) 300–305.

64 Stoddart, L. A., Smith, A. D., and Box, T. W., Range Management. McGraw Hill Book Co., New York 1975.

65 Strickland, R. C., Chaney, W. R., and Lamoreaux, R. J., Organic matter influences phytotoxicity of cadmium to soybeans. Plant Soil *52* (1979) 393–402.

66 U.S. Department of Agriculture, Residue Evaluation and Planning Staff, Heavy Metal Survey in Cattle. Unpublished data, 1971.

67 VanBruwaene, R., Kirchmann, R., and Impens, R., Cadmium contamination in agriculture and zootechnology. Experientia *40* (1984) 43–52.

68 Van Hook, R. I., and Yates, A. J., Transient behavior of cadmium in a grassland arthropod food chain. Envir. Res. *9* (1975) 76–83.

69 Viets, F. G. Jr, Chemistry and availability of micronutrients in soils. J. agric. Fd Chem. *10* (1962) 174–178.

70 Vinogradov, A. P., The Geochemistry of Rare and Dispersed Chemical Elements in Soils, 2nd edn, pp. 149–154. Consultants Bureau Inc, New York 1959.

71 White, D. H., Bean, J. R., and Longcore, J. R., Nationwide residues of mercury, lead, cadmium, arsenic, and selenium in starlings, 1973. Pestic. monit. J. *11* (1977) 35–39.

72 Williams, P. H., Shenk, J. S., and Baker, D. E., Cadmium accumulation by meadow voles (*Microtus pennsylvanicus*) from crops grown on sludge-treated soil. J. envir. Qual. *7* (1978) 450–454.

73 Williams, D. E., Vlamis, J., Pukite, A. H., and Corey, J. E., Trace element accumulation, movement, and distribution in the soil profile from massive applications of sewage sludge. Soil Sci. *129* (1980) 119–132.

Contamination and effects of cadmium in native plants

by R. M. Cox

Maritimes Forest Research Centre, P.O. Box 4000, Fredericton (New Brunswick E3B 5P7, Canada)

Introduction

Plants have no metabolic requirement for cadmium; however, owing to its similarity to, and its unique ubiquitous association with the nutrient element zinc, specific mechanisms to block its uptake into plant cells may be inherently difficult to evolve. Once inside the plant cell cadmium has a high affinity for sulphydryl groups and inhibits many key metabolic processes. Cadmium must rank among the more toxic heavy metals for plants because of its intrinsic toxicity and bioavailability. Cadmium in crop plants has been intensively studied recently, because of the effects of dietary intake of cadmium on health, but studies of native plants are fewer and are associated with biomonitoring of cadmium in the environment, or metal tolerance mechanisms, and phytotoxicity. These studies have contributed greatly to our knowledge of potential effects of the metal on individual plants and populations. However, little is known of the ecosystem effects of cadmium pollution, which is an area as yet largely neglected.

Contamination

Natural background exposure to cadmium can be estimated by its abundance in terrestrial rocks which, on average, approaches 0.3 $\mu g\ g^{-1}$. Its ratio to zinc, with which it always occurs, is 250. In the principal mineral ore[10], zincblend (Zn S) cadmium occurs at the 0.1–5% level[17], and it is obtained almost entirely as a by-product of this zinc deposit. Although there is little variation in the cadmium content of igneous rock, variation in cadmium content from 0.3 to 11.0 µg g^{-1} occurs in sedimentary rocks[65]. Some carbonaceous shales formed under reducing environments contain much cadmium[96]. Monterey shales along the Pacific coast of the U.S. may contain up to 90 µg Cd g^{-1} and weather to form soils with as much as 30 µg Cd g^{-1} [66]. Soils and glacial tills in areas close to ore bodies have been reported to contain up to 40 µg Cd g^{-1} along with high levels of zinc[10,91]. Some farmland reclaimed from old zinc mines near Shipham, Somerset, UK contain 30–800 µg Cd g^{-1}. This area has been mined for calamine (ZnCO$_3$) since the 16th century and much of the area has provided grazing[91]. Simon[84] has shown that cadmium levels in soils around zinc lead mines in Germany and Belgium contained up to 111.0 and 233.0 mg kg^{-1} total cadmium, respectively, and that the former site at Plombieres contained up to 1.0 mg Cd kg^{-1} soluble cadmium along with up to 54.8 mg Zn kg^{-1}.

Another source of cadmium in the environment is from coal-burning; coal contains 0.2–0.5 µg Cd g^{-1} whereas coal ash contains up to 50 µg Cd g^{-1} [51]. Industrial sources of cadmium are electroplating, pigments, alloys, batteries, tires and plastics of which the greatest percentage of cadmium emission to the atmosphere is from motor cars, oil, and tires[55]. Agricultural sources include fungicidal spraying[76] and sewage sludge application[1]. These diffuse sources together with major point sources have increased the cadmium in deposition in urban, rural, and remote areas; this is reviewed, along with the occurrence of other metals in precipitation, by Galloway et al.[29] and Williams and Harrison[100]. Even the background levels in the remote antarctic snow have shown a significant increase since industrialization[45].

Soil enrichment by deposition of particulate cadmium in the UK is reviewed by Parry et al.[69], who documented several sites near the Avon mouth smelters with soil concentrations exceeding 100 µg Cd g^{-1} dry soil with other notable enrichment near power stations and foundries. Previous studies of the Avon mouth smelter pollution by Little and Martin[53] showed increased accumulation of metals within soils of wooded sites as opposed to adjacent open grassland areas, i.e., Hallen wood open field site had 87.0, 21.0 and 1.0 ppm of acetic acid extractable zinc, lead, and cadmium, respectively, while in the forested area 1100.0, 33.0, and 5.0 ppm, respectively, had accumulated. These results may indicate that the forested area was more efficient at intercepting airborne particulate material than was the open grassland.

The relative importance and the contribution of dry and wet deposition of metals, including cadmium to the terrestrial ecosystem have been investigated by Lindburg et al.[50]. Both processes and rates of deposition of Cd, Mn, Pb, and Zn together with interactions with acid rain were studied in a deciduous forest site in the Tennessee valley, which was located within 22 km of three fossil fuel power plants. The study showed that the rates of metal deposition in rain events were orders of magnitude higher than in intervening periods of dry deposition. However, dry deposition occurs over longer periods and may supply about 20% of the total deposition of cadmium and zinc[49]. More important than rates of either wet deposition[9,29,36,80,82] or dry deposition[9,49] are the concentrations of these ions experienced in the canopy. Accumulated dry deposition together with the successive accumulation of evaporation deposits from rain events too light to produce appreciable leaf washing may produce concentrations of deposited ions 100 to 1000 times that of ambient rainfall. It was reported by Lindburg and Harris[49] that cadmium concentrations of 0.1 mg l^{-1} may be experienced in the canopy under certain conditions. Its high solubility in the acidic solutions caused by concurrent acid deposition enhances its potential for interaction with internal tissues and sensitive reproductive processes. Cadmium was shown to have no tendency to be retained in the canopy and eventual rates of net removal to the forest floor from the canopy exceed estimated dry deposition by factors of 5–10[50]. These processes plus the larger leaf to ground area in forests help to explain the higher Cd levels in the aerially polluted soil of Hallen wood as compared with an adjacent open area noted by Little and Martin[53].

Bioaccumulation and biomonitoring

Bioaccumulation and biomonitoring implies the accumulation within the plant of detectable increases of cadmium above that of ambient levels in the soil or air. Background levels in above-ground parts of plants are usually less than 1.0 µg Cd g^{-1} [30,43,53,57]. The cryptogams (mosses and lichens), however, owing to their high cation exchange capacity, are efficient at absorbing both dry and wet deposited metals including cadmium and have been used as good indicators of regional pollution[70,77–79]. Ruhling and Tyler[78] demonstrated that southern Scandinavia was significantly (p < 0.001) more polluted by various trace elements including cadmium than was the northernmost point of the peninsula. In this Scandinavian study, the level of cadmium in the moss *Hylocomium splendens* from the North was 0.18 µg g^{-1} whereas the moss from southern sites contained 0.99 µg Cd g^{-1}. Similar trends have been found in Finland by Pakarinen and Tolonen[68] using *Sphagnum* while Steinnes[87] likewise demonstrated the trend using bryophytes and lichens. In a Danish study, Pilegaard et al[71], using samples of the moss *Hypnum cupressiforme* and lichens *Hypogymnia physodes* and *Lecanora conizaeoides,* found Cd concentrations in the moss two and three times those found in the two lichens, respectively. In addition, it was found that the mean concentration of cadmium and other metals was lower in the northwest of the country than in the eastern and southeastern parts, indicating differential regional air pollution.

The sporophores (fruiting bodies) of 130 species of basidiomycete fungi in forest and pastures of south Sweden were analyzed for metal content by Tyler[92]. The sites were chosen to represent background conditions, i.e. were not affected by local sources of metal pollution. Ten of the species examined were considered to bioconcentrate cadmium to levels in excess of the median value for all samples (n = 200). The cadmium concentration in sporophore tissue ranged from 0.1 to 299 µg g^{-1} with a median concentration of 1.4 µg g^{-1}. The distribution of Cd concentration in the samples was highly positively skewed due to the high levels of bioaccumulation in a few species, namely *Agaricus macrosporus* (100–299 µg Cd g^{-1}) *Agaricus arvensis* (23–96 µg Cd g^{-1}) and *Amanita muscaria* (46–101 µg Cd g^{-1}).

The ability of higher plants to accumulate cadmium from metalliferous soils varies. Species that grow naturally on these soils may accumulate substantial amounts of cadmium with levels exceeding 100 µg g^{-1} dry weight in leaves of such species as *Ameria maritima* and *Campanula rotundifolia;* levels of 500 µg g^{-1} have been recorded in leaves of the metallophyte *Thalaspi alpestre*. However, these levels of Cd were accompanied by lead and zinc in excess of 1000 µg g^{-1} [26]. Levels of 11.0 and 22.0 mg Cd kg^{-1} have been reported by Simon[85] in shoots of *Euphrasia stricta* and *Hieracium pilosella,* respectively. Levels of cadmium in the shoots of the grasses *Agrostis tenuis* and *Festuca ovina* growing on the same soil, however, never exceeded 1.7 mg Cd kg^{-1} indicating an exclusion of the metal from the shoots. Accumulator and excluder strategies in the response of plants to heavy metals are discussed by Baker[6].

The ability of some plants to bioaccumulate elements from air and soil contaminated by aerial pollution has made them attractive as biological monitors not only of regional and background air pollution but to identify and monitor point sources. Examples of such studies together with levels of Cd bioaccumulation by the species used are given in table 1. These biological monitors offer increased cost effectiveness over mechanical air sampling equipment used in the identification and mapping of pollution distribution from point sources. However, differential aerosol capture and retention by plant species have to be considered[54]. Standardization of a biomonitoring technology has occurred with the development of the 'moss bag' which was pioneered by Goodman and Roberts[30] and Roberts[73] and later described by Parry et al[69] and Little and Martin[54]. This technique involves the exposure of a standard amount of previously mixed and washed moss, usually *Hypnum* or *Sphagnum* in nylon bags of various designs, to aerial pollution which is retained on the cation exchange sites on living or dead moss.

The longevity and stationary habit of trees makes them suitable biological archives of climate[18,28] and of heavy metal pollution[3,4,48,74,75,89,95]. However, few have studied cadmium in tree rings. Kardell and Larsson[40] studied the chronology of lead and cadmium in five successive years of wood in increment cores of *Quercus robur* and other trees at different distances from roads in Stockholm. Analysis of wood cadmium in one tree, 39 m from the road but within the discharge area of a gas works, revealed only 0.5–3.0 ppb of cadmium dry weight, which is considerably lower than levels normally found in test material from polluted areas[40]. Lead in the wood, however, showed a distinctive increase after 1956 when tetraethyl lead was added to motor fuel. The concentrations in the wood samples were correlated to the traffic density to which the tree was exposed at the time the lead was deposited. The mechanisms with which the different metals are deposited in the wood and their form while being transported within the plant is not known, nor is the degree of lateral transport of the metals known. These processes may vary according to the ligands used and how fixation in the xylem is influenced by the rate of transpiration. Baes III and McLaughlin[4] have shown the highest concentrations of trace elements including cadmium in the combined living phloem and cambium tissues of short leaf pine (0.47–7.5 µg Cd g^{-1}) in the Great Smoky Mountains National Park. Symeonides[89] tested the validity of tree ring analysis of *Pinus silvestris* in tracing the history of metal pollution around a copper zinc lead smelter at Rönnskär in northern Sweden. This

Table 1. Some biomonitoring studies of cadmium pollution from point sources using assimilation organs of native plants

Source	Bioaccumulation (µg · g^{-1})		Source of pollution	Reference
	Minimum	Maximum		
Woody plants				
Ulmus glabra	0.25	50.00	Zinc lead smelter	Little and Martin[53]
Crataegus monogyna	–	6.04	Zinc lead smelter	Little[52]
Salix alba	–	7.85	Zinc lead smelter	Little[52]
Quercus robur	–	6.82	Zinc lead smelter	Little[52]
Quercus robur	ND$^+$	38.00	Zinc lead smelter	Buchauer[13]
Picea abies	0.10	150.00	Aluminum plant	Mankovska[57]
Beech	0.92	4.20	Magnesite works	Mankovska[59]
Oak	0.93	9.00	Magnesite works	Mankovska[59]
Herbaceous plants				
Festuca spp.	0.80	40.00	Industrial complexes	Goodman and Roberts[30]
Taxacacum officinale	0.90	15.09	Non-ferrous plant	Kuleff and Djingova[43]
Capsella bursa-pastoris	0.08	0.92	Sakai City	Tatsumi et al.[90]
Poa annua	0.02	0.77	Sakai City	Tatsumi et al.[90]
Mosses				
Brachythecium rutabulum	0.49	0.77	City of Copenhagen	Andersen et al.[1]
Rhytidiadelphus squarrosus				
Hypnum spp.	1.80	9.50	Industrial complexes	Goodman and Roberts[30]
Sphagnum spp.	0.03	11.29	Zinc lead smelter	Little and Martin[54]
Dicranowisia cerrata	0.34	24.60	Steel works	Pilegaard et al.[71]
Lichens				
Parmelia spp.	20.00	90.00	Zinc lead smelter	Burkitt et al.[14]
Lecanora conizaeoides	0.70	17.80	Steel works	Pilegaard et al.[71]
Hypogymnia physodes	0.73	27.70	Steel works	Pilegaard et al.[71]
Lecanora conizaeoides	0.34	0.89	City of Copenhagen	Andersen et al.[1]

$^+$ ND; below detection limit.

study demonstrated that at the sites of high exposure, reduction in growth occurred and was correlated to metal content in the respective annual rings. Data showed clear evidence of lateral transport of both cadmium and zinc but not lead and copper. However, a general increase in cadmium was recorded in tree rings from the time the smelter started operations in 1930 (less than 1 μg g^{-1}) to a value of 6 μg g^{-1} in 1967–1977 and was significantly negatively correlated (p = 0.01) to growth. Multiple regression indicated that the order of importance in the contribution to the regression equation (F_3, 172 = 27.91) was copper > cadmium > lead and explained 56.5% of the variation. Cadmium in the bark of the trees in polluted and unpolluted sites was 4.4 ± 1.0 and 0.31 ± 0.08 μg g^{-1}, respectively, demonstrating that deposition in bark is also a good indication of accumulated pollution. Recently there has been much interest in metal and acidic deposition as causal agents of the decline of red spruce *Picea rubens* at high elevation on Camels Hump Mountain, Vermont[27,83]. These studies have prompted Scherbatskoy and Bliss[81] (unpublished report 1984, personal communications) to study certain metals including cadmium in the 10-year increments of wood of red spruce at different elevations at this site. The studies have documented a dramatic increase in Cd accumulation from 0.16 μg Cd g^{-1} in the wood during 1900–1910 to 0.42 μg g^{-1} during 1970–1980 at an elevation of 1160 m. A similar trend was also documented at 1040 m up to 1970, whereas no trend was apparent for cadmium in 10 successive years of annual rings at the lowest elevation of 910 m. This may provide evidence that interception of long range transported air pollution at high elevation may provide for better historical trends in Cd pollution in successive tree rings than at low elevations. However, lateral transport and rates of transpiration have to be considered and further investigated before firm conclusions can be made.

Physiology and toxicology

Effects of cadmium on native plants include studies with green algae which have shown variation between species in tolerance of cadmium in solution. Pakalne et al.[67] have shown stimulation of growth in *Chlorella* spp. when exposed to concentrations of Cd < 0.1 μg g^{-1} whereas 2.0 μg g^{-1} stopped growth and 3.0 μg g^{-1} proved lethal. The addition of Fe^{+3}, Mn^{+2} and Zn^{+2} ions was shown to reduce the toxicity of cadmium in solution. Hutchinson[32] determined that *Chlorella vulgaris* was more sensitive than *Scenedesmus acuminatus* which was in turn more sensitive than *Chlamydomonas eugametas* with thresholds of marked decline in cell division at cadmium concentrations of 0.05, 0.1 and > 0.2 μg g^{-1}, respectively. In addition, this study demonstrated that the addition of selenium to the media was antagonistic to Cd toxicity in *Chlorella* and *Scenedesmus,* whereas the selenium without Cd proved to be inhibitory. Burnison et al.[15] noted that 0.02 μg Cd g^{-1} inhibited photosynthesis in *Scenedesmus guadricauda* by 70%, whereas 0.1 μg g^{-1} was necessary to reduce primary productivity by 70% in a strain of *Chlorella vulgaris.* Ultrastructural changes such as vacuolation within the mitochondria and granule accumulation in the mitochondria, together with dilation of

endoplasmic reticulum, were also noted on exposure of the algae to Cd. The cadmium sensitive nature of the photoreaction rather than the dark reaction of photosynthesis in *Chlamydomonas reinhardii* was noted by Overnell[64]. Apart from these inhibitory effects of Cd, Bartlett et al.[7] found that Cd reduced copper toxicity in the alga *Selanastrum capricornutum.*

Most investigations on the availability of cadmium and its effects on higher plants involve crop plants, because of its importance as a contaminant of food and its potential effects on human health. Most of these crop studies are concerned with cadmium from fertilizers or sewage sludge applications to farmland, and will be dealt with elsewhere in this volume.

Studies by Markhinova and Gileva[60] using aquatic macrophytes have shown that the freshwater algae *Chara fragilis* and the water milfoil *Myrophyllum spicatum* can bioaccumulate cadmium 22,900 and 10,000 fold, respectively. This rate of bioaccumulation was reduced by 91.2% when EDTA was added to the water. Stanley[86] showed that 7.5 μg g^{-1} will inhibit root growth in this species by 50%. Hutchinson and Czyrska[33] studied the effects of Cd on ramit formation in *Lemna valdiviana* and *Salvinia nutans* placed in cultures where either one or both competing organisms were present. Concentrations of 0.01 μg Cd g^{-1} proved toxic to *Salvinia* and inhibited frond development in *Lemna*. Symptoms included chlorosis with necrotic patches on the *Salvinia* fronds. The combined culture of these plants resulted in the amelioration of toxicity to *Salvinia* and was later related to cadmium uptake. Bioaccumulation of cadmium in *Salvinia* and *Lemna* was 2000–9600 and 1500–6000-fold, respectively. The combined effects of copper and cadmium on the growth and flowering of *Lemna paucicostata* were investigated by Nasu et al[63]. It was found that cadmium only suppressed frond multiplication but not frond growth, as did copper. These differences in effects of the metals were also evident in combinations of the metals and related to their absorption. Copper absorption was not influenced by cadmium; however, Cd uptake was reduced in the presence of copper in the media. Copper ion induction of flowering in this plant was prevented by simultaneous addition of cadmium which suppresses neither copper absorption nor shortday flowering. Cowgill[21], also working with aquatic macrophytes, showed that Cd levels in *Ceratophylum* and *Potamogeton* could bioaccumulate levels 4000 times to reach tissue concentrations as high as 10.5 and 9.1 μg g^{-1}, respectively.

Investigating heavy metal effects on senescence of aquatic angiosperms, Jana and Choudhuri[35] found that 0.1 mM Cd reduced respiration rate by 7 and 5% in *Vallisneria spiralis* and *Hydrilla verticillata,* respectively, and 1.0 mM Cd reduced respiration by 53 and 50%. Immersion of isolated leaves in 1.0 mM Cd emulated senescence in *Potamogeton pectinatus* as well as in the above-mentioned species, in that reductions in chlorophyll, RNA, protein, alkaline phosphatase activity, and dry weight were induced compared with the control. Concomitant increases were noted in free amino acid and tissue permeability, together with increased RNase and acid phosphatase activity. The exception was acid phosphatase activity in *Hydrilla* which was increased rather than reduced when exposed to 1 mM Cd.

Cadmium effects on terrestrial plants have been studied mostly in the context of environmental contamination. Jordan[37] in her study of smelter emissions on a chestnut-oak woodland at Palmerton demonstrated that toxic foliar levels of Cd in *Quercus rubra* were between 5 and 30 $\mu g\ g^{-1}$, whereas levels in washed foliage reached 70 $\mu g\ g^{-1}$ in some tree leaves. Seedlings of *Q. rubra* and *Pinus strobus* planted in soil from close to the smelter showed inhibition of both root and shoot growth. In this study, seed germination of *Q. rubra*, *Betula populifolia* and *Populus tremuloides* was not inhibited by 10 $\mu g\ g^{-1}$ Cd in liquid culture but did show significantly reduced radicle elongation. In another study by Leavitt et al.[46], plant regulation of essential and nonessential heavy metals in an oak-hickory-pine forest growing on rock types known to include sulphide mineral deposits was investigated. This study concluded that, in general, these plants could not distinguish between Cd and Zn. Furthermore, concentration of these two metals in the tissue was correlated in the oak species which also accumulated more Cd than the other species. In fact, there was evidence from Zn/Cd ratios suggesting preferential uptake of Cd relative to zinc. Coughtrey and Martin[20] investigated Cd uptake in populations of *Holcus lanatus* growing at different distances from a lead smelter. The population on which the emissions had the most impact was inhibited in root growth by 2.0 $\mu g\ g^{-1}$ Cd in culture solution. However, this smelter population demonstrated both higher root Cd levels and reduced translocation of Cd to the shoots (42.7 $\mu g\ g^{-1}$) as compared with tillers of a population sampled from a more distant site. This control population accumulated up to 145 μg Cd g^{-1} in its shoots when exposed to the same 1.0 μg Cd g^{-1} in solution. The implications of this discovery are discussed in the tolerance section below.

McGrath[62], using populations of the grass *Holcus lanatus* from near a lead/zinc smelter and an aluminum tolerant race, indicated that Cd toxicity in the aluminum tolerant race is ameliorated by the addition of aluminum, whereas in the Hallen material (smelter population) the aluminum exerts an additive effect in depressing root elongation. Most of the bioaccumulation of both Al and Cd was in the roots; more Cd was accumulated here in the smelter population whereas more Al was accumulated here in the Al tolerant race than after reciprocal exposure of the two populations to the two metals. In addition, it was found that combination of the two metals in the culture media significantly (p < 0.001) reduced root concentrations of both metals in both populations in comparison with those found in roots treated with single metals. It was also suggested[62] that Al tolerant acid soil populations may be preadapted to low levels of Cd in the environment.

Enzyme sensitivity to cadmium and other metals has also been examined in terrestrial plants by Mathys[61]. Here the sensitivity of leaf enzymes to cadmium in zinc tolerant *Silene cucubalus* was examined in vitro. Nitrate reductase was found to be most sensitive, whereas glucose-6 phosphate dehydrogenase and malate dehydrogenase were in turn less sensitive with isocitrate dehydrogenase being the least sensitive enzyme. Levels of cadmium required to inhibit these enzymes by 50% were 0.001–0.01, 0.01–0.1, 0.1–1.0, and > 1.0 mM, respectively. The relative sensitivity of these enzymes may relate to the SH groups and other reactive sites on them[93] and may indicate potential physiological targets for cadmium.

The effect on leaf physiology of various concentrations of cadmium in solution supplied to excised leaves of *Acer saccharinum* through the petiole was examined by Lamoreaux and Chaney[44]. The cadmium-treated leaves exhibited reduced net photosynthesis and increased dark respiration up to 193% that of the control leaves. Reduced rates of net photosynthesis and transpiration were correlated to both solution and tissue concentrations of cadmium and were reduced by 18 and 21%, respectively, by 0.180 μM Cd as compared with the control treatment. Diffusive resistance of the leaves was also reduced with increasing Cd concentrations and time. A possible mode of action of cadmium was suggested, that of interference with the movement of potassium between guard cells and subsidiary cells of the stomata, as well as inhibition of a host of enzymic pathways.

Reproductive processes

Although little is known of the effects of heavy metals on the reproductive processes of higher plants, a few investigations have been carried out to determine the effects of cadmium on pollen function. In a study of cadmium effects on respiratory gas exchange in *Pinus resinosa* pollen, Strickland and Chaney[88] were able to demonstrate a stimulation of respiration after 1.5 h by 0.45 μM Cd, which declined sharply after 3.5 h in which CO_2 evolution was inhibited rather than O_2 uptake. In addition, pretreatment with Cd inhibited exogenous sucrose utilization which did not rectify the Cd inhibition of gas exchange. Furthermore, there was no correlation of Cd effects on gas exchange with either pollen germination or germ-tube length, which was severely affected by Cd levels (0.045 μM) that had no influence on gas exchange. The sensitivity of *Petunia alba* pollen to 0.2 μM Cd was previously shown by Kapur and Malik[39] who also examined other metabolic inhibitors on pollen tube growth and pollen germination.

The combined in vitro effect of cadmium (0, 1.0, 20, and 50 μM) and pH on *Trilllium grandiflorum* pollen was examined by Cox[23]. The response surfaces for pollen germination and tube growth are shown in figure 1. Analysis of variance revealed significant pH and cadmium effects (p < 0.01 and p < 0.001 respectively) in 50 μM cadmium. Cadmium was in general stimulatory at concentrations below 50 μM, however, the significant interaction between Cd and pH (p < 0.05) was in part the result of synergism at pH 3.6 and 50 μM Cd for both germination and tube growth. This indicated that one should not consider potential toxicity of metal pollutants in isolation from other pollutants. Although, in this study, concentrations are in excess of those expected in rain, the 1.0 μM Cd (112 $\mu g\ l^{-1}$ Cd) concentration does approach the levels expected to accumulate on vegetation from the combined effects of dry and wet deposition[50]. The lowest Cd concentration used (1.0 μM) was slightly stimulatory to pollen tube growth and slightly inhibitory to the pollen germination response. These responses indicated that at the levels of regional deposition both pH and Cd have the potential to affect reproduction in plants and that these

106

effects may be increased in areas local to Cd emissions. The sensitivity of *Pinus resinosa* pollen to cadmium was confirmed by Chaney and Strickland[16] who estimated the ED_{10} dosage of cadmium (that which inhibits by 10%) to be as low as 0.27 and 1.00 µmol l[-1] for germination and germ-tube growth, respectively.

Tolerance

Variation in tolerance to cadmium may be expected in natural populations, as any physiological process will reflect the natural genetic variation that exists between genetically distinct individuals. Evidence of such variation in cadmium tolerance within species or between individuals has been documented in various grass species by Simon[85] for *Agrostis tenius* and *Festuca ovina,* by Coughtrey and Martin[19] for *Holcus lanatus,* and by Cox and Hutchinson[22] for *Deschampsia cespitosa.* These species' tolerances were examined using relative root growth of tillers of each individual in the presence and absence of cadmium (1.0–2.0 ppm) in water culture, a technique in common use for determining plant tolerances to a variety of trace metals[25,98,99].

Differential survival using soil spiked with cadmium has also been used by White and Rolfe[97] to identify differences in cadmium sensitivity between varieties of cottonwood. Given this diversity within species it becomes important to determine whether populations differ from each other and if these differences have a genetic basis. The significance of genetic differentiation between populations is that it may represent the accumulated judgments of natural selection on existing variation over many generations. The evolution of metal tolerance can occur in a few generations[94,101] given the requisite genetic variation and sufficient selection pressure. Differentiation in cadmium tolerance of populations exposed to different levels of cadmium contamination has been examined by Simon[85] in a zinc mine habitat, and by

Coughtrey and Martin[19] in a population contaminated by aerial pollution. Both studies concluded that population differentiation had occurred with regard to root cadmium tolerance in water culture. However, the cadmium tolerance reported in these two studies is of a similar magnitude to those reported in individual clones of another grass, *Deschampsia cespitosa,* sampled from two populations neither of which was contaminated with cadmium[24]. The combined frequency distribution of cadmium tolerance in these *D. cespitosa* populations seemed continuous rather than distinct or bimodal as in the case of combined distributions of tolerances to other metals from contaminated and control populations[11,24,38]. Distinct bimodal distribution of tolerances from tolerant and control populations may be attributed to major gene effects[56], whereas no such effect has been conclusively shown for cadmium tolerance as one is unable to ascertain this distribution of tolerance among individuals in bulk population samples of tillers in which the number of different genotypes or individuals is neither given nor estimated.

Cadmium contamination also presents a problem to the ecologist in that it never occurs alone in the environment under natural exposure conditions; it inevitably co-occurs with greater quantities of Pb and Zn. Possible evolutionary response to Cd must then be examined in the light of similar or greater responses to the co-occurring metals. Likewise, the cadmium-tolerance mechanism must be viewed in operation in the presence of excess and competing ions of these other metals. Although the former considerations were taken into account by Simon[85], he attributed the difference in the mean cadmium tolerance of the species populations studies to genetic adaptation to cadmium toxicity in the contaminated population. In these experiments, it was documented that the highest indices of tolerance to Cd were exhibited by plants that also had high tolerances to Zn. Tolerance of *Holcus lanatus* to the co-deposited metals from the Avon mouth smelters were

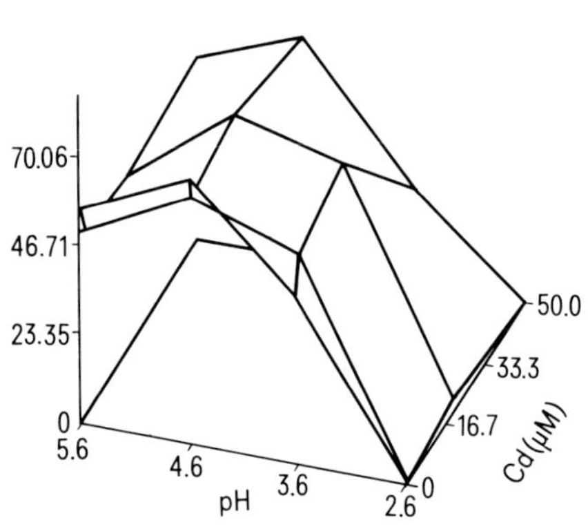

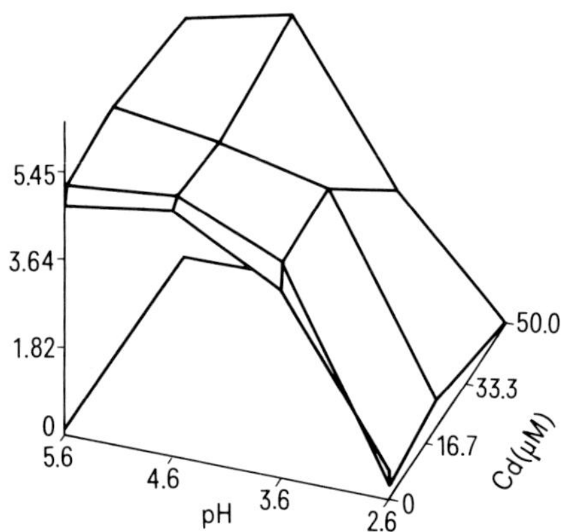

Figure 1. The effects of acidity of liquid culture media on *Trillium grandiflorum* pollen in the presence of different cadmium concentrations. (Means derived from 5 replicates).

Table 2. The interrelationships of tolerance indices (ti) for seven metals in clones of *Deschampsia cespitosa* samples from Coniston and from Hay Bay. Values are Spearman correlation coefficients (r_s)

	Cu	Ni	Al	Co	Pb	Cd	Zn
Cu		0.63**	− 0.43	− 0.38	0.08	0.06	0.03
Ni	0.48		0.08	− 0.03	− 0.03	0.10	0.12
Al	− 0.46	− 0.22		0.71**	− 0.19	− 0.14	0.08
Co	− 0.37	− 0.29	0.49		− 0.27	− 0.28	− 0.18
Pb	− 0.11	0.06	0.23	0.40		0.78**	0.75**
Cd	− 0.17	0.09	0.16	0.33	0.88**		0.88**
Zn	− 0.04	0.13	0.14	0.46	0.93**	0.94**	

(Upper triangle: *Smelter population*; lower triangle: *Control population*.)

* p < 0.05; ** p < 0.01.

examined by Coughtrey and Martin[20]. It was found that the *H. lanatus* populations previously studied for cadmium tolerance also had slightly elevated tolerances to Pb and Zn compared with an uncontaminated population. Factorial combinations of metals used for tolerance testing indicated significant increased tolerance of the contaminated populations to all permutations of Pb, Zn, and Cd except for that of Zn alone. The tolerances for the Cd/Zn treatment was higher than for Cd alone and it was suggested that there was antagonistic interaction of these metals on toxicity. The Pb, Zn, Cd permutation showed a higher tolerance index (Ti) than did any of the paired combinations, which indicated that the antagonistic interactions on root toxicity dominate where the three metals are mixed.

In the light of these observations, one may ask if cadmium toxicity is ever experienced in field conditions in populations that have adapted to overwhelming excesses of Zn and Pb ions. The population differentiation noted in cadmium tolerance may be an incidental increase or co-tolerance brought about by the evolution of a lead- and zinc-tolerance mechanism. Co-tolerances in *Deschampsia cespitosa* were examined by Cox and Hutchinson[22] in relation to population differentiation in Cu/Ni tolerance. Lead and zinc tolerance in the smelter population was incidentally increased with no increase in concentration of these elements in the populations' soil of origin. Correlations of the various tolerance indices examined in the smelter and control population of *D. cespitosa* (table 2) indicate the tolerances to Pb, Zn, and Cd are significantly (p < 0.01) positively intercorrelated in both populations. Scatter plots of the combined population for these intercorrelations are shown in figure 2, which suggest a greater increase in cadmium tolerance per unit increase in zinc tolerance in the control population than in the smelter population. In general, these observations demonstrate that a similar mode of toxicity and tolerance may exist for Cd, Pb, and Zn in both populations of *D. cespitosa*. Although cadmium tolerance was not incidentally increased along with those of Pb and Zn with the evolution of nickel and copper tolerances in the smelter population it is not hard to envisage such an incidental increase with the evolution of a zinc and lead tolerance mechanism. Such a co-tolerance has

been determined in animals[47] and was discussed by Cox[22] and Brown and Martin[12]. Furthermore, McGrath et al.[62] indicated that cadmium tolerance in *H. lanatus* may be increased in populations adapted to increased available aluminum. Co-tolerances in terms of phenotypic plasticity are discussed by Humphreys and Nicholls[31].

These observations indicate that population differentiation in cadmium tolerance may not necessarily be due to a genetic adaptation in response to increases in cadmium in the environment, it may be a result of preadaptation or co-tolerance. This argument also points to the dangers of circularity in logic in assuming that all population differentiation, in particular traits that appear to correlate with environmental gradients, is the direct result of natural selection.

Phenotypic plasticity of *Holcus lanatus* was investigated by Brown and Martin[12] who found evidence that low-level cadmium pretreatment of roots of tillers, from cadmium contaminated and control populations in water culture, increased the tolerance of tillers of both populations to higher cadmium dosages. This indicated that increased tolerance to Cd was induced. This is supported by the work of Baker[5] who reported that plants from the same tolerant population of *H. lanatus* lost their tolerance over time when kept in uncontaminated soil. However, it remains to be seen whether the increased cadmium tolerance in this species can be induced by pretreatment with Zn or Pb or Zn and Pb. A cadmium detoxification metallothionein mechanism has been shown to be induced by Zn in animals[47]. The discovery of similar proteins in plants by Rauser and Curvetto[72] supplies an attractive biochemical theory for the mechanism of metal tolerance and co-tolerance in plants.

More startling evidence of phenotypic plasticity in cadmium tolerance was found by Jackson et al.[34]. They also found cadmium tolerance to be induced in cell cultures of *Datura innoxia* by preexposure to the metal. Furthermore, cadmium tolerant cell lines were produced that tolerated 250 μM Cd in media by selection in successively higher concentrations of the metal. Such cloned cell lines remained tolerant through 400 generations of cells. Similarly, Bennetzen and Adams[8] selected cell lines in suspension culture which were tolerant to 1500 μM Cd, from relatively nontolerant progenitor cells of the wild tomato *Lycopersicon peruvianum*. Here the production of the cadmium-binding protein could be induced by copper ions as well as by cadmium. Thes observations may demonstrate that phenotypic plasticity may be reflected in the varia-

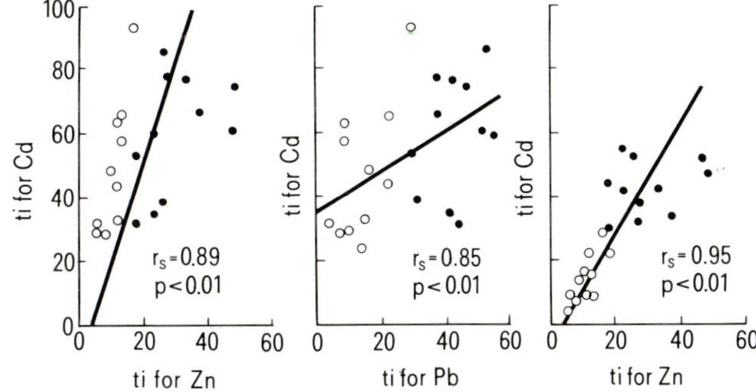

Figure 2. Scattergrams showing overall relationships between lead, cadmium and zinc tolerances in clones of *D. cespitosa* from Coniston (●) and clones from Hay Bay (○).

tion of tolerance in individual component cells of the plant which not only includes possible variation due to induction but also possible variation due to somatic mutation. The expression of adaphic tolerance of plants would then rely on the availability of tolerant or inducible cell lines in the exposed and growing root meristems, as, unlike cell suspension cultures, not all the cells in the plant tissues are equally exposed to the metal. Fixation of selectively neutral cell lines and selective loss of disadvantageous cell lines in the meristems of plants have been discussed by Klekowski and Kozarinova-Fukshansky[41,42]. The selective loss of cell lines in meristems presented in the array of adventitious root initials in grasses may explain the loss of cadmium tolerance of *H. lanatus* individuals noted by Baker[5] and may provide evidence that cadmium tolerant cell lines are a 'cost' to the plant in the absence of the metal.

1 Andersen, A., Hovmand, M. F., and Johnson, I., Atmospheric heavy metal deposition in the Copenhagen area. Envir. Pollut. *17* (1978) 133–151.

2 Andersson, A., and Nilsson, K. O., Influence of lime and soil pH on Cd availability to plants. Ambio *3* (1974) 198–200.

3 Ault, W. U. Senechal, R. G., and Erlebach, W. E., Isolopic composition as a natural tracer of lead in our environment. Envir. Sci. Technol. *4* (1970) 305–313.

4 Baes, III, C. F., and McLaughlin, S. B., Trace elements in tree rings: Evidence of recent and historical air pollution. Science *224* (1984) 494–497.

5 Baker, A. J. M., Environmentally-induced cadmium tolerance in the grass *Halcus lanatus* L. Chemosphere *13* (1984) 585–589.

6 Baker, A. J. M., Accumulations and excluders – strategies in response of plants to heavy metals. J. Plant Nutr. *3* (1981) 643–654.

7 Bartlett, L., Rabe, F. W., and Funk, W. H., Effects of copper, zinc and cadmium on *Selanastrum capricornutum*. Water Res. *8* (1974) 179–185.

8 Bennetzen, Z. L., and Adams, T. L., Selection and characterization of cadmium-resistant suspension cultures of wild tomato *Lycopersicon peruvianum*. Plant Cell Rep. *3* (1984) 258–261.

9 Bormann, F. H., Air pollution stress and energy policy, in: New England Prospects: Critical choice in a time of change, pp. 85. Ed. C. H. Reidel. University Press of New England, Hanover, New Hampshire 1982.

10 Boyle, R. M., and Jonasson, I. R., Geochemistry of cadmium, in: Effects of Cadmium in the Canadian Environment, pp. 15–21. Nat. Res. Counc. Can. NRCC No. 16743. 1979.

11 Bradshaw, A. D., McNeilly, T. S., and Gregory, R. P. G., Industrialization, evolution and the development of heavy metal tolerance in plants, in: Ecology and the Industrial Society, pp. 327–343. Br. Ecol. Soc. Sym. No. 5, 1965.

12 Brown, H., and Martin, M. H., Pretreatment effects of cadmium on the root growth of *Holcus lanatus* L. New Phytol. *89* (1981) 621–629.

13 Buchauer, M. J., Contamination of soil and vegetation near a zinc smelter by zinc cadmium and lead. Envir. Sci. Technol. *7* (1973) 131–135.

14 Burkitt, A., Lester, P., and Nickless, G., Distribution of heavy metals in the vicinity of an industrial complex. Nature, Lond. *238* (1972) 327–328.

15 Burnison, G., Wang, P. I. S., Chau, Y. K., and Silvergerg, B., Toxicity of cadmium to fresh water algae. Proc. 18th Ann. Meeting Can. Fed. Biol. Soc., Winnipeg, Manitoba, June 1975, p. 46.

16 Chaney, W. R., and Strickland, R. C., Relative toxicity of heavy metals to red pine pollen germination and germ tube elongation. J. envir. Qual. *13* (1984) 391–394.

17 Chizhikov, D. V., Cadmium, p. 263. Pergamon Press, New York 1966.

18 Cook, E. R., and Jacoby G. C. Jr, Tree-ring-drought relationships in the Hudson Valley, New York. Science *198* (1977) 399.

19 Coughtrey, P. J., and Martin, M. H., Cadmium tolerance of *Holcus lanatus* from a site contaminated by aerial fall out. New Phytol *79* (1977) 273–280.

20 Coughtrey, P. J., and Martin, M. H., Cadmium uptake and distribution in tolerant and non-tolerant populations of *Holcus lanatus* grown in solution culture. Oikos *30* (1978) 555–560.

21 Cowgill, U. M., The hydrogeochemistry of Linsle pond, North Branford, Connecticut II. The chemical composition of the aquatic macrophytes. Arch. Hydrobiol. Suppl. *45* (1974) 1–119.

22 Cox, R. M., Multiple tolerance relations in native plants and their application to reclamation. Proc. Int. Conf. Heavy Metals in the Environment, London, pp. 202–205. CEP Consultants, Edinburgh EH1 3QH, UK 1979.

23 Cox, R. M., Sensitivity of forest plant reproduction to acid rain, in: Proc. Int. Conf. Acid Rain and Forest Resources. Eds P.J. Rennie and G. Robitaille. Quebec City.

24 Cox, R. M., and Hutchinson, T. C., Metal co-tolerances in the grass *Deschampsia cespitosa*. Nature, Lond. *279* (1979) 231–233.

25 Cox, R. M., and Hutchinson, T. C., Multiple metal tolerances in the grass *Deschampsia cespitosa* (L.) Beauv. from the Sudbury smelting area. New Phytol. *84* (1980) 631–647.

26 Ernst, W. H. O., Physiology of heavy metal resistance in plants, in: Proc. Int. Conf. Heavy Metals in the Environment, vol. II, pp. 121–136. University Press Toronto, Canada 1975.

27 Friedland, A. J., Johnson, A. H., and Siccama, T. G., Trace metal content of the forest floor in the Green Mountains of Vermont. Spatial and temporal patterns. Water Air Soil Pollut. *21* (1984) 161–170.

28 Fritts, H. C., Tree rings and climate. Academic Press, London 1976.

29 Galloway, J. N., Thornton, J. D., Norton, S. A., Volchok, H. L., and McLean, R. A. N., Trace metals in the atmospheric deposition. A review and assessment. Atmos. Envir. *16* (1982) 1677–1700.

30 Goodman, G. T., and Roberts, T. M., Plants and soils as indicators of metals in the air. Nature, Lond. *231* (1971) 287–292.

31 Humphreys, M. O., and Nicholls, M. K., Relationships between tolerance to heavy metals in *Agrostis capillaris* L. (*A. tenuis* sibth.) New Phytol. *98* (1985) 177–190.

32 Hutchinson, T. C., Comparative studies of the toxicity of heavy metals to phytoplankton and their synergistic interactions. Water Pollut. Res. Can. *8* (1973) 68–90.

33 Hutchinson, T. C., and Czyrska, H., Cadmium and zinc toxicity and synergism to floating aquatic plants. Water Pollut. Res. Can. *7*, (1972) 59–65.

34 Jackson, P. J., Roth, E. J., McClure, P. R., and Naranjo, C. M., Selection, isolation and characterization of cadmium-resistant *Datura innoxia* suspension cultures. Plant Physiol. *75* (1984) 914–918.

35 Jana, S., and Choudhuri, M. A., Senescence in submerged aquatic angiosperms: effects of heavy metals. New Phytol. *90* (1982) 477–484.

36 Jeffries, D. S., and Snyder, W. R., Atmospheric deposition of heavy metals in central Ontario. Water Air Soil Pollut. *15* (1981) 127–152.

37 Jordan, M. J., Effects of zinc smelter emissions and fire on a chestnut-oak woodland. Ecology *56* (1975) 78–91.

38 Jowett, D., Some aspects of the genecology of resistance to heavy metal toxicity in the Genus *Agrostis*. Ph. d. Thesis, University College North Wales, Bangor UK 1959.

39 Kapur Arvind, and Malik, C. P., Effects of metabolic inhibitions on pollen germination and pollen tube growth of *Petunia alba*. Plant Sci. (Lucknow) *8* (1976) 26–27.

40 Kardell, L., and Larsson, J., Lead and cadmium in oak tree rings (*Quercus robur* L.). Ambio *7* (1978) 113–121.

41 Klekowski, E. J., Jr, and Kazarinova-Fukshansky, N., Shoot apical meristems and mutation: fixation of selectively neutral cell genotypes. Am. J. Bot. *71* (1984) 28–34.

42 Klekowski, E. J. Jr, and Kazarinova-Fukshansky, N., Shoot apical meristems and mutation: selective loss of disadvantageous cell genotypes. Am. J. Bot. *71* (1984) 28–34.

43 Kuleff, I., and Djingova, R., The dandelion *(Taraxacum officinale)* – a monitor for environmental pollution? Water Air Soil Pollut. *21* (1984) 77–85.

44 Lamoreaux, R. J., and Chaney, W. R., The effects of cadmium on net photosynthesis, transpiration and dark respiration of excised silver maple leaves. Physiol. Plant. *43* (1978) 231–236.

45 Landry, M. P., and Peel, D. A., Short term fluctuations in heavy metal concentrations in Antarctic snow. Nature, Lond. *291* (1981) 144–146.

46 Leavitt, S. W., Dueser, R. D., and Goodell, H. G., Plant regulation of essential and non-essential heavy metals. J. appl. Ecol. *16* (1979) 203–212.

47 Leber, A. P., and Miya, T. S., A mechanism for cadmium-and-zinc-induced tolerance to cadmium toxicity: Involvement of metal-lothionein. Toxic. appl. Pharmac. *37* (1976) 403–414.

48 Lepp, N. W., and Dollard, G. J., Studies in the behavior of lead in wood: Binding of free and complexed Pb to xylem tissue. Oecologia *16* (1974) 369–373.

49 Lindberg, S.E., and Harriss, R.C., The role of atmospheric deposition in an eastern U.S. deciduous forest. Water Air Soil Pollut. *16* (1981) 13–31.

50 Lindberg, S.E., Turner, R.R., Shriner, D.S., and Huff, D.D., Proc. Int. Conf. Heavy metals in the Environment, Amsterdam, September 1981, pp. 306–309. CEP Consultants Ltd., Edinburgh EH1 3QH, UK 1981.

51 Lisk, D.J. Trace metals in soil, plants and animals. Adv. Agron. *24* (1972) 267–325.

52 Little, P., A study of heavy metal contamination of leaf surfaces. Environ. Pollut. *5* (1973) 159–172.

53 Little, P., and Martin, M.H., A survey of zinc, lead and cadmium in soil and natural vegetation around a smelting complex. Envir. Pollut. *3* (1972) 241–254.

54 Little, P., and Martin, M.H., Biological monitoring of heavy metal pollution. Envir. Pollut. *6* (1974) 1–19.

55 Lymburner, D.B., The production, use and distribution of cadmium in Canada. Environmental Contaminants Control Study No. 2. Report Series No. 39. Canada Centre for Inland Waters Directorate, Burlington, Ontario, Canada 1974, pp. 71.

56 Macnair, M.R., The genetics of copper tolerance in the yellow monkey flower *Mimulus guttatus* 1:Crosses to non-tolerant. Genetics *91* (1979) 553–563.

57 Mankovska, B., The pollution of spruce *Pica abies* Karst. by emission of F, As, Pb, Cd, and S from an aluminium plant. Biologia (Bratislava) *34* (1979) 563–570.

58 Mankovska, B., The natural content of F, As, Pb, and Cd inforest trees. Biologia (Bratislava) *35* (1980) 267–274.

59 Mankovska, B., Contamination of beech and oak by Mg, S, F, Pb, Cd and Zn near a magnesite works. Biologia (Bratislava) *36* (1981) 489–496.

60 Marklinova, G. I., and Gileva, E.A., Accumulation of zinc[65] cadmium[115] and mercury[203] by freshwater plants and the effect of EDTA on accumulation coefficient of these radiosotopes. Tr. Inst. Ekol. Rast. Zhivotn., Ural. Filial, Akad, Nauk. SSSK *61* (1968) 72–78.

61 Mathys, W., Enzymes of heavy-metal-resistant and non-resistant populations of *Silene cucubalus* and their interaction with some heavy metals in vitro and in vivo. Physiol. Plant *33* (1975) 161–165.

62 McGrath, S.P., Baker, A.J.M., Morgan, A.N., Salmon, W.J., and Williams, M., The effect of interactions between cadmium and aluminium on the growth of two metal tolerant races of *Holcus lanatus* L. Envir. Pollut. (series A) *23* (1980) 267–277).

63 Nasu Y., Kugimoto, M., Tanaka, O., Yanaka, O., and Takimoto, A., Effects of cadmium and copper co-existing in the medium on the growth and flowering of *Lemna paucicostata* in relation to their absorption. Envir. Pollut. (series A) *33* (1984) 267–274.

64 Overnell, J., The effect of some heavy metal ions on photosynthesis in a fresh water algae. Pest. Biochem. Physiol. *5* (1974) 19–26.

65 Page, A.C., and Bingham, F.I., Cadmium residues in the environment. Res. Rev. *48* (1973) 1–44.

66 Page, A.L., Bingham, F.I., and Chang, A.C., Cadmium, in: Effects of Heavy Metal Pollution on Plants, pp. 77–110. Ed. N.W. Lepp. Applied Science Publ., London and New Jersey 1981.

67 Pakalne, D., Nallendorf, A.F., and Upites, V., Little investigated trace elements in *Chorella* cultures: cadmium. Chem. Abstr. *74* (1970) 31772c.

68 Pakarinen, P., and Tolonen, K., Regional survey of heavy metals in peat moss *(Sphagnum)*. Ambio *5* (1976) 38–40.

69 Parry, G.D.R., Goodman, G.T., and Smith, S., A simple technique for the monitoring of airborne heavy metals prior to revegetation, in: Environmental Management of Mineral Wastes, pp. 273–295. Eds G.T. Goodman and M.J. Chadwick. Sijthoff and Noordhoff International Publishers, BV Alphen aan. den Rijn, The Netherlands 1978.

70 Percy, K., Heavy metal and sulphur concentrations in *Sphagnum magellanicum* Brid. in the Maritime Provinces, Canada. Water Air Soil Pollut. *19* (1983) 341–350.

71 Pilegaard, K., Rasmussin, L., and Gydesen, H., Atmospheric background deposition of heavy metals in Denmark monitored by epiphytic cryptogams. J. appl. Ecol. *16* (1979) 843–853.

72 Rauser, W.E., and Curvetto, N.R., Metallothionein occurs in roots of *Agrostis* tolerant to excess copper. Nature *287* (1980) 563–564.

73 Roberts, T.M., Plants as monitors of airborne metal pollution. J. Envir. Plant Pollut. Control *1* (1972) 43–54.

74 Robitaille, G., Heavy metal accumulation in the annual rings of balsam fir *Abies balsamea* (L.) Mill. Envir. Pollut. (series B) *2* (1981) 193–202.

75 Rolfe, G.L., Lead distribution in tree rings. Forest Sci. *20* (1974) 283–286.

76 Ross, R.G., and Stewart, D.K.R., Cadmium residues in apple fruit and foliage following a cover spray of cadmium chloride. Can. J. Plant Sci. *49* (1969) 49–52.

77 Ruhling, A., and Tyler, G., Ecology of heavy metals – a regional and historic study. Bot. Notiser *122* (1969) 248–259.

78 Ruhling, A., and Tyler, G., Regional differences in the deposition of heavy metals over Scandinavia. J. appl. Ecol. *8* (1971) 497–507.

79 Ruhling, A., and Tyler, G., Heavy metal deposition in Scandinavia. Water Air Soil Pollut. *2* (1973) 445–455.

80 Scherbatskoy, T., The effect of acidic deposition in forest ecosystems in the Green Mountains of Vermont. Proc. Forest Issues Conferences, Pennsylvania State University 1984 (in press).

81 Scherbatskoy, T., and Bliss, M., Studies on tree cores and metal uptake. Unpubl. Rep., University of Vermont, 1982.

82 Schlesinger, W.H., Reiners, W.A., and Knopman, D.S., Heavy metal concentrations and deposition in bulk precipitation in montane ecosystem of New Hampshire, USA. Envir. Pollut. *6* (1974) 39–47.

83 Siccama, T.G., Bliss, M., and Vogelmann, H.W., Decline of red spruce in the Green Mountains of Vermont. Bull. Torrey Bot. Club *109* (1982) 162–168.

84 Simon, E., Heavy metal in soils, vegetation development and heavy metal tolerance in plant populations from metalliferous areas. New Phytol. *81* (1978) 175–188.

85 Simon, E., Cadmium tolerance in populations of *Agrostis tenuis* and *Festuca ovina*. Nature, Lond. *265* (1977) 328–330.

86 Stanley, R.A., Toxicity of heavy metals and salts to Eurasian water milfoil (*Myriophyllum spicatum* L.). Arch. envir. Contam. Toxic. *2* (1974) 331–341.

87 Steinnes, E., Atmospheric deposition of trace elements in Norway studied by means of moss analysis. Kjeller Report, 1977, KR 154 Institute for Atomenergi Kjeller, Norway.

88 Strickland, R.C., and Chaney, W.R., Cadmium influence on respiratory gas exchange of *Pinus resinosa* pollen. Physiol. Plant. *47* (1979) 129–133.

89 Symeonides, C., Tree-ring analysis for tracing the history of pollution. Application to a study in Northern Sweden. J. envir. Qual. *8* (1979) 484–486.

90 Tatsumi, Y., Yoda, K., and Ikeda, A., Effects of soil pollution by heavy metals on annual plants in Sakai City. Jap. J. Ecol. *33* (3) (1983) 293–303.

91 Thornton, I., Abrahams, P. and Mathews, H., Some examples of the environmental significance of heavy metal anomalies disclosed by the Wolfson Geochemical Atlas of England and Wales. Proc. Int. Conf. Heavy metals in the Environment, London, pp. 218–221. CEP Consultants Ltd., Edinburgh, EH13QH, U.K. 1979.

92 Tyler, G., Metals in sporophores of Basidiomycetes. Trans. Br. mycol. Soc. *74* (1980) 41–49.

93 Vallee, B.L., and Ulmer, D.D., Biochemical effects of mercury cadmium and lead. A. Rev. Biochem. *41* (1972) 91–128.

94 Walley, K.A., Khan, M.S.I., and Bradshaw, A.D., The potential for evaluation of heavy metal tolerance in plants. I. Copper and zinc tolerance in *Agrostis tenuis*. Heredity *32* (1974) 309–319.

95 Ward, N.I., Brooks, R.R., and Reeves, R.D., Effect of lead from motor-vehicle exhausts on trees along a major thoroughfare in Palmerton North New Zealand. Envir. Pollut. *6* (1974) 149–158.

96 Wedpole, K.L., Chemical fractionation in sedimentry environments, in: Origin and Distribution of Elements, p. 997. Ed. L.H. Ahrens. Pergamon Press, New York 1968.

97 White, T.A., and Rolfe, G.L., Differing effects of cadmium on two varieties of cottonwood *Populus deltoides* Bartr. Envir. Pollut. (series A) *22* (1980) 29–38.

98 Wilkins, D.A., A technique for the measurement of lead tolerances in plants. Nature, Lond. *180* (1957) 37–39.

99 Wilkins, D.A., The measurement of tolerance to edaphic factors by means of root growth. New Phytol *80* (1978) 623–633.

100 Williams, C.R., and Harrison, R.M., Cadmium in the atmosphere. Experientia *40* (1984) 29–36.

101 Wu, L., and Bradshaw, A.D., Aerial pollution and the rapid evolution of copper tolerance. Nature, Lond. *238* (1972) 167–169.

Part III: Cadmium and Human Health

Cadmium in foods and the diet

by J. C. Sherlock[1]

Ministry of Agriculture, Fisheries and Food, Great Westminster House, Horseferry Road, London SW1P 2AE (England)

Summary. Information on the sources of cadmium in food are presented and the effects of raised environmental levels of cadmium on the concentration of cadmium in plant based foods, fish and shellfish, meat and offals, and dairy produce are discussed. Information is also presented on normal dietary intakes of cadmium and how these intakes may be elevated by environmental pollution or atypical dietary habits. The estimation of dietary intakes of cadmium using data about extreme intakes of specific foods is described.

Introduction

Cadmium is naturally present in all parts of the environment; it is present in all soils and sediments at concentrations which are generally < 1 mg/kg[31] and its total concentration in unpolluted seawater, where it exists mainly as chlorocomplexes, is generally < 1 µg/kg[27]. The concentration of cadmium in air in non-industrialized areas rarely exceeds 2.5 ng/m³ which is equivalent to 3 ng/kg[5]. Consequently, all food, whether it be of plant or animal origin, is exposed to and contains cadmium. Unlike many other metals, cadmium has come to be used by man only relatively recently. It was identified as an element in 1817; its large scale use dates from the 1940s; and it is only in the last 3 decades that serious consideration has been given to cadmium as a food contaminant.

The use of cadmium may increase, in several ways, the extent to which it is found in foods: 1. deposition from the atmosphere onto crops growing near sources of cadmium emissions, for example smelters[2]; 2. discharge and deposition of cadmium in water and subsequent uptake by animals and other food grown in water[37]; 3. the use on land of phosphatic fertilizers which contain high concentrations of cadmium[22]; 4. the disposal to land or to sea of cadmium contaminated sewage sludge[17]; 5. the use of cadmium containing food contact materials, such as decorated glazed ceramic ware[14,16]; 6. the dispersion through mining activities of material richer in cadmium than the surrounding environment[4].

This paper discusses the concentrations of cadmium normally present in food and presents information showing how these concentrations are elevated by environmental cadmium pollution. Finally, information is presented about the dietary intake of cadmium with particular emphasis on the estimation of extreme intakes.

Cadmium in food and the impact of environmental pollution

For the purposes of simplicity, consideration of normal concentrations of cadmium in food and the impact that environmental pollution has on them has been divided into 4 sections. Each section deals with a specific class of food.

Plant based foods. Individual samples of plant-based foods grown in uncontaminated environments rarely contain more than 0.2 mg/kg on a fresh weight basis[26] (unless otherwise stated all cadmium concentrations in food are reported on a fresh weight basis) and average values for cadmium in specific foods are unlikely to exceed 0.1 mg/kg. Some root crops, such as carrot and parsnip, and some leafy crops, such as lettuce and spinach, tend to contain more cadmium than do other plant foods. The same is true of cereals, perhaps by virtue of their relatively high dry matter content. In contrast, fruits and fruit juices contain uniformly low concentrations of cadmium. Evidence in support of this is presented in table 1 which compares normal concentrations of cadmium in plant-based foods with those found in foods grown in soils containing more than 1 mg/kg of cadmium. It has been assumed that the concentration of cadmium in the soils in which the normal crops were grown was < 1 mg/kg. The data demonstrate that, unlike the case with lead[6], plants tend to take up and translocate cadmium from soil. The process by which this happens is not well understood and much effort is currently being expended on research into this problem.

Cadmium pollution of the soil often arises because of

Table 1. Cadmium in crops grown on land containing elevated concentrations of cadmium compared with cadmium in normal crops

Crop	Mean concentration of cadmium in crop (mg/kg, fresh weight basis)	
	Crops grown on land with elevated cadmium concentration[a]	Normal crops
Lettuce	0.18 (50)[b]	0.06 (17)
Cabbage	0.04 (45)	0.01 (22)
Spinach	0.63 (11)	0.08 (4)
Carrot	0.15 (2)	0.05 (13)
Potato	0.14 (7)	0.03 (20)
Plums	0.01 (7)	< 0.02 (5)
Corn grain	0.06	0.03

[a] Soils containing greater than 1 mg/kg of cadmium.
[b] Number of samples in parenthesis.

excessive applications of sewage sludge or because of the dispersion of mine spoil to the environment. There is also some evidence[22] to show that the use of phosphatic fertilizers, all of which contain some cadmium, has caused an increase in the cadmium content of cereal crops. Analysis of aged samples of cereals did not indicate that there had been a marked change in the cadmium content of cereals grown in the United Kingdom (UK) over the last century[18]. The current awareness about the risks to health from cadmium has done much to reduce emissions to the air and to reduce the dispersal of cadmium-contaminated industrial waste to the environment. However, cadmium-containing effluents can still be discharged into the sewers and the consequent treatment of sewage results in the concentration of cadmium in sewage sludge. Much of the sewage sludge produced in the UK is desposed of on land where it provides a source of organic matter and plant nutrients. Many countries have recommended maximum limits to regulate the addition of cadmium in sewage sludge to agricultural land, and in the UK the level is not to exceed more than 5 kg/ha over 30 years[10]. It is only by controlling the discharge of cadmium into the sewers that limitation of future increases in the cadmium content in the soil and, therefore, in food, can be achieved.

Fish and shellfish. One of the problems encountered in monitoring fish for cadmium is that most species contain so little cadmium that it is difficult to determine accurately the concentration of cadmium without resorting to sophisticated and time-consuming analytical techniques and without taking great care to avoid adventitious contamination. Table 2 presents information gathered as part of a routine monitoring exercise carried out in the UK. The samples of fish were taken from both distant water and coastal water about the UK. The limits of detection varied according to the analytical techniques employed, but it is clear that the average concentration of cadmium is certainly less than 0.2 mg/kg and there is evidence[19] to indicate that the cadmium concentrations in fish are often < 0.005 mg/kg. Shellfish contain higher concentrations of cadmium than do most other foods. With exceptions of lobster, whelks and

crabs, shellfish taken from unpolluted waters rarely contain an average cadmium concentration in excess of 1 mg/kg, although individual samples sometimes contain more than this. In contrast, whelks, the body meat of lobsters and, especially, brown body meat of crabs often contain an average cadmium concentration of more than 1 mg/kg. Pollution of the marine environment, for example by discharge of cadmium-containing effluents to rivers and estuaries, appears to have resulted in increased concentrations of cadmium in shellfish[36,37] but not fish.

Meat and offal. The concentration of cadmium in meat, other than offal, is uniformly low, average concentrations being < 0.05 mg/kg. Animal offal, especially liver and kidney, generally contains an average cadmium concentration in excess of 0.05 mg/kg; individual samples of kidney often contain more than 0.5 mg/kg of cadmium. This is not surprising since the kidneys and, to a lesser extent, the liver of animals, including man, accumulate about 65% of the cadmium absorbed by the body and the major area of concern about the effects of cadmium on man is because of its nephrotoxicity[32].

Animals grazing on land contaminated by cadmium, or consuming fodder grown on contaminated land yield meat which contains normal or slightly elevated concentrations of cadmium[17,33]. However, the liver and kidneys from animals consuming elevated amounts of cadmium contain significantly more cadmium than is usual and it is prudent that this offal should not be consumed.

Dairy produce and other foods. Milk, cheese, butter, oils and fats, and eggs contain uniformly low concentrations of cadmium. Average concentrations of cadmium in butter, cheese, lard and margarine do not exceed 0.05 mg/kg and individual samples rarely contain more than 0.1 mg/kg[26]. Normal average concentrations of cadmium in cow milk are generally less than 0.005 mg/l[12] and in eggs the average concentration is 0.01 mg/kg[21]. Milk from cows whose intake of cadmium is high does not appear to contain elevated levels of cadmium[33]. It has been reported that eggs are unlikely to contain more than 0.05 mg/kg of cadmium unless the feed contains more than 13 mg/kg[9]. Supporting evidence for this was found at the English village of Shipham, where previous mining activity has caused localized elevated cadmium levels in soil. Eggs from chickens and ducks at Shipham contained < 0.01 mg/kg of cadmium[34].

Dietary intakes of cadmium

Normal exposure to dietary cadmium. Before the dietary intake of any chemical can be estimated, it is essential to know the weight of the diet and the concentration of the chemical in the diet. In some countries the weight and types of food in the average diet are determined each year by extensive surveys among the population. In Great Britain, for example, the weight and composition of an average persons diet are derived from data gathered by the National Food Survey[30] in which the weekly food acquisitions of more than 7000 households are determined. In other

Table 2. Cadmium in fish and shellfish

Fish	Mean cadmium concentration (mg/kg)
Bass	0.2 (14)[a]
Cod	0.16 (343)
Herring	0.18 (171)
Monkfish	0.02 (10)
Whiting	0.17 (360)
Sea trout	0.03 (68)
Salmon	0.03 (17)
Pink shrimps	0.2 (7)
Winkles	0.2 (1)[b]
Whelks	1.8 (4)
Crabs (brown meat)	4.3 (43)
Crabs (white meat)	0.2 (44)

[a] Number of samples in parenthesis.
[b] Analysis made on 250 samples bulked together.

countries a hypothetical diet may be designed for a person who is considered likely to have the highest intake of food and, therefore, a high intake of food contaminants. This is done, for example, in New Zealand[11]. The concentrations of the contaminants in the diet may be determined by the analysis of individual foods[28], or by the analysis of groups comprising like foods, for example leafy vegetables[20], or by the analysis of whole diets prepared by institutions such as hospitals[3].

The weight of adult diets used in the estimation of intakes of contaminants in various countries usually lies between about 1.5 and 3.5 kg; this figure refers to the weight of food and beverages consumed in 1 day. The weight taken as representative of total intake can account for at least a 2-fold variation in the estimated contaminant intake from country to country. Furthermore, both the accuracy of the analysis of the diets and the limit of determination of the analytical methodology is critical if intakes are to be reliably determined. If the limit of determination for cadmium was as high as 0.1 mg/kg and findings at or below this level were considered to be 0.1 mg/kg for the purpose of estimation, then the minimum estimated daily intake might be 0.15 mg/kg (0.1 mg/kg × 1.5 kg). Generally, limits of determination for cadmium in food, milk and beverages lie between 0.002 and 0.05 mg/kg depending upon the analytical method and the food matrix; the lower limit of determination is usually achieved for milk and beverages but not for other food. Table 3 presents typical information on the dietary intakes of cadmium in various countries but must be viewed within the constraints outlined above. The WHO/FAO Joint Expert Committee on Food Additives has recommended a provisional tolerable weekly intake (PTWI) for cadmium which, for adults, is 0.4–0.5 mg (0.06–0.07 mg/day), based on a tolerable intake of 1 µg/kg b.wt/day for a 60–70 kg adult[38]. For children the tolerable intake is less. The PTWI concept is akin to the acceptable daily intake (ADI) concept for a chemical which was defined as the 'daily intake which, during an entire lifetime, appears to be without appreciable risk of the basis of all the known facts at the time'[38]. In respect of the PTWI, the word 'provisional' expresses the tentative nature of the evaluation and the word 'tolerable' signifies permissibility rather than acceptability since the intake of the contaminant is unavoidably associated with the consumption of otherwise wholesome and nutritious foods. All of the average daily intakes presented in table 3 are below the PTWI, although four[3,15,23,29] appear to be close to it.

It may be inferred from the data in table 3 and from the meaning of the PTWI that the 'average' consumer in the countries for which data are presented is not endangered by cadmium in the diet. Nevertheless it is prudent to try to keep exposure as low as possible by minimizing the output to the environment from controllable sources.

Abnormal exposure to dietary cadmium. Some individuals have consistently higher intakes of cadmium because they have atypical dietary habits or because the food they eat is produced or grown in areas suffering from cadmium pollution or because of both of these reasons. When such circumstances have been revealed special dietary studies or other estimates of intake have often been undertaken. Where practical, these studies have been coupled with a medical examination of the exposed population and entail, for example, measurements of blood cadmium and B_2-microglobulin excretion. These studies are used to provide information on which a judgement can be made about the risks to health of the exposed population from dietary cadmium. In the UK the dietary intake of cadmium by people consuming food grown in cadmium-polluted areas is generally assessed by means of a duplicate diet study[7]. This involves the collection of exact replicates of all the food consumed in the course of 1 week by a sample of individuals from the exposed population. These studies are difficult and costly to undertake, and rely heavily on the ability of the participants to ensure that there will be no discrepancy between the amount of food they actually eat and the amount of food claimed to be exact replicates of their diet. Practical experience in the UK shows that the average weight of duplicate diets is about 25% less than would be expected on basis of data gathered from the National Food Survey. Instances of severe cadmium pollution are rare although recent investigations have been made at the village of Shipham in Somerset, England[34]. Some samples of Shipham soil contained more than 100 mg/kg of cadmium, and some vegetable samples contained more than 1 mg/kg of cadmium. The cadmium intakes of the study sample as determined by a duplicate diet study were about double the estimated UK intake and some individuals exceeded the PTWI for cadmium. There was no evidence from the medical checks that any of the present residents of Shipham had suffered adverse health effects related to cadmium.

Estimation of cadmium intakes by individuals who have atypical dietary habits is extremely difficult because, in contrast to people consuming food grown on contaminated land, the individuals cannot be easily identified. For example, regular consumption of liver will increase the dietary intake of cadmium, but individuals who consume abnormal amounts of liver are not easy to identify by, for example, their place of residence. In these circumstances it is usual to estimate intake by assuming a high consumption of the food and multiplying this by the average concen-

Table 3. Typical dietary intakes of cadmium

Country	Intake (mg/week)	Reference
Germany	0.40	3
Germany	0.20[a]	35
Poland	0.13[a]	23
Japan	0.27	28
New Zealand	0.11	11
Australia	0.15	8
Belgium	0.35	15
Denmark	0.21	24
Italy	0.38	29
USA	0.23	20
Great Britain	< 0.15	25

[a] Intakes by children.

tration of cadmium (or any other contaminant) in the food. Investigations in the UK have shown[7] that high consumptions of individual foods may be related to the average consumption by the consuming population using simple equations. Thus the 90th percentile consumption (y in g/week) of a food has been found to be related to the average consumption (x) by:

$$y = 1.6\,x + 60.5$$

If this relationship holds widely it means that, once an extreme consumption is defined in terms of the percentile of the consuming population, and the average consumption of the food is known, then a high intake of the contaminant may be estimated using data on the concentration of cadmium in the food and the calculated extreme consumption. More recent investigations by the author[33*] indicate that the frequency distribution of consumption of individual foods for a wide range of foods is approximately lognormal. The logarithmic standard deviation of the distributions are similar for a wide range of foods as are the ratios of the arithmetic to geometric mean for each distribution. This finding explains to some extent the relationships described by Coomes et al.[7]. For example the logarithmic mean consumption of liver in the UK is equivalent to 0.1 kg/week and the logarithmic standard deviation of the frequency distribution is 0.257. Using these data it may be shown that the 97.5th percentile consumption in any one week is 0.32 kg. The mean cadmium content of liver sold in the UK is about 0.1 mg/kg, so that cadmium intakes from liver in the UK will rarely regularly exceed 0.03 mg/week (0.32 mg × 0.1 mg/kg = 0.03 mg), which is well within tolerable limits. Obviously, more liver may be consumed in other countries where the population has different dietary habits and similar differences will apply to other foods.

Conclusions

In general, the concentrations of cadmium in food are low although some foods of minor dietary importance, such as shellfish or kidney, often contain cadmium concentrations greater than 0.5 mg/kg. The concentration of cadmium in many foods can be increased due to environmental cadmium pollution. At present, dietary intakes of cadmium in most countries are generally less than the PTWI but emissions of cadmium into the environment are continuing. While control of cadmium emissions may decrease the rate of emission, the dispersion of cadmium into the environment will inevitably continue as long as emissions take place. For this reason there is a clear need for continued monitoring of cadmium in food and the environment, coupled with research into the effects on man of low level exposure to cadmium. This research should take into account other important sources of exposure to cadmium. There is, for example, very strong evidence[13] to show that the total body burden of cadmium and the concentration of cadmium in the liver and kidneys of smokers is significantly higher than for non-smokers.

1 The author wishes to thank the Ministry of Agriculture, Fisheries and Food for permission to publish this paper.

2 Averman, E., and Dassler, H.G., Cadmium content of vegetable foods in the effective range of a lead smelting plant. Die Nahrung *23* (1979) 875.

3 Barchet, R., and Wilk, G., Untersuchung zur täglichen Aufnahme von Cadmium durch die Nahrung mit Hilfe der Emissionsspektralanalyse. Dt. Lebensmitt. Rdsch. *76* (1980) 348–351.

4 Carruthers, M., and Smith, B., Evidence of cadmium toxicity in a population living in a zinc mining area. Lancet *1* (1979) 845–847.

5 Cawse, P.A., A survey of atmospheric trace elements in the UK results for 1976. United Kingdom Atomic Energy Authority, Report No. AERE R 8869, 1977.

6 Chumbley, C.G., and Unwin, R.J., Cadmium and lead content of vegetable crops grown on land with a history of sewage sludge application. Envir. Pollut. *4* (1982) 231–237.

7 Coomes, T.J., Sherlock, J.C., and Walters, C.B., Studies in dietary intake and extreme food consumption. R. Soc. Hlth J. *102* (1982) 119–23.

8 Commonwealth of Australia, Report of the 87th Session of the National Health and Medical Research Council; Market Basket (noxious substances) Survey 1977. Australian Government Publishing Service, Canberra A.C.T. 1979.

9 Crossmann, G., and Seifert, D., Contamination of soil in agriculture areas by sewage sludge and aspects of transfer and carryover to plants and farm animals, Cadmium 81. Proceedings, Third International Cadmium Conference, Miami, USA. Cadmium Association, London 1982.

10 Department of the Environment, Report of the sub-committee on the disposal of sewage sludge to land. Department of the Environment, National Water Council, Standing Technical Committee Reports No. 20. London 1981.

11 Dick, G.L., Hughes, J.T., Mitchell, J.W., and Davidson, F., Survey of trace elements and pesticide residues in the New Zealand Diet. N.Z. Jl Sci. *21* (1978) 57–69.

12 Dorn, C.R., Pierce, J.O., Chase, G.R., and Phillips, P.E., Cadmium, copper, lead and zinc in blood, milk, muscle and other tissues in cattle from an area of multiple source contamination. Trace Substances in Environmental Health; Proc. of the 7th Annual Conference. Ed. D.D. Hemphill, University of Missouri, 1973.

13 Ellis, K.J., Vartsky, D., Zanzi, I., Cohn, S.H., and Yusumura, S., Cadmium: in vivo measurements in smokers and non-smokers. Science *205* (1979) 323–325.

14 Engberg, A., and Bro-Rasmussen, F., Study on information already available in the literature on food contamination caused by lead and cadmium in ceramic household containers. CEC Document No. V/F/3799/74e, Luxembourg, 1975.

15 Fouassin, A., and Fondu, M., Evaluation de la teneur moyenne en plomb et en cadmium de la ration alimentaire en Belgique. Arch. belg. Med. soc. *38* (1980) 453–467.

16 Friberg, L., Piscator, M., Nordberg, G., and Kjellstrom, T., Cadmium in the Environment. 2nd ed. Chem. Rubber Co., Press, Cleveland, Ohio, 1974.

17 Heffron, C.L., Reid, J.T., Elfving, D.C., Stoewsand, G.S., Hascheck, W.M., Telford, J.H., Furr, A.K., Parkinson, T.F., Bache, C.A., Gutenmann, W.H., Wszolek, P.C., and Lisk, D.J., Cadmium and zinc in growing sheep fed silage corn grown on municipal sludge amended soil. J. agric. Fd Chem. *28* (1980) 58 61.

18 Hubbard, A.W., and Sherlock, J.C., Legislation and surveillance for heavy metals in the UK with special reference to cereals. Proceedings of the 10th International Cereals Congress, Vienna, May 7–9, 1980.

19 International Council for the Exploration of the Sea (ICES) Coop. The ICES co-ordinated monitoring programmes 1975 and 1976. Res. Report *72*, 1977.

20 Johnson, R.D., Manske, D.D., and Podrebara, D.S., Pesticide, metal and other chemicals residues in adult total diet samples, August 1975–July 1976. Pestic. monit. J. *15* (1981) 54–69.

21 Kirkpatrick, D.C., and Coffin, D.E., Trace metal content of chicken eggs. J. Sci. Fd Agric. *26* (1975) 99–103.

22 Kjellstrom, T., Lind, B., Linnaman, L., and Elinder, C.G., Variation of cadmium concentration in Swedish Wheat and Barley. Archs envir. Hlth *30* (1975) 321–328.

23 Koktysz, N., and Bulinski, R., Lead, mercury and cadmium

content in daylong meals for children. Bromat. Chem. Toksy-kol, *13* (1980) 215.

24 Miljøministeriet: Cadmium forurening. En redegotelse om and vendelse, forekomst og skad evirkninger at cadmium i Dan-mark. Miljøministeriet: miljostyrelsen, Strandgade 29, 1401 Kobenhavn K, 1980.

25 Ministery of Agriculture, Fisheries and Food: A survey of cadmium in food: first supplemantary report. 12th Report of the Steering Group on Food Surveillance, HMSO, London 1983.

26 Ministery of Agriculture, Fisheries and Food: Survey of cadmium in food. Working Party on the Monitoring of Foodstuffs for Heavy Metals, 4th Report, HMSO, London 1973.

27 Mohlenberg, F., and Jensen, A., The ecotoxicology of cadmium in fresh and seawater, and water pollution with cadmium in Denmark. The National Agency of Environmental Protection, Denmark, 1980.

28 Nishihara, T., Shimamoto, T., Kando, M., Ando, Y., Kushida, I., Iio, T., Onosaka, T., Tanaka, K., and Yokoyama, T., Cadmium contents in foods and the estimation of cadmium uptake from foods. Eisei Kagaku *25* (1979) 346.

29 Orlando, P., Perdelli, F., Franco, J., and Pietrini-Pallotta, A., Content of trace elements in prepared foods and total diets. IV Cadmium. G. Igiene Med. prevent. *18* (1977) 76–84.

30 Peattie, E. M., Buss, D. H., Lindsay, D. G., and Smart, G. A., Reorganisation of the British Total Diet Study for monitoring food constituents from 1981. Fd chem. Toxic. 1983, to be published.

31 Peterson, P. J., and Alloway, B. J., in: The chemistry, biochemistry and biology of cadmium, chapt. 2, pp.45–92. Ed. M. Webb. Elsevier/North-Holland Biomedical Press, Amsterdam 1979.

32 Piscator, M., in: The chemistry, biochemistry and biology of cadmium, chapt. 2. Ed. M. Webb. Elsevier/North-Holland Biomedical Press, Amsterdam 1979.

33 Sharma, R. P., Street, J. C., and Verma, M. P., Cadmium uptake from feed and its distribution to food products of livestock. Envir. Hlth Perspect. *28* (1979) 59–66.

33* Sherlock, J. C., and Walters, C. B., Dietary intake of heavy metals and its estimation. Chem. Ind. 4th July 1983, pp.505–508.

34 Shipham Survey Committee, Soil contamination at Shipham; report on studies completed in the village and advice to residents, December 1980. Department of the Environment, London 1980.

35 Stolley, H., Kersting, M., and Droese, W., Intake of trace elements and heavy metals with the diet of 2–14 year old children. Mschr. Kinderheilk. *129* (1981) 233–238.

36 Ward, T. J., The distribution of cadmium in shallow marine sediments, flora and fauna near a lead smelter, Spencer Gulf, South Australia. Cadmium 81. Proceedings, 3rd International Cadmium Conference, Miami, USA. Publ. Cadmium Association, London 1982.

37 Weidow, M. A., O'Connor, J. M., Hazen, R., Sloan, R., and Korcher, R., Cadmium concentrations in tissues of Hudson River blue crabs (Calinectes sapides). Bull envir. Contam. Toxic. 1982 (in press).

38 World Health Organisation, Evaluation of mercury. lead, cadmium and the food additives amaranth, diethylpyrocarbonate and octyl gallate. WHO Food Additive Series No. 4. World Health Organisation, Geneva. 1972.

Cadmium in human population

by A. Bernard[*] and R. Lauwerys

Département de Médecine et Hygiène du Travail, Université Catholique de Louvain, 30.54 Clos Chapelle-aux-Champs, B–1200 Bruxelles (Belgium)

Introduction

Due to man's activity, trace metals are slowly being redistributed in the environment. During the last decades, concentrated metal deposits that are confined in the earth's crust and which are usually harmless to living beings, have been exploited at an increasing rate and discharged partly in the environment. Among these metals, cadmium has raised the most concern because of its high toxicity coupled with an exceptional tendency to accumulate in the human organism.

Although the acute toxicity of cadmium on the lung and the gastrointestinal tract has been known for a long time[91,102], it was only in the 1950's that one first became really aware of the danger resulting from long term exposure to this metal.

In 1942, Nicaud et al.[66], when examining French workers in an alkaline accumulator factory, found several cases of severe osteoporosis with pseudofractures of the bones, associated with an impairment of the general health. A few years later, in Sweden, Friberg[25,26] conducted a toxicological study on workers exposed to CdO dust in an electrical battery plant. He reported emphysema and renal damage characterized by a proteinuria rich in low molecular weight proteins.

However, the greatest concern over cadmium pollution was triggered when it appeared that chronic cadmium poisoning was not restricted to industrial workers, but could constitute a health hazard to the general population. In Japan, contamination of water and rice by cadmium was responsible, in combination with other nutritional factors, for the outbreak of a severe bone disease (Itai-Itai disease)[30]. In industrialized countries, the cadmium body burden of the general population has increased during the 20th century[18,94]. In some population groups, the mean concentration of cadmium in the kidney cortex at the age of 50 is only 2–5 times lower than the critical level for renal damage[27,100]. The potential health effects of the current environmental pollution by cadmium remain largely to be clarified. Whether cadmium plays a role in the occurrence of frequent diseases, such as hypertension or various types of cancer remains a matter of controversy.

The aim of this review is to summarize our current knowledge on the metabolism and toxicity of cadmium in man. Because this field is so large and continues to develop rapidly, a comprehensive review of all published data is impossible. For more details, the reader is referred to several monographs on this subject[27,51,69,101].

Exposure

Working environment

In industry, cadmium enters the organism mainly by inhalation. However, oral intake, namely through contaminated hands, may sometimes be significant. Smoking at one's place of work may also increase the cadmium intake.

Exposure to cadmium may occur in conjunction with a number of industrial processes: production of cadmium and its compounds, fabrication of alloys and solders, plating of metals, use of cadmium in pigments and stabilizers for plastics, fabrication of Cd-Ni batteries, etc. Because of the relatively high volatility of this metal (boiling point: 765 °C), the hazard is particularly high for people welding cadmium-plated materials.

The total airborne cadmium concentration may vary considerably depending on the type of industry and also on the efforts made to reduce the emissions. But, usually, values currently observed in cadmium-using or producing factories range from 5 to 200 µg Cd/m^3. In the past, and also in plants where cadmium emissions are not yet controlled, values as high as 10,000 µg Cd/m^3 have been reported[26,55].
It is likely that cadmium constituted a health hazard long before its industrial use and even before its discovery. Since cadmium occurs in nature together with zinc, lead and copper, the use of these metals over several thousand years, has necessarily been a source of human exposure to cadmium. Some toxic effects (such as proteinuria and emphysema) attributed in the past to lead[87], were probably caused by cadmium. Occupational exposure to cadmium is limited in space and is normally amenable to adequate control by well-designed preventive measures. Furthermore, it concerns a limited number of adult workers, who are in good clinical health and are kept under medical surveillance. The situation is different for the general population, which is continuously exposed to moderate doses of a widely dispersed contaminant, and certain groups may be more susceptible than others to the toxicity of cadmium.

General environment

For non-smokers, food constitutes the principal environmental source of cadmium. Basic foodstuffs such as wheat or rice can accumulate relatively high amounts of cadmium when grown on a cadmium-polluted soil. Internal organs of mammals, such as liver and kidneys, may also contain high amounts of cadmium. The daily intake depends on both the dietary habits and the concentration of cadmium in foodstuffs. In the United States, it has recently been calculated that, even assuming an increase of cadmium levels in food between 1945 and 1975, the intake of cadmium has remained fairly constant because of profound changes in eating habits[98]. On the other hand, in Japan, where rice represents the basic staple food, the cadmium intake has, at least in some areas, drastically increased as a result of cadmium pollution.

Friberg et al.[27] have estimated the daily intake of cadmium in an uncontaminated area to be 25–60 µg for a 70 kg person. Values of 10–30 µg/day in the United Kingdom[64] and of 26–61 µg/day in the United States[60] have been reported.
The more recent estimates are lower. A mean daily intake of 16 µg Cd was found in Sweden[20] and a very similar value was obtained in Belgium (15 µg/day)[9].
The most dramatic example of detrimental health effects resulting from environmental exposure to cadmium is the Itai-Itai disease, which broke out in Japan during the early half of the 20th century. This unusual disease was characterized by pain and bone fractures (hence the name of the disease 'Ouch-Ouch'), proteinuria, aminoaciduria and severe osteomalacia. It was first diagnosed among the inhabitants in the Jintzu River basin in Toyama Prefecture[30]. The patients had been exposed to cadmium as a result of contamination of water and rice by discharges from a zinc mine. The average daily intake of cadmium in the endemic area was estimated at 500–800 µg whereas outside the area, the intake amounted to about 30–50 µg[28]. Several other areas in Japan have been found polluted by cadmium[49,67,83]. In these areas, the incidence of tubular protenuria in the general population is significantly higher than in non-polluted areas.

Excessive environmental exposure to cadmium is, however, not restricted to Japan. In 1979, a national geological survey carried out in England indicated a substantial contamination of soil by cadmium at Shipham, a Somerset village. Cadmium originated from nearby extinct calamine workings. The cadmium level in the liver of inhabitants was found to be on the average 5 times higher than in control subjects[33]. After examining 32 residents of the village, Carruthers and Smith[14] concluded that some of the abnormal biological findings observed (namely at the kidney level) could be attributed to cadmium. Their conclusion is certainly premature since the study was based on volunteers and did not include a control group. At about the same time, studies were carried out on the general population in Belgium; these suggested that industrial pollution by cadmium in one area of this country may exacerbate the age-related decline of the renal function, and even increase the proportional mortality rate from nephrosis and nephritis[52,53].

In a recent mortality study carried out at Shipham[39], slight increases in the standardized mortality ratios from nephritis, genito-urinary disease, hypertensive and cardiovascular disease were found by comparison with a nearby control village. However, no definitive conclusion can be drawn from this study, because of the small size of the population examined and of the low number of deaths from specific causes.

Tobacco

Tobacco smoking, too, may significantly increase man's exposure to cadmium. Cigarettes contain generally 1–2 µg cadmium, of which about 10% is absorbed by the lung. Hence, heavy smokers have at age 50–60 a cadmium body burden on the average twice higher than that of non-smokers[21,57].

Metabolism

The main feature of cadmium metabolism is its very long biological half-life and hence its progressive accumulation in the organism. Evolution seems to have provided man with no effective homeostatic mechanism to deal with an increasing intake of this heavy metal. No metabolic pattern allowing its elimination from the organism has evolved. Although the synthesis of metallothionein – an intracellular cadmium-binding protein – corresponds to a defence mechanism, it is also responsible for the selective accumulation of cadmium in the kidney and thus indirectly for its toxic effect on that organ.

Absorption

Inhalation. Depending on the size of the particles, 10–50% of the inhaled cadmium is deposited in the alveolar compartment of the lungs. A part of the deposited cadmium is eliminated by the lung clearance mechanisms. The absorption of the retained cadmium depends on its chemical form. For CdO, it has been estimated at about 60%[51].

Ingestion. In man, the average oral absorption of cadmium amounts to about 5%[63,76]. However, the absorption by the gastrointestinal tract – the main route of exposure for the general population – is influenced by several nutritional factors.

Animal experiments have shown that a low intake of calcium and protein may considerably increase the intestinal absorption of cadmium[50,93]. It is very likely that the low calcium and protein intake in Japan at the end of World War II contributed to the accumulation of cadmium in Itai-Itai disease patients. The study of Flanagan et al.[24] has demonstrated that subjects with low iron stores (low ferritin levels in serum) absorb significantly more cadmium (up to 15% of ingested cadmium) than persons with normal iron stores. Oral absorption of cadmium is therefore higher in females than in males.

It should be stressed that, independent of the absorption rate, age may also influence the cadmium intake through food. Since the greatest caloric intake occurs between the ages of 10 to 20 years, it can be expected that the rate of cadmium accumulation in the organism is the greatest in this age group. The amount of cadmium absorbed via food range from about 0.2 to 5 μg/day[51,69]. However, in view of the nutritional factors mentioned above, it is clear that the intake may be much higher in certain groups of the general population.

Distribution. In blood, more than 90% of cadmium is found in cells. In vitro experiments[35] have shown that red blood cells accumulate little cadmium, whereas lymphocytes – which synthetize metallothionein – can concentrate cadmium at a level 3000-fold greater than in the cadmium medium.

The two main sites of storage of cadmium are the liver and kidneys. The amount of cadmium accumulated in the liver and kidneys represents about 50% of the total body burden. In the kidney, the concentration of cadmium in the cortex is about 1.5 times higher than in the whole organ[58].

The concentration of cadmium in the liver increases continuously with age; that in the kidney increases also with age at least until age 50–60. From that age, cadmium concentration in kidney levels off and/or decreases. Several explanations have been proposed for this phenomenon, e.g. the effect of age-related functional and morphological changes in the renal cortex[98], an effect due to a lower environmental pollution by cadmium in the past and a decreased food intake with age.

The cadmium body burden of non-occupationally exposed adults in Europe and the United States ranges from 5 to 40 mg[51]. Smokers have on the average a higher cadmium body burden than non-smokers. Ellis et al.[21] have measured by the neutron activation technique the absolute amounts of cadmium stored in the kidney and the liver of 20 adult male Americans. They estimated that the average body burden of cadmium at age 50 is 19.3 mg for non-smokers and 35.5 mg for smokers.

Many studies were carried out to measure the concentration of cadmium in the human kidney cortex. The highest values at age 50 were observed in Japan (60–120 ppm)[99]. In Europe and in the USA, mean values between 15 and 50 ppm were reported[31,74,86].

Recently, a study[100] was carried out to assess the human exposure to cadmium in different areas of the world. The concentration of cadmium was measured in samples of kidney cortex from deceased teachers. The highest values were observed in Japan in the age group of 50 (geometric mean = 67.0 ppm). Values obtained in other countries using the same age group were 2–4 times lower, with geometric means ranging from 15.7 ppm (India) to 39.3 ppm (Belgium). Since these values are average estimates, it can be assumed that in certain countries as in Japan and Belgium, a small fraction of the population has already reached the critical level (around 200 ppm; see toxic effects on the kidney).

In industrial workers, the concentration of cadmium in the kidney cortex may be much higher (e.g. up to 300 ppm)[27,81]. However, when renal damage has occurred, the cadmium level in kidney decreases. This explains why, in most severely poisoned workers and also in patients with Itai-Itai disease, the concentration of cadmium in kidney cortex may be relatively low and sometimes not markedly different from that observed in normal subjects. However, in these persons, the concentration of cadmium in liver remains elevated.

The levels of cadmium in the liver are generally much lower than those in renal cortex. Because of its weight, however, the liver contains a significant fraction of the total body burden. In non-occupationally exposed adults, the concentration of cadmium in the liver ranges approximately from 0.5 to 5 ppm[69]. In cadmium-exposed workers, hepatic cadmium levels may exceed 150 ppm[81]. The ratio between cadmium concentration in liver and in kidney increases with the intensity of exposure[27]; it is therefore much higher for occupationally exposed persons than in the general population. A significant accumulation of cadmium also takes place in other tissues, namely in the pancreas, the thyroid and the salivary glands. The placenta barrier is effective against the transfer of cadmium to the fetus.

In tissues, cadmium is mainly bound to metallothionein, a low molecular weight protein (mol. w 6600) rich in cystein residues but deficient in aromatic aminoacids. The synthesis of this protein corresponds very probably to a defence mechanism against the toxic cadmium ion. One has also hypothesized[27] that metallothionein is involved in the transport of cadmium from the liver to the kidneys. Because of its small size, the cadmium-metallothionein complex released from the liver is freely filtered through the glomeruli to be reabsorbed by the tubular cells. Such a mechanism could explain the selective accumulation of cadmium in kidney cortex.

The cadmium stored in tissues is released very slowly. The biological half-life of cadmium in the liver has been estimated to be 5–10 years and that in the kidney to be about twice as long as in the liver[99]. The total biological half-life of cadmium in the body has been estimated up to 30 years[27].

Excretion

As can be expected from a cumulative toxic, only a small part of absorbed cadmium is excreted. This excretion occurs mainly via the urine. An important fraction of urinary cadmium is probably bound to metallothionein. A closed relationship between both parameters has been reported[15,80] (fig. 1).

On a group basis, the urinary excretion of cadmium is proportional to the body burden and thus increases with age, at least up to 50–60 years[19]. When renal damage has occurred, cadmium excretion by urine increases drastically and the relationship with body burden disappears.

Evaluation of cadmium exposure

The determination of cadmium concentration in liver and kidney samples collected at autopsy or by biopsy provides a direct assessment of the internal integrated dose of cadmium of population groups. Another approach, more directly useful on an individual basis, consists in measuring in vivo by a neutron activation technique the amount of cadmium accumulated in the liver[97] or the kidney[1]. This technique has been applied with occupationally exposed workers[81] and with the general population[21]. For various practical reasons, its routine use cannot be envisaged.

So far, in the health surveillance programs of workers in industry, or in large scale studies on the general population, the exposure to cadmium has usually been evaluated indirectly by measuring the metal in easily accessible biological samples such as urine or blood.

It is now well established that the concentration of cadmium in blood reflects mainly the average intake of the past few months[47,55]. In non-occupationally exposed people, the cadmium level in whole blood is normally below 5 µg/l.

The significance of urinary cadmium is somewhat

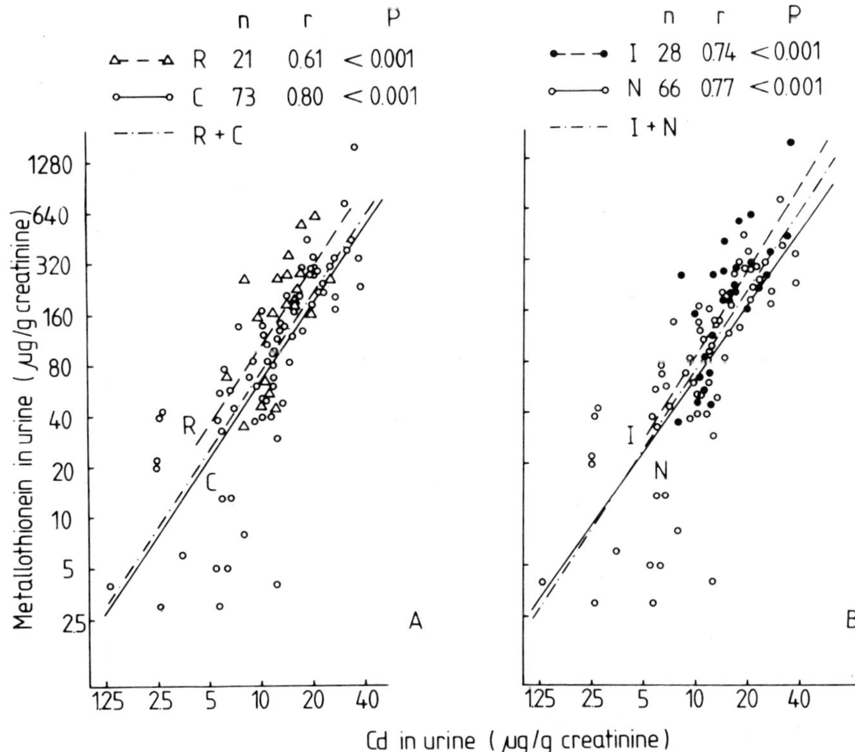

Figure 1. Relationship between urinary metallothionein and urinary cadmium. *A* Workers currently exposed to Cd (n = 73, group C) and workers removed from Cd exposure (n = 21, group R). *B* Workers with normal (n = 6, group N) and increased (n = 28, group I) β_2-microglobulinuria. The regressive line for the total population is shown (reproduced from Roels et al.[80]).

more complex[55]. Under moderate exposure (environmental or moderate industrial exposure) and as long as the cadmium-binding sites in the kidney are not saturated, the amount of cadmium excreted in urine is a relatively good indicator of the amount of cadmium stored in the organism. In the absence of occupational exposure, the concentration of cadmium in urine is usually lower than 1 µg/g creatinine.

A urinary excretion of cadmium higher than 10 µg/g creatinine indicates that the saturation of cadmium-binding sites of the organism is occurring and hence that renal dysfunction, if not already present, may develop if exposure is pursued. At this stage, urinary cadmium may also be significantly influenced by recent exposure.

There is now sufficient evidence to suggest that metallothionein level in urine has the same significance as urinary cadmium. As shown in figure 1, good correlations have been observed between both parameters whatever the status of renal function.

Toxic effects

Acute exposure. Ingestion. Acute oral intoxication by cadmium has been observed after the ingestion of acidic food or beverage stored in cadmium-plated containers. For instance, a poisoning incident among Swedish school children was caused by fruit juice contaminated in a vending machine which had a cadmium-plated reservoir[68]. Cases of acute oral intoxication have also been reported in workers exposed to cadmium dust who had eaten or smoked at their place of work. The use of cadmium-plated cooking utensils, currently prohibited in many countries, was in the past frequently a cause of acute cadmium poisoning.

Following acute ingestion of cadmium the target organ is the gastro-intestinal tract. Symptoms are severe nausea, vomiting, salivation, abdominal cramps and diarrhea. In the case of fatal intoxication, these symptoms are followed up by shock due to the loss of liquid or by acute renal failure and cardiopulmonary depression. The no-effect level of cadmium administered at a single oral dose to adults has been estimated at 3 mg and the lethal doses range from 350 to 8900 mg[51].

Inhalation. Acute inhalation of CdO fume generated by heating cadmium metal (in welding, smelting or soldering operations) has been responsible for most fatal or near-fatal accidents in the industry. Between 1958 and 1976, 24 cases of acute exposure to CdO were reported in the literature, involving 100 workers of whom 17 died[59]. It is likely that many more cases were not reported. Symptomatology is that of severe bronchial and pulmonary irritation appearing several hours after exposure. The lethal concentration of CdO for man has been estimated to be around 5 mg/m³ for a 8-h exposure[51].

Although acute – and sometimes fatal – cadmium poisoning still constitutes a hazard in industry, it is much less frequent than the chronic intoxication.

Chronic effects

Following long term exposure, cadmium may exert various toxic effects, such as renal disturbances, lung insufficiency, osteomalacia, anemia and anosmia. The hypothesis has also been put forward that cadmium is carcinogenic for man and plays a role in the development of cardiovascular diseases, in particular hypertension. Although the lung may be the organ first affected after repeated subacute exposure to cadmium fume, there are presently sufficient data to conclude that under present exposure conditions in industry, the kidney is usually the critical organ. This is also the case when exposure occurs by ingestion.

Effects on the kidney

Cadmium nephropathy was first described in occupa-

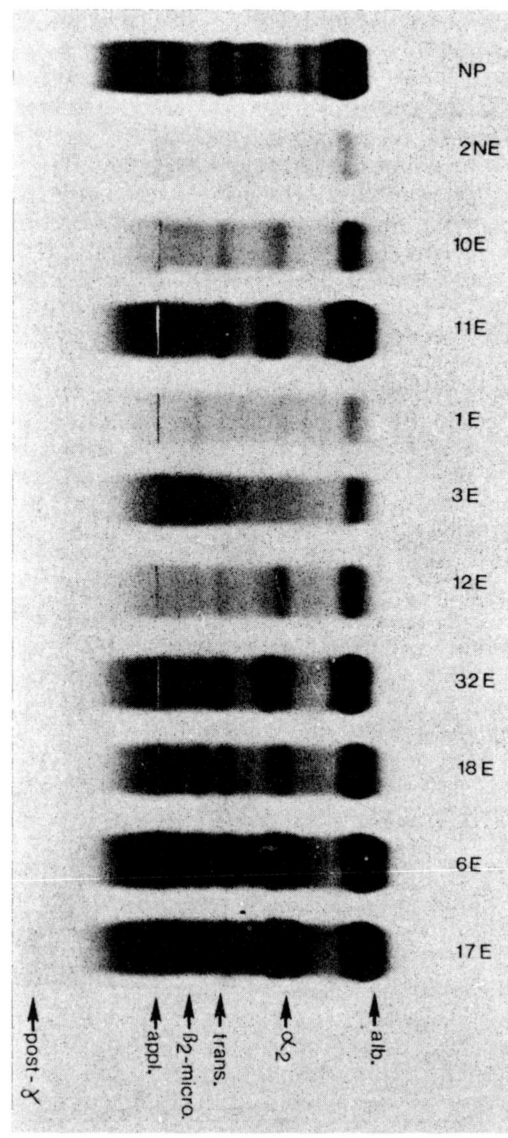

Figure 2. Electrophoresis on agarose gel of about 100-fold concentrated urinary proteins from non-exposed (NE) and cadmium-exposed (E) workers. 2NE: normal pattern; 10E and 11E: glomerular type patterns; 1E and 3E: tubular type pattern; 12E, 32E, 18E, 6E and 17E: mixed type patterns. NP: reference plasma (reproduced from Bernard et al.[5]).

tionally exposed workers[25,26]. Classically, the functional disturbances induced by cadmium involve the proximal tubule, giving rise to a tubular type proteinuria as in the Fanconi syndrome[12,44,72,90]. The urinary excretion of low molecular weight proteins (mol. wt below 40,000) such as β_2-microglobulin, lysozyme and retinol-binding protein, is markedly increased (up to 1000-fold). Cadmium affects the tubular reabsorption of these proteins which are freely filtered through the glomeruli and taken up by the proximal tubule cells. Currently, the determination of urinary β_2-microglobulin is the most widely used test for the early detection of cadmium proteinuria. It has been shown, however, that because of its greater stability in urine, retinol-binding protein is a more practical and reliable parameter[4].

In cadmium-exposed workers, the low molecular weight proteinuria is often accompanied by an increased urinary excretion of high molecular weight proteins like albumin, transferrin or IgG (mixed type proteinuria)[5,6,32,54]. In some workers, only a glomerular type proteinuria with an enhanced excretion of high molecular weight proteins may be observed (fig. 2). Therefore, an increased urinary excretion of both low and high molecular weight proteins should be regarded as critical effects and thus should be looked for concurrently in order to detect at an early stage renal damage induced by cadmium.
Increased levels of β_2-microglobulin and creatinine in serum, generally accompanied by a reduction in creatinine clearance, have been observed in cadmium-exposed workers, which is suggestive of a glomerular dysfunction[5,73]. Cadmium proteinuria may be also associated with enzymuria, aminoaciduria, glucosuria, impaired acid excretion, decreased concentration capacity of the kidneys and increased excretion of calcium and phosphates. The disturbances in calcium and phosphorus metabolism may lead to the formation of kidney stones and/or to a demineralization of the bones.

These signs of renal damage, observed in cadmium workers, are commonly found in Itai-Itai patients. A higher incidence of proteinuria, glucosuria and β_2-microglobulinuria has been observed in the Jintzu river basin in Toyama Prefecture, where Itai-Itai disease was first seen, and also in other areas where high concentrations of cadmium were found in rice. In the endemic area of Toyama, the increased urinary excretion of β_2-microglobulin was strongly related to the residence time in that area as well as to the purposes for which contaminated river water was used[48]. β_2-Microglobulin concentration in urine was also correlated with the cadmium level in urine and blood. It has been suggested that farmers who usually eat rice containing 0.6 ppm Cd may develop renal damage in their fifties.
It seems however that Japan is not the only country where environmental exposure to cadmium may adversely affect the health of the general population. Recently, a study was carried out in Belgium, suggesting that environmental exposure to cadmium in an industrialized area polluted by this metal may exacerbate the age-related decline of renal function in elderly resident groups[53,78]. The renal function of a group of 60 women over 60 years who had spent the major part of their lives in the cadmium-polluted area, was compared with that of 2 groups of aged women from areas less polluted by this heavy metal. The group of aged women from the contaminated area had, on the average, a higher cadmium body burden (as reflected by a higher urinary excretion of cadmium) than the groups from the other areas. The parameters selected for evaluating renal function (total protein, aminoacids, β_2-microglobulin and albumin in urine) followed the same trend. Significant correlations were observed between the urinary excretion of cadmium and that of the 4 renal function parameters. Furthermore, a retrospective mortality study carried out shortly after these observations[52] revealed that the standardized mortality ratio from nephritis and nephrosis was significantly higher in the cadmium-polluted area than in other areas. The increase was observed for both males and females, which tends to confirm the influence of an environmental factor. An autopsy study is presently under way in our laboratory, comparing the cadmium body burden of inhabitants in Liège area with that of persons living in other parts of Belgium. Preliminary results indicate that the cadmium concentration in the kidney cortex in the age group 40-60 years is, on the average, nearly twice as high in the Liège area (geometric mean 37 ppm) as in other areas of the country.

The latency period before the onset of renal damage depends on the intensity of exposure and generally exceeds 10 years. After removal from exposure, the signs of renal dysfunction may not be reversible[79]. On the basis of the results of cadmium analysis in renal tissues collected from about 30 autopsies or biopsies in persons who had been exposed to cadmium, Friberg et al.[27] has suggested that the critical level of cadmium in renal cortex for the appearance of tubular proteinuria is about 200 ppm. In 1977, a WHO task group[104] concluded that the critical level of cadmium in the kidney cortex may be between 100 and 300 ppm with 200 ppm as the most likely estimate. With the development of neutron activation analysis, the critical levels of cadmium in the human organism have been more precisely estimated. Roels et al.[77,81,82] have measured by this technique the concentrations of cadmium in the liver and the kidney cortex of 309 male workers in 2 Belgian zinc-cadmium plants. The renal function of these workers was evaluated at the same time by determining the urinary excretion of β_2-microglobulin, albumin and of total protein. The critical concentration of cadmium in the kidney cortex has been estimated between 215 and 385 ppm (assuming a mean kidney depth of 8 cm), which corresponds to concentrations in liver from 30 to 60 ppm. The minimal critical body burden of cadmium has been estimated to be around 180 mg. On the basis of the relationship between urinary and renal cortex cadmium, the authors estimated the critical concentration of cadmium in urine between 10-15 µg/g creatinine, a value in agreement with that derived previously from the relationships between urinary cadmium and prevalences of signs of renal dysfunction[4,6,10]. The conclusions of Roels et al.[77,81,82] were confirmed by Ellis et al.[22] who performed a similar study on American cadmium workers.

By using an 8-compartment kinetic model, Kjellström and Nordberg[47] have calculated that a daily intake via food of 440 µg was necessary for a European-American population to reach an average renal cortex concentration of 200 ppm. In 1979, a WHO task group[103] concluded that a daily food intake of cadmium of 200–400 µg may be associated with the appearance of renal critical effects.

It should be stressed, however, that these estimates are derived from data obtained in cadmium-exposed workers. They do not necessarily apply to the general population in which some groups may be more susceptible to the toxic effects of cadmium.

On the other hand, the health significance of an increased urinary excretion of specific low or high molecular height proteins (the effect regarded as critical) is still a matter of discussion. The Sub-committee on Cadmium of the British Occupational Hygiene Society's Committee on Hygiene Standards[8] does not consider as an adverse effect an increase in the excretion of proteins which is not detectable by classical tests for proteinuria (clinical proteinuria). More information is required on the disturbances before it can be concluded whether they must be considered adverse health effects. Meanwhile, it is reasonable to try to indicate at which levels of exposure these disturbances do not occur.

Effects on the lungs

Impairment of lung function has been described only among workers exposed to cadmium by inhalation. Studies performed in the 1950's[2,7,26,43,44] have reported emphysema in workers exposed to cadmium dust and fume. However, in these investigations, the influence of tobacco was not always taken into account so that it is likely that all the pulmonary changes described were not due to cadmium alone. More recent investigations[54,92] on workers currently exposed to cadmium indicate that because of improved working conditions, the pulmonary changes are mild and occur with a much lower incidence and probably at a later stage than the biological signs of renal damage.

Effect on the bones

Bone lesions are usually a late manifestation of a severe chronic cadmium poisoning. They are characterized by osteomalacia, osteoporosis and spontaneous fractures. The patient complains of pain in the back and in the extremities, difficulties in walking and pain on bone pressure. It is generally thought that these bones changes are secondary to renal tubular dysfunction (increased urinary loss of calcium and phosphorus) eventually associated with an altered metabolism of vitamin D. Indeed, it has been shown experimentally that in the chicken kidney, the hydroxylation of vitamin D was inhibited by cadmium, which leads to a deficiency in the active form of the vitamin[23].

An extreme manifestation of the toxic effect of cadmium on the bones, is the Itai-Itai disease, which broke out at the end of World War II in the cadmium-polluted area of Toyama in Japan. The etiology of this disease is not yet fully elucidated. A few authors[46,96] have suggested that it was not related to cadmium-pollution but resulted from vitamin D-deficiency. However, a task group of international experts convened by WHO in 1977 concluded, after a review of the epidemiological data regarding Itai-Itai disease, that the close epidemiological association between cadmium exposure and the disease, the accumulation of cadmium in tissues, the finding of osteomalacia in cadmium-exposed workers[66] and the occurrence of renal tubular dysfunction among both Itai-Itai patients and cadmium-exposed persons indicate that cadmium was a necessary factor in the development of Itai-Itai disease.

Itai-Itai disease struck mainly aged women who have had several children and had suffered from calcium and phosphorus deficiency. It is now believed that this severe bone disease resulted from the combination of nutritional deficiencies and cadmium-induced nephropathy.

Carcinogenic effects

Several epidemiological studies have suggested that inhalation of cadmium might increase the incidence of prostate and probably lung cancer in occupationally exposed workers[36,37,45,75]. No firm conclusion can yet be drawn because of the relatively small number of workers examined and also because of the difficulty, in many studies, to evaluate the possible role of other carcinogenic agents such as tobacco. Conflicting results have also been obtained regarding the induction by cadmium of chromosome anomalies in lymphocytes from cadmium-exposed workers (see other effects).

In animals, the injection of cadmium can induce local sarcoma. However, the significance of these tumors can be questioned since it is well known that tumors can be produced by the repeated injection of normally noncarcinogenic substances. Animal experiments have not shown any incidence of prostatic carcinoma. The only indirect tumors which have been induced in different animal species are Leydig cell tumors, but such tumors have not been observed with a higher incidence in cadmium-exposed persons.

There is presently no clinical or experimental indication that exposure to cadmium via food favors the development of cancer, even in Japan where the intake is relatively high.

The association between cancer and exposure to cadmium in man thus appears to be tenuous. Nevertheless, in 1976, the International Agency for Research on Cancer[40] classified cadmium among chemicals which are probably carcinogenic for man. However, new concern about cadmium carcinogenicity has been generated recently by the report of Takenaka et al.[95] on lung cancer in rat chronically exposed to cadmium chloride by inhalation. The animals were continuously exposed to cadmium concentrations in air (aerosols) of 12.5, 25 and 50 µg/m^3 for a period of 18 months. 13 months after the end of the exposure histopathological examination revealed a dose-dependent incidence of lung carcinoma (71% in the highest, 53% in the middle and 15% in the lowest concentration group).

Hypertension

Schroeder[84] was the first to suggest that cadmium might contribute to the development of essential hypertension in man. He showed that long-term oral administration of cadmium to rat may induce a systolic hypertension which persists throughout their life. He reported also that patients dying from cardiovascular diseases had higher concentrations of cadmium or higher cadmium to zinc ratios than those dying for other causes. This association between hypertension and cadmium levels in tissues was confirmed by some authors[42,56], but not by others[62,65]. Similarly[27,101], the hypertensive action of cadmium in animals was reproduced by some investigations while others failed to show any effect. Carroll[13] and later Hickey et al.[34] found a correlation between the cadmium concentration in the air of more than 20 American cities and the death rate from cardiovascular diseases. However, Hunt et al.[38], in a study involving 77 cities in the USA, failed to show a correlation between cadmium fallout and mortality from cardiovascular diseases. Furthermore, reanalyzing Carroll's data, they were able to demonstrate a higher correlation between population density and hypertension than between hypertension and the cadmium level in the air.
A significant negative correlation between total water hardness and cardiovascular diseases has been fequently reported[61]. Since hard water usually contains fewer heavy metals – in particular, cadmium – than soft water, this explains why one may find a correlation between cadmium in water and prevalence of hypertension[85]. However, with regard to this relationship between hypertension and cadmium in water and also that between hypertension and cadmium in air, it should be remembered that the cadmium intake by the general population through air and water is lower than through food. So far, there is no report of hypertension in industrially exposed workers nor in Itai-Itai disease patients. In Japan, the determination of blood pressure in more than 10,000 persons living in cadmium-polluted areas did not reveal any difference in blood pressure with the degree of pollution[41]. A recent epidemiological study[88] confirms this observation.
These negative findings under conditions of relatively high exposure to cadmium do not disprove that cadmium, at a low environmental level, plays a role in the etiology of hypertension. As suggested by Perry and Erlanger[71], the relationship between cadmium and blood pressure may not be linear and the potential action of cadmium on blood pressure may be divergent according to the intensity of exposure.

Indeed, these authors have observed in animals that at low doses cadmium exhibits a pressor action while at higher doses, it decreases systolic blood pressure. There also have been some speculations that factors such as age, sex, smoking habits, individual susceptibility or mineral content of the diet interfere with the hypertensive action of cadmium. In summary, the role of cadmium in the development of hypertension remains uncertain but the failure, after more than 15 years of epidemiological and experimental studies, to show a clear relationship between cadmium and hypertension casts some doubt on the validity of Schroeder's hypothesis.

Other effects

Among other effects reported in persons chronically exposed to cadmium, one can mention: anosmia, ulceration of the nasal mucosa, yellowing of dental necks, mild anemia and occasionally some signs of liver damage. So far, no report on teratogenic effects in populations exposed to cadmium has been published. In a study on cadmium-exposed women in USSR[16], congenital malformations were not observed, but the birth weights of newborns were lower than those of newborns from control mothers. Data on genetic effects of cadmium are conflicting. One study[89] on Itai-Itai patients showed some chromosomal aberrations, but this was not confirmed by another study[11]. Bui et al.[11] did not find any chromosome anomaly in 5 cadmium-exposed workers. By contrast, Deknudt and Leonard[17] and Bauchinger et al.[3] found a slight but significant increase of the chromosomal anomalies in workers exposed to cadmium. However, in these 2 studies, the workers had also been exposed to other metals, such as lead, zinc or arsenic so that the anomalies observed can not be firmly attributed to cadmium exposure. O'Riordan et al.[70] found no increased aberration yields in blood lymphocytes of cadmium workers. They suggest that the aberrations described previously result from a synergistic action of heavy metal with some environmental mutagens. Recently, Gasiorek and Bauchinger[29] studied the effect of lead, cadmium and zinc separately and in combination on the incidence of chromosomal aberrations in human lymphocytes. They observed a significantly increased incidence of anomalies exclusively with cadmium, an observation the authors relate to the metabolism of cadmium in lymphocytes (synthesis of metallothionein).

1 Al-Haddad, I.K., Chettle, D.R., Fletcher, J.G., and Fremlin J.H., A transportable system for measurement of kidney cadmium in vivo. Int. J. appl. Radiat. Isot. *32* (1981) 109–112.
2 Baader, E.W., Gesundheitsfürsorge und Arbeitsmedizin: die chronische Kadmium Vergiftung. Dt. Med. Wschr. *76* (1951) 484–487.
3 Bauchinger, M., Schmid, E., Einbrodt, H.J., and Dresp, J., Chromosome aberrations in lymphocytes after occupational exposure to lead and cadmium. Mutation Res. *40* (1976) 57–62.
4 Bernard, A., Moreau, D., and Lauwerys, R., Comparison of retinol-binding protein and β_2-microglobulin determination in urine for the early detection of tubular proteinuria. Clinica chim. Acta *126* (1982) 1–7.
5 Bernard, A., Buchet, J.P., Roels, H., Masson, P., and Lauw-

erys, R., Renal excretion of proteins and enzymes in workers exposed to cadmium. Eur. J. clin. Invest. *9* (1979) 11–22.

6 Bernard, A., Roels, H., Hubermont, G., Buchet, J.P., Masson, P.L., and Lauwerys, R., Characterization of the proteinuria in cadmium-exposed workers. Int. Archs occup. envir. Hlth *38* (1976) 19–30.

7 Bonnell, J.A., Emphysema and proteinuria in men casting copper-cadmium alloys. Br. J. ind. Med. *12* (1955) 181–197.

8 British Occupational Hygiene Society Committee on Hygiene Standards: Sub-Committee on Cadmium. Ann. occup. Hyg. *20* (1977) 215–228.

9 Buchet, J.P., Lauwerys, R., Vandevoorde, A., and Pycke, J.M., Oral daily intake of cadmium, lead, manganese, copper, chromium, mercury, calcium, zinc and arsenic in Belgium: a duplicate meal study. Fd chem. Toxic. *21* (1983) 19–24.

10 Buchet, J.P., Roels, H., Bernard, A., and Lauwerys, R., Assessment of renal function of workers exposed to inorganic lead, cadmium or mercury vapour. J. occup. Med. *22* (1980) 741–750.

11 Bui, T.H., Lindsten, J., and Nordberg, G.F., Chromosome analysis of lymphocytes from cadmium workers and Itai-Itai patients. Envir. Res. *9* (1975) 187–195.

12 Butler, E.A., Flynn, F.V., Harris, H., and Robson, E.B., A study of urine proteins by two dimensional electrophoresis with special reference to the proteinuria of renal tubular disorders. Clinica chim. Acta *7* (1962) 34–41.

13 Carrol, R.E., The relationship of cadmium in the air to cardiovascular disease death rates. J. Am. med. Assoc. *198* (1966) 267–269.

14 Carruthers, M., and Smith, B., Evidence of cadmium toxicity in a population living in a zinc-mining area. Pilot survey of Shipham residents. Lancet *1* (1979) 845–847.

15 Chang, C.C., Lauwerys, R., Bernard, A., Roels, H., Buchet, J.P., and Garvey, S.J., Metallothionein in cadmium exposed workers. Envir. Res. *23* (1980) 422–428.

16 Cvetkova, R.P., Materials on the study of the influence of cadmium compounds on the generative function. Gig. Tr. Prof. Zabol. *14* (1970) 31–33.

17 Deknudt, G., and Leonard, A., Cytogenetic investigations on leucocytes of workers from a cadmium plant. Envir. Physiol. Biochem. *5* (1975) 319–327.

18 Elinder, C.G., and Kjellström, T., Cadmium concentration in samples of human kidney cortex from the 19th century. Ambio *6* (1977) 270.

19 Elinder, C.G., Kjellström, T., Linnman, L., and Pershagen, G., Urinary excretion of cadmium and zinc among persons from Sweden. Envir. Res. *15* (1978) 432–440.

20 Elinder, C.G., Kjellström, T., Friberg, L., Lind, B., and Linnman, L., Cadmium Concentration in kidney cortex, liver and pancreas from Swedish autopsies. Archs envir. Hlth *31* (1976) 292–302.

21 Ellis, K.J., Vartsky, D., Zanzi, I., Cohn, S.H., and Yasumara, S., Cadmium: in vivo measurement in smokers and non-smokers. Science *205* (1979) 323–335.

22 Ellis, K.J., Morgan, W.D., Zanzi, I., Yasumura, S., Vartsky, D., and Cohn, S.H., Critical concentration of cadmium in human renal cortex: dose-effect studies in cadmium smelter workers. J. toxic. envir. Hlth *7* (1981) 691–703.

23 Feldman, S.L., and Cousins, R.J., Influence of cadmium on the metabolism of 25-hydroxycholecalciferol in chicks. Nutr. Rep. int. *8* (1973) 251–259.

24 Flanagan, P.R., McLellan, J.S., Haist, J., Cherian, G., Chamberlain, M., and Valberg, L., Increased dietary cadmium absorption in mice and human subjects with iron deficiency. Gastroenterology *74* (1978) 841–846.

25 Friberg, L., Proteinuria and kidney injury among workmen exposed to cadmium and nickel dust. J. ind. Hyg. Toxic. *30* (1948) 32–36.

26 Friberg, L., Health hazards in the manufacture of alkaline accumulators with special reference to chronic cadmium poisoning. Acta med. scand. *138*, suppl. 240 (1950) 1–124.

27 Friberg, L., Piscator, M., Nordberg, G.F., and Kjellström, T., Cadmium in the environment, 2nd edn. CRC Press Inc., Cleveland Ohio, 1974.

28 Fukushima, M., Ishizaki, A., Sakamoto, M., and Kobayashi, E., Cadmium concentration in rice eaten by farmers in the Jinzu river basin. Jap. J. Hyg. *28* (1973) 406–415.

29 Gasiorek, K., and Bauchinger, M., Chromosomes changes in human lymphocytes after separate and combined treatment with divalent salts of lead cadmium and zinc. Envir. Mut. *3* (1981) 513–518.

30 Hagino, N.J., About investigations on Itai-Itai disease. J. Toyama med. Assoc., Suppl. 21 (1957).

31 Hammer, D.I., Coalocci, A.V., Hasselblad, V., Williams, M.E., and Pinkertson, C., Cadmium and lead in autopsy tissues. J. occup. Med. *15* (1973) 956–963.

32 Hansen, L., Kjellström, T., and Vesterberg, O., Evaluation of different urinary proteins excreted after occupational cadmium exposure. Int. Archs. occup. envir. Hlth *40* (1977) 273–282.

33 Harvey, T.C., Chettle, D.R., Fremlin, J.H., Al-Haddad, I.K., and Downey, S.P.M.J., Cadmium in Shipham. Lancet *1* (1979) 551.

34 Hickey, R.J., Schoff, E.P., and Clelland, R.C., Relationship between air pollution and certain chronic disease death rates. Archs envir. Hlth *15* (1967) 728–738.

35 Hildebrand, C.E., and Cram, L.S., Distribution of cadmium in human blood culture in low levels of $CdCl_2$. Accumulation of cadmium in lymphocytes and preferential binding to metallothionein. Proc. Soc. exp. Biol. Med. *161* (1979) 438–443.

36 Holden, H., Cadmium toxicology. Lancet *2* (1969) 57.

37 Humperdinck, K., Kadmium und Lungenkrebs. Med. Klin. *63* (1968) 948–951.

38 Hunt, W.F., Pinkerton, C., McNulty, O., and Creason, J., A study in trace element pollution of air in 77 midwestern cities. Trace substances in environmental health, pp.55–68. Ed. D. Hemphill, University of Missouri Press, Columbia, Missouri, 1971.

39 Inskip, H., and Beral, V., Mortality of Shipham residents: 40-year follow up. Lancet *1* (1982) 896–899.

40 International Agency for Research on Cancer (IARC) Monographs on the evaluation of carcinogenic risk of chemicals to man. Cadmium and cadmium compounds, vol.11, p.3974 (1976).

41 Japan Public Health Association, Effects of cadmium on human health: a review on studies mainly performed in Japan, Tokyo 1976.

42 Karlicek, V., and Topolcan, O., Cadmium in kidneys of patients with essential and renal hypertension. Cas. Lek. Cesk. *113* (1974) 41–43.

43 Kazantzis, G., Respiratory function in men casting cadmium alloys. Part I: Assessment of ventilatory function. Br. J. ind. Med. *13* (1956) 30–36.

44 Kazantzis, G., Flynn, F.V., Spowage, J.S., and Trott, D.G., Renal tubular malfunction and pulmonary emphysema in cadmium pigment workers. Q. Jl Med. *32* (1963) 165–192.

45 Kipling, M.D., and Waterhouse, J.A.H., Cadmium and prostatic carcinoma. Lancet *1* (1967) 730–731.

46 Kitamura, S., Etiology of Itai-Itai disease. Proc. First Int. Cadmium Conference, Metal Bulletin Ltd, London 1978.

47 Kjellström, T., and Nordberg, G.F., A kinetic model of cadmium metabolism in the human being. Envir. Res. *16* (1978) 248–266.

48 Kjellström, T., Shiroishi, K., and Evrin, P.E., Urinary β_2-microglobulin excretion among people exposed to cadmium in the general environment. An epidemiological study in cooperation between Japan and Sweden. Envir. Res. *13* (1977) 318–344.

49 Kojiama, S., Haga, Y., Kurihara, T., Yamawaki, T., and Kjellström, T., A comparison between fecal cadmium and urinary β_2-microglobulin, total protein and cadmium among Japanese farmers. An epidemiological study in cooperation between Japan and Sweden. Envir. Res. *14* (1977) 436–451.

50 Larsson, S.E., and Piscator, M., Effect of cadmium on skeletal tissue in normal and calcium deficient rats. Israel J. med. Sci. *7* (1971) 495–498.

51 Lauwerys, R., Criteria (dose/effect relationships) for cadmium, C.E.C., Pergamon Press, Oxford, 1978.

52 Lauwerys, R., and De Wals, Ph., Environmental pollution by cadmium and mortality from renal diseases. Lancet *1* (1981) 383.

53 Lauwerys, R., Roels, H., Bernard, A., and Buchet, J.P., Renal response to cadmium in a population living in a nonferrous smelter area in Belgium. Int. Archs occup. envir. Hlth *45* (1980) 271–274.

54 Lauwerys, R., Buchet, J.P., Roels, H., Brouwers, J., and Stanescu, D., Epidemiological survey of workers exposed to cadmium. Archs envir. Hlth *28* (1974) 145–148.

55 Lauwerys, R., Roels, H., Regniers, M., Buchet, J.P., Bernard, A., and Goret, A., Significance of cadmium concentration in blood and in urine in workers exposed to cadmium. Envir. Res. *20* (1979) 375–394.

56 Lener, J., and Bibr, B., Cadmium and hypertension. Lancet *1* (1971) 970.

57 Lewis, G.P., Coughlin, L., Jusko, W., and Hartz, S., Contribution of cigarette smoking to cadmium accumulation in man. Lancet *2* (1972) 291–292.

58 Livingston, H.D., Measurement and distribution of zinc cadmium and mercury in human kidney tissue. Clin. Chem. *18* (1972) 67–72.

59 Macfarland, H., Pulmonary effects of cadmium, in: Cadmium Toxicity, vol. 15, p. 113. Ed. J.D., J.D. Mennear. Marcel Dekker, Inc., New York 1979.

60 Mahaffey, K.R., Corneliussen, P.E., Jelinek, C.F., and Fiorino, J.A., Heavy metal exposure from foods. Envir. Hlth Perspect. *12* (1975) 63–68.

61 Masiromi, R., Trace elements in relation to cardiovascular diseases. WHO, Geneva 1974.

62 McKenzie, J.M., Tissue concentration of cadmium, zinc and copper from autopsy samples. New Zealand med. J. *79* (1974) 1016–1019.

63 McLellan, J.S., Flanagan, P.R., Chamberlain, M.J., and Valberg, L.S., Measurement of dietary cadmium absorption in human. J. Toxic. envir. Hlth *4* (1978) 131–138.

64 Ministery of Agriculture, Fisheries and Food of United Kingdom. Survey of cadmium in food, in: Working Party on the Monitoring of Foodstuffs for heavy metals, 4th report, London 1973, p. 314.

65 Morgan, J.M., 'Normal' lead and cadmium content of the human kidney. Archs envir. Hlth *24* (1972) 364–368.

66 Nicaud, P., Lafitte, A., and Gros, A., Les troubles de l'intoxication chronique par le cadmium. Archs Mal. prof. Méd. trav. *4* (1942) 192–202.

67 Nogawa, K., Ishizaki, A., Fukushima, M., Shibata, I., and Hagino, N., Studies on the women with acquired Fanconi syndrome observed in the Ichi River basin polluted by cadmium. Envir. Res. *10* (1975) 280–307.

68 Nordberg, G.F., Slorach, S., and Stenström, T., Cadmium poisoning caused by a cooled soft drink machine. Läkartidn. *70* (1973) 601–604.

69 Nriagu, J.O., Cadmium in the environment. Part II: Health Effects. John Wiley & Sons, New York 1981.

70 O'Riordan, M.L., Hughes, E.G., and Evans, H., Chromosomal studies on blood lymphocytes of men occupationally exposed to cadmium. Mutation Res. *58* (1978) 305–311.

71 Perry, H.M., and Erlanger, M.W., Metal-induced hypertension following chronic feeding of low doses of cadmium and mercury. J. Lab. clin. Med. *83* (1974) 541–547.

72 Piscator, M., Proteinuria in chronic cadmium poisoning. I. An electrophoretic and chemical study of urinary and serum proteins from workers with chronic cadmium poisoning. Archs envir. Hlth *4* (1962) 607–622.

73 Piscator, M., Serum β_2-microglobulin in cadmium-exposed workers. Path. Biol. *26* (1978) 321–323.

74 Piscator, M., and Lind, B., Cadmium, zinc, copper and lead in human renal cortex. Archs envir. Hlth *24* (1972) 426–431.

75 Potts, C.L., Cadmium proteinuria. The health of battery workers exposed to cadmium oxide dust. Ann. occup. Hyg. *8* (1965) 55–61.

76 Rahola, T., Aaran, R.K., and Miettinen, J.K., Half-time studies of mercury and cadmium by whole-body counting, in: Assessment of radioactive organ and body burdens. I.A.E.A., Vienna 1972.

77 Roels, H., Lauwerys, R., and Dardenne, A.N., The critical level of cadmium in human renal cortex: a reevaluation. Toxic. Lett. *15* (1983) 357–360.

78 Roels, H., Lauwerys, R., Buchet, J.P., and Bernard, A., Environmental exposure to cadmium and renal function of aged women in three areas of Belgium. Envir. Res. *24* (1981) 117–130.

79 Roels, H., Djubgang, J., Buchet, J.P., Bernard, A., and Lauwerys, R., Evolution of cadmium-induced renal dysfunction in workers removed from exposure. Scand. J. envir. Hlth *8* (1982) 191–200.

80 Roels, H., Lauwerys, R., Buchet, J.P., Bernard, A., Garvey, J.S., and Linton, H.J., Significance of urinary metallothionein in workers exposed to cadmium. Int. Archs occup. envir. Hlth *52* (1983) 159–166.

81 Roels, H., Lauwerys, R., Buchet, J.P., Bernard, A., Chettle, D., Harvey, T.C., and Al-Haddad, I.K., In vivo measurement of liver and kidney cadmium in workers exposed to this metal. Envir. Res. *26* (1981) 217–240.

82 Roels, H., Bernard, A., Buchet, J.P., Goret, A., Lauwerys, R., Chettle, D.R., Harvey, T.C., and Al-Haddad, I.K., The critical concentration of cadmium in the renal cortex and in urine in man. Lancet *1* (1979) 221.

83 Saito, H., Shioji, R., Furukawa, Y., Nagai, K., Arikawa, T., Saito, T., Sasaki, Y., Furuyama, T., and Yoshinaga, K., Cadmium-induced proximal tubular dysfunction in a cadmium-polluted area. Nephrology *6* (1977) 1–12.

84 Schroeder, H.A., Cadmium as a factor in hypertension. J. chron. Dis. *18* (1965) 647–656.

85 Schroeder, H.A., Cadmium, chromium and cardiovascular disease. Circulation *35* (1967) 570–582.

86 Schroeder, H.A., Nason, A.P., Tipton, I.H., and Balassa, J.J., Essential trace metals in man: zinc. Relation to environmental cadmium. J. chron. Dis. *20* (1967) 179–210.

87 Seiffert, Die Erkrankungen der Zinkhüttenarbeiter und hygienische Massregeln dagegen. Dt. Vjschr. öff. GesundhPflege *29* (1897) 419–454.

88 Shigematsu, I., Minowa, M., Yoshida, T., and Miyamoto, K., Recent results of health examinations on the general population in cadmium-polluted and control areas in Japan. Envir. Hlth Perspect. *28* (1979) 205–217.

89 Shiraishi, Y., Cytogenetic studies in 12 patients with Itai-Itai disease. Humangenetik *27* (1975) 31–44.

90 Smith, J.C., Wells, A.R., and Kench, J.E., Observations on the urinary protein of men exposed to cadmium dust and fume. Br. J. ind. Med. *18* (1961) 70–77.

91 Sovet, Empoisonnement par une poudre à récurer l'argenterie. Presse med. Belge *10* (1858) 69–71.

92 Stanescu, D., Veriter, C., Frans, A., Goncette, L., Roels, H., Lauwerys, R., and Brasseur, L., Effects on lung of chronic occupational exposure to cadmium. Scand. J. Resp. Dis. *58* (1977) 289–303.

93 Suzuki, S., Taguchi, T., and Yokohashi, G., Dietary factors influencing upon the retention rate of orally administered 115 m $CdCl_2$ in mice, with special reference to calcium and protein concentration in diet. Ind. Hlth *7* (1969) 155–162.

94 Takata, F., Nakamura, K., Kajizuka, E., and Nakano, J., Concentrations of some heavy metals in preserved human tissues, Pb, Cd, Cu and Zn. Jap. J. Hyg. *191* (1978) 691–697.

95 Takenaka, S., Oldiges, H., König, H., Hochrainer, D., and Oberdörster, G., Carcinogenicity of cadmium aerosols in Wistar rats. J. natl. Cancer Inst. *70* (1983) 367–373.

96 Takeuchi, J., Etiology of Itai-Itai disease. Proc. First Int. Cadmium Conference, San Francisco. Metal Bulletin Ltd, London 1978.

97 Thomas, B., Harvey, T.C., Chettle, D., McLellan, J.S., and Fremlin, J.H., A transportable system for the measurement of liver cadmium in vivo. Phys. med. Biol. *24* (1979) 432–437.

98 Travis, C.C., and Etnier, E.L., Dietary intake of cadmium in the United States: 1920–1975. Envir. Res. *27* (1982) 1–9.

99 Tsuchiya, K., Seki, Y., and Sugita, M., Organ and tissue cadmium concentrations of cadavers from accidental death. Proc. 17th Int. Cong. Occup. Health, Buenos Aires, Argentina, 1972.

100 Vather, M., Assessment of human exposure to lead and cadmium through biological monitoring. Report prepared for United Nations Environment Programme and World Health Organisation by National Swedish Institute of Environmental Medicine and Karolinska Institute, Stockholm 1982.

101 Webb, M., ed., The Chemistry, biochemistry and biology of cadmium Topics in environmental health. Elsevier/North-Holland, Biomedical Press, Amsterdam 1979.

102 Wheeler, G.A., A case of poisoning by bromide of cadmium. Boston med. surg. J. *95* (1876) 434–436.

103 W.H.O. Trace elements in human nutrition. Geneva 1973.

104 W.H.O. Task group on Environmental Health Criteria for Cadmium. Geneva 1977.

Biological indicators of cadmium exposure and toxicity

by Z. A. Shaikh* and L. M. Smith

Department of Pharmacology and Toxicology, University of Rhode Island, Kingston (Rhode Island 02881, USA)

Summary. The increasing environmental and occupational exposure of populations to cadmium creates the need for biological indicators of cadmium exposure and toxicity. The advantages and disadvantages of monitoring blood cadmium, urinary, fecal, hair, and tissue cadmium, serum creatinine, β_2-microglobulin, α_1-antitrypsin and other proteins, and urinary amino acids, enzymes, total proteins, glucose, β_2-microglobulin, retinol-binding protein, lysozyme, and metallothionein are discussed. It is concluded that urinary cadmium, metallothionein and β_2-microglubulin may be used together to assess cadmium exposure and toxicity.

Introduction

Cadmium is a heavy metal of increasing prevalence in our environment, due to its industrial production and usage and its emissions. from fossil fuel combustion. Specific populations exposed to high levels of cadmium have been studied intensively in order to determine not only the mechanisms of cadmium toxicity, but also biological indicators which can be used to prevent its toxic manifestations by giving an estimate of body burden.

There now exists a great awareness of cadmium's potential health hazard and measures have been taken to reduce the exposure both in the environment and in industry. However, the continuous life-long accumulation of cadmium in the body creates the need for a monitor of body burden and not recent exposure. Also, the severe and possibly irreversible renal tubular dysfunction caused by cadmium necessitates the development of a therapeutic index which is sensitive enough to detect cadmium body burden before irreparable damage is done. Therefore, a search is needed to discover the most sensitive parameters which accurately reflect body burden and also exhibit a measurable change after cadmium exposure.

Many schools of thought exist concerning which parameter would be the best indicator of exposure and of toxicity. A review of the advantages and disadvantages of each one suggested will aid in the development of a rational approach for monitoring cadmium toxicity in susceptible populations. Some of the more commonly used parameters are listed in the table.

Blood cadmium

There is general agreement that although blood cadmium levels do increase with exposure, this increase mainly reflects recent exposure rather than body burden of the metal. Lauwerys et al.[26] showed that blood cadmium correlated with recent exposure in occupationally exposed workers. Also, it was found that blood cadmium reflected recent exposure after an equilibrium was reached at 4 months[28]. Bernard et al.[5] found a similar relationship in rats, in which blood cadmium reached an equilibrium at 3 months, after which it reflected recent exposure. In agreement with this data, blood cadmium was found to be a good indicator of acute exposure to cadmium[3]. Contrary to this evidence, however, Elinder et al.[14] found a signifi-

cant correlation between blood cadmium and hepatic cadmium determined in biopsy material.

Cadmium in urine

The mechanism of urinary excretion of cadmium was studied by Nomiyama et al.[37]. Most cadmium is reabsorbed at the proximal tubules and tubular damage may cause the increased excretion observed. However, pretreatment with urinyl acetate, causing tubular dysfunction, did not subsequently increase the excretion of cadmium, suggesting that the cadmium excretion is not related to renal function.

In the same study Nomiyama et al.[37] also analyzed the relationship between cadmium in urine and in the renal cortex. It was found that urinary cadmium increased with renal cortex cadmium in rabbits that had been given 1.5 mg cadmium/kg/day for up to 5 weeks. This suggests that a close relationship exists between the two. However, they also found a proportional increase in urinary cadmium with plasma cadmium and most data show that cadmium in blood relates more to recent exposure.

Parameters used for determining cadmium exposure and renal dysfunction

Parameter	Possible indicator of	References
Blood cadmium	Recent exposure	3, 5, 14, 26, 28
Urinary cadmium	Recent exposure; body burden	5, 8, 26, 28, 37, 45, 52, 64
Fecal cadmium	Dietary intake; recent exposure	17, 25, 56, 58, 65
Hair cadmium	Body burden	3, 10, 16, 29, 42, 44
Tissue cadmium	Body burden	15, 34, 38, 39, 52
Serum creatinine	Glomerular dysfunction	4, 20, 21, 48
Serum β_2-microglobulin	Glomerular dysfunction	4, 20, 21, 48, 66
Serum enzymes	Liver dysfunction	30, 33
Plasma α_1-antitrypsin	Cadmium toxicity	7, 12
Other serum proteins	Liver dysfunction	33, 46, 47, 49
Aminoaciduria	Renal dysfunction	2, 18, 35, 41
Enzymuria	Renal dysfunction	9, 18, 35, 41, 60
Glucosuria	Renal dysfunction	18, 32, 36, 39, 54
Mixed proteinuria	Glomerular and tubular dysfunction	4, 6, 51, 53
Urinary β_2-microglobulin	Tubular dysfunction	1, 4, 21-24, 30, 45, 55, 59, 66
Urinary retinol-binding protein	Tubular dysfunction	30, 31, 33, 39, 40, 50
Urinary lysozyme	Tubular dysfunction	33, 50
Urinary metallothionein	Chronic exposure; body burden; tubular dysfunction	11, 13, 19, 43, 57, 61-64

In vivo measurement of liver and kidney cadmium and its significance to cadmium in urine was studied by Roels et al.[52]. It was found that the concentration of cadmium in urine followed body burden until this reached levels where renal dysfunction was evident. After this, urinary cadmium increased greatly. It was concluded that the probability of getting renal dysfunction was low when urinary cadmium levels of 10 µg/g creatinine were not regularly exceeded[52]. This supported studies done earlier with animals[5,8,45], showing that cadmium excretion was directly related to dosage, but then increased markedly upon renal damage.

In 2 different studies, Lauwerys et al.[26,28] concluded that cadmium in urine reflected body burden when exposure was low, and recent exposure when exposure was high. This finding was in agreement with the data presented thus far. They postulated that when exposure is high enough to saturate all binding sites, cadmium is excreted in relation to recent exposure. After renal dysfunction, excess cadmium excretion results from 2 sources: from the damaged tubular epithelium, and as a result of decreased reabsorption.

The above data support the use of urinary cadmium as a biological indicator for cadmium exposure in the absence of renal dysfunction. However, contamination of the urine during collection and final analysis makes it difficult to obtain accurate cadmium values. Furthermore, on the basis of cadmium in urine, differences between people with and without renal dysfunction cannot always be found[64], suggesting that cadmium determination in urine should be carried out in combination with other tests.

Cadmium in feces

From a study on the daily excretion of cadmium in normal populations, it was concluded that Japanese adults excrete 25–80 µg cadmium daily in feces. These values agreed with daily cadmium intake via food[65]. Kojima et al.[25] also found that people living in a cadmium-polluted prefecture excreted significantly more cadmium in feces than in control populations; 16.1% of the cadmium-exposed group excreted ≥ 300 µg cadmium/day, while none of the control group excreted this large an amount.

Fecal cadmium has been used as an estimate of daily intake in studying the environmentally exposed people in Japan[25]. The rationale for this practice was derived from evidence that cadmium has low gastrointestinal absorption[17,56] and a low intestinal secretion[17]. However, this method is valid only when the values obtained are used as an estimate of daily dietary intake. Also, large variations in fecal cadmium content probably occur daily. Because of this, it was suggested[25] that data from one day's fecal excretion should be analyzed on a group basis only. Even then, it may not be an accurate indicator of total cadmium exposure (i.e., gastrointestinal, pulmonary and dermal), especially in occupationally exposed populations.

Hair cadmium

Measurement of cadmium in hair has also been suggested as an indicator of body burden. Nordberg and Nishiyama[44] found that hair cadmium correlated well with body burden in mice. Brancato et al.[10] also found a qualitative relationship between hair cadmium and kidney and liver cadmium in rats. It was suggested that it could be used as a biological indicator in the future, however, a quantitative relationship between hair and body cadmium did not exist. Ellis et al.[16] also found a qualitative, but not quantitative, relationship between hair cadmium and body burden in exposed workers.

Another problem exists concerning whether hair cadmium only measures endogenous metal if the exposure is from the air. Significant amounts of cadmium have been shown to adsorb to hair, with individual variability, based on hair acidity[29].

Reviewing diagnostic procedures for subacute cadmium toxicity in jewelry workers, hair was considered a poor indicator because of extrinsic contamination with cadmium[3]. The inability to remove the extrinsic cadmium by washes could have been due to its adsorption and incorporation into the hair's protein matrix by binding to sulfhydryl groups. The possibility of using hair cadmium cannot be ruled out completely, however, since under conditions of solely oral intake of cadmium, it may very well reflect body burden[42].

Tissue cadmium

The introduction of a mobile facility for determining tissue metal levels by in vivo neutron activation provides a significant and unique contribution to the field of heavy metal toxicology[52]. There now exists a means of measuring cadmium in the kidney and liver of exposed individuals. Recent studies using this technique have proposed new critical concentrations of cadmium in the human renal cortex. In cadmium smelter workers, 300–400 µg/g renal cortex was designated as a critical concentration above which renal dysfunction may occur[15]. However, Roels et al.[52] suggest a lower range of critical cadmium concentration in the renal cortex of 160–285 µg/g, beyond which a high percentage of the subjects develop renal dysfunction.

Nomiyama[34] suggests a possible critical cadmium concentration of > 300 µg/g wet weight based on animal data. In rhesus monkeys exposed to cadmium through the diet, the critical concentration was estimated as 380 µg/g for the appearance of low molecular weight proteinuria, and 450 µg/g for proteinuria, glucosuria and amino acid uria[39].

The use of measuring cadmium solely in the renal cortex has questionable value, however. Roels et al.[52] found that the cadmium concentration is higher in the renal cortex of exposed individuals who still have normal renal function. This is in agreement with data on monkeys[38] which showed a decrease in renal cortex cadmium with the appearance of protein and glucose in the urine.

Therefore, renal cadmium alone is not an acceptable indicator of cadmium exposure or toxicity, and other parameters would have to be examined concomitantly. Furthermore, in humans, due to the potential radiation hazard of the neutron activation analysis, the procedure has limited usefulness in monitoring the same population routinely.

Serum β_2-microglobulin and creatinine

Serum β_2-microglobulin and creatinine, which exhibit changes in concentration with cadmium exposure, have been suggested as biological indicators. For example, in cadmium-exposed workers, serum β_2-microglobulin increases along with an increase in blood cadmium[48,66]. This increase is independent of serum creatinine and occurs in the presence of normal creatinine values[48]. This suggests that it may be a result of increased synthesis of β_2-microglobulin in addition to decreased glomerular filtration rate. This parameter may, then, give false positive results if used for testing glomerular function. Therefore, serum creatinine, which rises when glomerular filtration decreases, is probably a better estimate of glomerular function in exposed people. Knowledge of glomerular status is relevant because several studies from Belgium indicate that cadmium causes both glomerular and tubular damage[4,27,53]. This will be discussed in greater depth when reviewing urinary parameters. Another study on occupationally exposed workers found no changes in serum β_2-microglobulin even with high β_2-microglobulin in urine. Thus, they concluded that it was not a good indicator of toxicity[20].

In conclusion, increased serum creatinine or β_2-microglobulin are non-specific indicators of glomerular function, with secondary effects also causing increases in serum β_2-microglobulin. Alone they cannot be used as a monitor of cadmium toxicity, unless it is well established that no other disease state is present[21].

Serum enzymes

In studying chronic cadmium poisoning manifested as Itai-Itai disease, high blood levels of glutamic oxaloacetic transaminase, alkaline phosphatase and lactate dehydrogenase were observed[33]. These data are representative of an advanced picture of cadmium poisoning, however, and studies which show the most sensitive of these enzymes have yet to be done.

When comparing these to urinary values, high alkaline phosphatase and low phosphorus in serum correlate significantly with proteinuria or combined proteinuria and glucosuria in inhabitants of an Itai-Itai endemic district[30]. Also, the prevalence rates of these indices rise proportionately with increasing cadmium concentrations in urine. Therefore, the possibility exists that serum alkaline phosphatase and phosphorus levels could give an indication of cadmium induced renal toxicity.

Serum α_1-antitrypsin

Finally, there has been an interest in monitoring α_1-antitrypsin levels in plasma as a biological indicator of cadmium toxicity. One study shows a dose-dependent decrease in α_1-antitrypsin and its respective tryp-

sin inhibitory capacity in human plasma in vitro, with cadmium administration[12]. This finding caused Bernard et al.[7] to measure plasma α_1-antitrypsin levels in cadmium-exposed workers. However, in this population, they did not find a significant change in antitrypsin levels.

Other serum proteins

Piscator studied the electrophoretic patterns of both urinary and serum proteins extensively[46,47]. In one study, cadmium-exposed workers were found to have a significantly higher γ-globulin fraction, causing a statistical difference in total protein content between control and exposed workers. Similarly, total hexose and seromucoid hexose contents were significantly different between the two groups[46]. However, a later study, also done on cadmium-exposed workers, gave generally normal results in the exposed group, with increases in α- and γ-globulins in some cases[47].

Piscator and Axelsson[49] then studied serum proteins in rabbits, 7 months after a 6-month exposure to cadmium. There was no difference found between those rabbits which were exposed and the controls. This suggests that liver function normalized in the exposed rabbits, even though much greater hepatic cadmium concentrations were found. This evidence indicates that a) serum proteins are not good indicators of cadmium body burden, and b) large amounts of cadmium can be tolerated in the kidney and liver without developing pathological abnormalities.

Lastly, total protein content in serum has been shown to decrease in patients with Itai-Itai disease[33]. Whether this is an effect of advanced renal damage is unknown.

Amino aciduria

There are opposing reports regarding the value of using amino aciduria as an indicator of cadmium toxicity. Nomiyama[35,41] found that it was one of the earliest signs of cadmium toxicity in rabbits, having appeared in the urine after 16 weeks of exposure, whereas proteinuria and glucosuria occurred after 27 weeks. However, Axelsson and Piscator[2] found that amino aciduria occurred later than proteinuria and glucosuria in cadmium-exposed workers. Similarly, amino aciduria in Japanese workers in lead and cadmium industries was found to be a less sensitive parameter than proteinuria[18]. In summary, there is still some question about the validity of amino aciduria as a sensitive indicator of cadmium exposure.

Enzymuria

There is more agreement that enzymuria may be indicative of renal damage, especially if particular interest is given to where the enzymes originate from, and, therefore, where the damage has been done.

In analyzing the effects of dietary cadmium (300 ppm for 54 weeks) on rabbits, Nomiyama[35,41] found that, in addition to amino aciduria, enzymuria was one of the earliest signs of toxicity. While enzymes (alkaline phosphatase, glutamic oxaloacetic transaminase, glutamic pyruvic transaminase) appeared in the urine after 16–28 weeks of exposure, protein and glucosuria were observed only after 27 weeks. Enzyme clearance studies also showed that acid phosphatase, alkaline phosphatase, glutamic oxaloacetic transaminase, and glutamic pyruvic transaminase were of renal rather than serum origin. However, in this study, acid phosphatase levels were independent of alkaline phosphatase, even though both originate from the proximal tubules.

Two-phase increases in alkaline phosphatase were also seen in rats after parenteral administration of cadmium. An initial increase was seen within 48 h, and was followed by a persistent enzymuria after 15 days[9]. This is further evidence for the use of alkaline phosphatase in urine as a non-specific indicator of renal damage.

Other specific enzymes have been observed which increase in urine with cadmium exposure. Specifically, carbonic anhydrase C increases in urine 10–250 times in cadmium-exposed workers compared to normal subjects[60]. This is particularly relevant since this enzyme originates from damaged tubular cells.

In summary, enzymuria is a possible indicator for non-specific renal damage. However, the absence of any other possible disease state must be assured before renal damage in a clinical setting can be recognized as a result of cadmium toxicity. Yet, the observation that general enzymuria occurs before changes in tubular function or glomerular filtration rate[35,41] suggests that this parameter may be indicative of very early renal dysfunction. This gives it potential value as an index of function which reflects renal status before the development of nephropathy. Further work, however, is needed to identify the most suitable enzyme to be analyzed.

Glucosuria

Glucosuria has been observed in cadmium-exposed animals[39,54], cadmium workers[54], and inhabitants of cadmium-polluted areas[32]. After subcutaneous administration of cadmium in rabbits, the critical concentration of cadmium in the renal cortex was observed to be 120 µg/g for the appearance of glucosuria[36]. Using oral administration of cadmium in rabbits, however, Nomiyama found 300 µg/g to be the critical concentration at which glucosuria develops[36]. In other studies on rabbits and monkeys, Nomiyama[18,39] observed that glucosuria was indicative of a later stage in cadmium toxicity than low molecular weight proteinuria. It is generally not considered an early sign of renal effects of cadmium in exposed populations, and therefore would only be a biological indicator of advanced cadmium toxicity.

Mixed proteinuria

Proteinuria gives great insight into the nature of the dysfunction present. The excretion of low molecular weight proteins in the urine reflects a problem with tubular reabsorption, while high molecular weight proteins in urine reflect problems with glomerular filtration. Although classically cadmium is thought to cause only proximal tubular damage, there is evidence that glomerular damage also occurs. Bernard et al.[4] found, in exposed workers, increased high molecular weight proteins (orosomucoid, immunoglobulin G, albumin, transferrin) in the urine, and a decreased creatinine clearance. This, in addition to increased serum creatinine and β_2-microglobulin, suggests glomerular impairment. There is even evidence that glomerular dysfunction may precede classical tubulopathy since workers with less than 20 years exposure showed only high molecular weight proteinuria, and workers with more than 20 years exposure showed a mixed proteinuria[53]. This theory is confirmed in cadmium-exposed rats which developed a similar mixed proteinuria and not just low molecular weight proteinuria[6]. However, Roels et al.[51] concluded that low molecular weight proteinuria, especially β_2-microglobulinuria, is more sensitive for cadmium toxicity than high molecular weight proteinuria because increases of β_2-microglobulin are significantly higher than increases in albumin at a urine cadmium concentration of 10 µg/g creatinine. Also, since their excretion takes place by 2 separate mechanisms, conclusions cannot be made that one systematically occurs before the other.

β_2-Microglobulin in urine

The significance of low molecular weight proteinuria, particularly β_2-microglobulinuria, has been studied extensively. The level of this protein in urine was shown to increase in long term cadmium exposure in mice[45] and also in environmentally or occupationally exposed populations[4,30,55,59,66]. This low molecular weight proteinuria was much higher (100–300 ×) than total proteinuria (7–17 ×) in Itai-Itai patients, and was, therefore, considered a more sensitive indicator[59]. β_2-Microglobulin values greater than 280 µg/g were not found in occupationally exposed workers until after 4 years of exposure. Similarly, in another group of exposed workers values greater than 1000 µg/l were not found unless 10 years or more of exposure had occurred[1]. Therefore, duration, and not intensity, of exposure appears to be important. The correlation between years of exposure to cadmium and urinary β_2-microglobulin was reinforced by various studies[4,22,24]. Tsuchiya[66] found that β_2-microglobulin in urine closely correlated with age, again indicating an association with the duration of cadmium exposure, since cadmium accumulates with age. He also postulates that increased urinary β_2-microglobulin could be due to increased serum β_2-microglobulin. Furthermore, the levels of the protein in serum correlated with cadmium levels in blood, suggesting the possibility that cadmium stimulates its synthesis. He concluded that it is difficult to assume that increased urinary β_2-microglobulin is solely an effect of tubular dysfunction.

Kazantzis[21] suggested that urinary β_2-microglobulin be used for cumulative cadmium exposure and not short-term exposure, indicating that it does reflect body burden and toxicity. Kjellström and Piscator[23], however, reasoned that cadmium-induced tubular damage has progressed severely when β_2-microglobulin increases in the urine, and using it as an indicator would only serve to determine that irreversible damage has occurred.

In conclusion, most evidence supports the use of urinary β_2-microglobulin as an indicator of renal dysfunction. Yet, the possibility does exist that increased synthesis of the protein causes increased levels in blood and subsequent increases in urine. Also, it should be pointed out that the ability of other disease states (i.e., congenital Fanconi syndrome, chronic pyelonephritis, upper urinary tract infection, etc.) to increase urinary β_2-microglobulin levels, makes it a non-specific indicator, more related to renal function than cadmium exposure.

Retinol-binding protein and lysozyme in urine

Various low molecular weight proteins, besides β_2-microglobulin, are excreted in the urine after cadmium exposure. Retinol-binding protein appears to be the most evident and increases in its excretion are seen in exposed animals[39] and cadmium-exposed populations in Japan[30,31,33,40].

Nomiyama[39] found that the highest exposure group of rhesus monkeys (300 ppm, diet) excreted increased retinol-binding protein after 12 weeks and increased protein, glucose, and amino acids in urine only after 16 weeks. Nogawa et al.[30,31,33] used urinary retinol-binding protein as one of the indices of renal effects of cadmium in various Japanese studies and reported that the retinol-binding protein increased significantly in urine along with other low molecular weight proteins.

The excretion of this protein is commonly thought to be a result of decreased tubular reabsorption[50]. This is also true in the case of lysozyme, another protein which has been shown to increase in urine with chronic cadmium poisoning[33]. Both of these, however, are non-specific indices of renal function and are not directly related to cadmium. Furthermore, among all substances used as indices of renal damage, β_2-microglobulin in urine showed the highest prevalence rate for both men and women environmentally exposed to cadmium[30]. For this reason, β_2-microglobulin may be a better low molecular weight protein to measure in urine as a non-specific indicator of renal function.

Metallothionein in urine

Metallothionein is a low molecular weight protein which is induced by cadmium[13] and which binds cadmium ions[19]. Small amounts of metallothionein are detected in plasma after cadmium exposure[61]. It is also shown that metallothionein is filtered and reabsorbed by the kidney. Its relationship with cadmium and renal function makes it potentially ideal to study cadmium's effect on the kidney.

In both occupationally and environmentally exposed people, metallothionein excretion increases[62–64] and appears to be related to cadmium concentrations in the kidney and liver[62]. Such a relationship in cadmium-exposed workers was shown when cadmium was measured by in vivo neutron activation analysis[15] and metallothionein by a radioimmunoassay[61]. The

logarithm of urinary metallothionein concentration of the workers showed a linear relationship with the logarithm of the liver cadmium concentration. Similarly, in workers which had normal renal function, cadmium in the kidney showed a significant correlation with urinary metallothionein[62] (fig. 1). Studies in rats support this observation[62]. One difference found, however, was the appearance of a critical concentration of cadmium in rat kidney and liver. Once this threshold was reached, urinary metallothionein increased greatly.

When analyzing metallothionein as a potential index of renal dysfunction in environmentally and occupationally exposed populations, a significant correlation between metallothionein in urine and cadmium in urine was found[57,64] (fig. 2). Furthermore, it was possible to separate the cadmium-exposed population into groups of normal or abnormal renal function on the basis of urinary metallothionein values. This suggests that metallothionein not only has a definitive relationship with cadmium, but also with renal function.

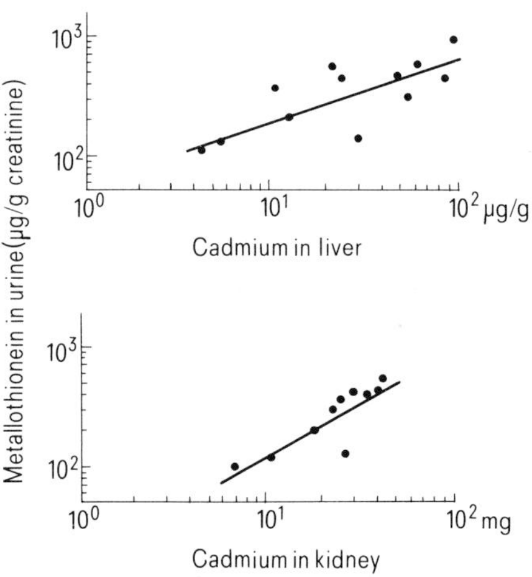

Figure 1. The relationship of metallothionein in urine to cadmium in liver (r = 0.75, p < 0.01) and in left kidney (r = 0.85, p < 0.01) of cadmium smelter workers. Modified from Tohyama et al.[62].

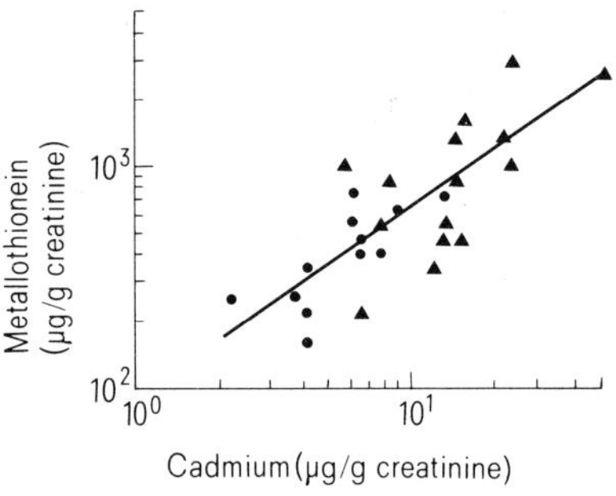

Figure 2. The relationship of metallothionein and cadmium in urine of Japanese women (60–90 years old) from the cadmium-polluted (▲) and non-polluted (●) areas (r = 0.80, p < 0.001). Adapted from Tohyama et al.[64].

The relationship between urinary cadmium and urinary metallothionein was also observed by Chang et al.[11] in cadmium-exposed workers. They suggest that metallothionein in urine is more related to cadmium exposure than renal function because no difference was found in subjects with and without renal dysfunction in this study.
Nordberg et al.[43] suggest that because β_2-microglobulin and metallothionein in urine are not always directly related in cadmium-exposed workers (the highest β_2-microglobulin corresponded to the lowest metallothionein value), the type of information derived from both may be different. This is in agreement with the view that β_2-microglobulin is a better indicator of general, non-specific renal status, and metallothionein a better indicator of cadmium body burden.
In summary, although metallothionein is also induced by other heavy metals, and not just by cadmium, it correlates well with the body burden of cadmium. This, in addition to its correlation with cadmium in urine, suggests that it is a good indicator of cadmium exposure. Because the subjects with and without renal dysfunction can be separated by urinary metallothionein and β_2-microglobulin values, and not by cadmium in urine[64], the former two parameters may be more sensitive for determining cadmium-induced renal damage.

Conclusion

Although there are many parameters suggested for use as potential biological indicators for cadmium toxicity, none appears to be suitable by itself. Based on the existing knowledge, the measurement of urinary cadmium, metallothionein, and β_2-microglobulin appears to be the most rational strategy to use in detecting and monitoring cadmium toxicity.

Acknowledgment. This work was supported by US Public Health Service grant No ES03187.

* To whom all correspondence should be addressed.

1 Adams, R., β_2-Microglobulin levels in nickel-cadmium battery workers, in: Occupational Exposure to Cadmium, Report on Seminar, London, 20 March 1980, pp.41–42. Cadmium Association, London 1980.
2 Axelsson, B., and Piscator, M., Renal damage after prolonged exposure to cadmium. Archs envir. Hlth 12 (1966) 360–373.
3 Baker, E.L., Peterson, W.A., Holtz, J.L., Coleman, C., and Landrigon, P.J., Subacute cadmium intoxication in jewelry workers: an evaluation of diagnostic procedures. Archs envir. Hlth 20 (1979) 173–177.
4 Bernard, A., Buchet, J.P., Roels, H., Masson, P., and Lauwerys, R., Renal excretion of proteins and enzymes in workers exposed to cadmium. Eur. J. clin. Invest. 9 (1979) 11–22.
5 Bernard, A., Goret, A., Buchet, J.P., Roels, H., and Lauwerys, R., Significance of cadmium levels in blood and urine during long-term exposure of rats to cadmium. J. Toxic. envir. Hlth 6 (1980) 175–184.
6 Bernard, A., Goret, A., Roels, H., Buchet, J.P., and Lauwerys, R., Experimental confirmation in rats of the mixed type proteinuria observed in workers exposed to cadmium. Toxicology 10 (1978) 369–375.
7 Bernard, A., Roels, H., Buchet, J.P., Masson, P.L., and Lauwerys, R., α_1-Antitrypsin levels in workers exposed to cadmium. In Clinical Chemistry and Chemical Toxicology of Metals, pp.161–164. Ed. S.S. Brown. Elsevier/North Holland Biomedical Press, Amsterdam 1977.
8 Bonner, F.W., King, L.J., and Parke, D.V., The tissue disposition and urinary excretion of cadmium, zinc, copper, and iron following repeated parenteral administration of cadmium to rats. Chem. biol. Interact. 27 (1979) 343–351.
9 Bonner, F.W., King, L.J., and Parke, D.V., The urinary excretion of alkaline phosphatase after the repeated parenteral administration of cadmium to rats given a high dietary supplement of zinc. Toxic. Lett. 6 (1980) 369–372.
10 Brancato, D.J., Picchioni, A.L., and Chin, L.J., Cadmium levels in hair and other tissues during continuous cadmium intake. J. Toxic. envir. Hlth 2 (1976) 351–359.
11 Chang, C.C., Lauwerys, R., Bernard, A., Roels, H., Buchet, J.P., and Garvey, J.S., Metallothionein in cadmium-exposed workers. Envir. Res. 23 (1980) 422–428.
12 Chowdhury, P., and Louria, D.B., Influence of cadmium and other trace metals on human α_1-antitrypsin: an *in vitro* study. Science 191 (1976) 480–481.
13 Durnam, D., and Palmiter, R.J., Transcriptional regulation of the mouse metallothionein-I gene by heavy metals. J. biol. Chem. 256 (1981) 5712–5716.
14 Elinder, C.G., Kjellström, T., Lind, B., Molander, M.-L., and Silander, T., Cadmium concentration in human liver, blood, and bile: comparison with a metabolic model. Envir. Res. 17 (1978) 236–241.
15 Ellis, K.J., Morgan, W.D., Zanzi, I., Yasumura, S., Vartsky, D., and Cohn, S.H., Critical concentration of cadmium in human renal cortex: dose-effect studies in cadmium smelter workers. J. Toxic. envir. Hlth 7 (1981) 691–703.
16 Ellis, K.J., Yasumura, S., and Cohn, S.H., Hair cadmium content: is it a biological indicator of the body burden of cadmium for the occupationally exposed worker? Am. J. ind. Med. 2 (1981) 323–330.
17 Friberg, L., Piscator, M., Nordberg, G., and Kjellström, T., Cadmium in the Environment, 2nd edn, pp.23–91. CRC Press, Cleveland, Ohio, 1974.
18 Goyer, R.A., Tsuchiya, K., Leonard, D.L., and Kahyo, H., Aminoaciduria in Japanese workers in the lead and cadmium industries. Am. J. clin. Path. 57 (1972) 635–642.
19 Jacobson, K.B., and Turner, J.E., The interaction of cadmium and other metal ions with proteins and nucleic acids. Toxicology 16 (1980) 1–37.
20 Jeandot, R., and Doisneau, R., Epidemiologic study of β_2-microglobulin in cadmium-exposed workers in a nickel-cadmium battery plant, in: Occupational Exposure to Cadmium, Report on Seminar, London, 20 March 1980, pp.37–41. Cadmium Association, London 1980.
21 Kazantzis, G., Significance of β_2-microglobulin as a measure of cadmium exposure, in: Occupational Exposure to Cadmium, Report on Seminar. London, 20 March 1980, pp.55–56. Cadmium Association, London 1980.
22 Kjellström, T., Evrin, P.-E., and Rahnster, B., Dose-response analysis of cadmium-induced tubular proteinuria: a study of urinary β_2-microglobulin excretion among workers in a battery factory. Envir. Res. 13 (1977) 303–317.
23 Kjellström, T., and Piscator, M., Phadedoc Diagnostic Communications, No.1, pp.4–20. Pharmacia Diagnostics, Uppsala, Sweden 1977.
24 Kjellström, T., Shiroishi, K., and Evrin, P.-E., Urinary β_2-microglobulin excretion among people exposed to cadmium in the general environment: an epidemiological study in cooperation between Japan and Sweden. Envir. Res. 13 (1977) 318–344.
25 Kojima, S., Haga, Y., Kurihara, T., Yamawaki, T., and Kjellström, T., A comparison between fecal cadmium and urinary β_2-microglobulin, total protein, and cadmium among Japanese farmers. An epidemiological study of cooperation between Japan and Sweden. Envir. Res. 14 (1977) 436–451.
26 Lauwerys, R.R., Buchet, J.P., and Roels, H., The realationship between cadmium exposure or body burden and the concentration of cadmium in blood and urine in man. Int. Archs occup. envir. Hlth 36 (1976) 375–385.
27 Lauwerys, R.R., Buchet, J.P., Roels, H.A., Brouwers, J., and Stanescu, D., Epidemiological survey of workers exposed to cadmium, Archs envir. Hlth 28 (1974) 145–148.
28 Lauwerys, R.R., Roels, H., Regniers, M., Buchet, J.P., Bernard, A., and Goret, A., Significance of cadmium concentration in blood and urine in workers exposed to cadmium. Envir. Res. 20 (1979) 375–391.
29 Nishiyama, K., and Nordberg, G.F., Adsorption and elution of cadmium on hair. Archs envir. Hlth 25 (1972) 92–96.

30 Nogawa, K., Ishizaki, A., and Kobayashi, E., A comparison between health effects of cadmium and cadmium concentration in urine among inhabitants of the Itai-Itai disease endemic district. Envir. Res. *18* (1979) 397–409.

31 Nogawa, K., Ishizaki, A., and Kawano, S., Statistical observation of the dose-response relationships of cadmium based on epidemiological studies in the Kakehaski River Basin. Envir. Res. *15* (1978) 185–198.

32 Nogawa, K., Kobayashi, E., Inaoka, H., and Ishizaki, A., The relationship between the renal effects of cadmium and cadmium concentrations in urine among the inhabitants of cadmium-polluted areas. Envir. Res. *14* (1977) 391–400.

33 Nogawa, K., Kobayashi, E., and Ishizaki, A., Clinico-chemical studies on chronic cadmium poisoning (Part II). Results of blood examinations. Jap. J. Hyg. *34* (1979) 415–419.

34 Nomiyama, K., Does a critical concentration of cadmium in human renal cortex exist? J. Toxic. envir. Hlth *3* (1977) 607–609.

35 Nomiyama, K., Toxicity of cadmium-mechanism and diagnosis, in: Progress in Water Technology. vol. 7, pp. 15–23. Ed. P. A. Krenkel. Pergamon Press, Oxford, 1975.

36 Nomiyama, K., Experimental studies on animals: in vivo experiments, in: Cadmium Studies in Japan, a review, pp. 47–86. Ed. K. Tsuchiya. Elsevier/North Holland Biomedical Press, New York and Kodansha, Tokyo 1978.

37 Nomiyama, K., and Nomiyama, H., Mechanism of urinary excretion of cadmium: experimental studies in rabbits, in: Effects and Dose-Response Relationships of Toxic Metals, pp. 371–379. Ed. G. F. Nordberg, Elsevier Scientific Publishing Company, Amsterdam 1976.

38 Nomiyama, K., Nomiyama, H., Akahori, F., and Masaoki, T., Further studies on effects of dietary cadmium on rhesus monkeys. (5) Renal effects, in: Recent Studies in Health Effects of Cadmium in Japan, pp. 59–104. Environment Agency, Office of Health Studies 1981.

39 Nomiyama, K., Nomiyama, H., Nomura, Y., Taguchi, T., Matsui, K., Yotoriyama, M., Akahori, F., Iwao, S., Koizumi, N., Masaoka, T., Kitamura, S., Tsuchiya, K., Suzuki, T., and Kobayashi, K., Effect of dietary cadmium on rhesus monkeys. Envir. Hlth Perspect. *28* (1979) 223–243.

40 Nomiyama, K., Sugata, Y., Murata, I., and Nakagawa, S., Urinary low molecular weight proteins in Itai-Itai disease. Envir. Res. *6* (1974) 373–81.

41 Nomiyama, K., Sugata, Y., Yamamoto, A., and Nomiyama, H., Effects of dietary cadmium on rabbits. I. Early signs of cadmium intoxication. Toxic. appl. Pharmac. *31* (1975) 4–12.

42 Nordberg, G. F., Cadmium metabolism and toxicity. Experimental studies on mice with special reference to the use of biological materials as indices of retention and the possible role of metallothionein in transport and detoxification of cadmium. Envir. Physiol. Biochem. *2* (1972) 7–36.

43 Nordberg, G. F., Garvey, J. S., and Chang, C. C., Metallothionein in plasma and urine of cadmium workers. Envir. Res. *28* (1982) 170–182.

44 Nordberg, G. F., and Nishiyama, K., Whole body and hair retention of cadmium in mice. Archs envir. Hlth *24* (1972) 209–214.

45 Nordberg, G. F., and Piscator, M., Influence of long term cadmium exposure on urinary excretion of protein and cadmium in mice. Envir. Physiol. Biochem. *2* (1972) 37–49.

46 Piscator, M., Proteinuria in chronic cadmium poisoning. I. An electrophoretic and chemical study of urinary and serum proteins from workers with chronic cadmium poisoning. Archs envir. Hlth *4* (1962) 607–621.

47 Piscator, M., Proteinuria and chronic cadmium poisoning. III. Electrophoretic and immunoelectrophoretic studies on urinary proteins from cadmium workers, with special reference to the excretion of low molecular weight proteins. Archs envir. Hlth *12* (1966) 335–344.

48 Piscator, M., Serum β_2-microglobulin in cadmium exposed workers. Path. Biol. *26* (1978) 321–323.

49 Piscator, M., and Axelsson, B., Serum proteins and kidney function after exposure to cadmium. Archs envir. Hlth *21* (1970) 604–608.

50 Price, R. G., Urinary enzymes, nephrotoxicity, and renal disease. Toxicology *23* (1982) 99–134.

51 Roels, H., Bernard, A., Buchet, J. P., Lauwerys, R., and Masson, P., Urinary excretion of β_2-microglobulin and other proteins in workers exposed to cadmium, lead or mercury. Path. Biol. *26* (1978) 329–331.

52 Roels, H., Lauwerys, R. R., Buchet, J. P., Bernard, A., Chettle, D. R., Harvey, T. C., and Al-Haddad, I. K., In vivo measurement of liver and kidney cadmium in workers exposed to this metal: its significance with respect to cadmium in blood and urine. Envir. Res. *26* (1981) 217–240.

53 Roels, H., Lauwerys, R., Materne, D., and Buchet, J. P. Study on cadmium proteinuria. Glomerular dysfunction: an early sign of renal impairment. Proceedings, Recent Advances in the Assessment of the Health Effects of Environmental Pollution *2* (1974) 631–641.

54 Sakurai, H., Occupational exposure, in: Cadmium Studies in Japan, a review, pp. 133–144. Ed. K. Tsuchiya. Elsevier/North Holland Biomedical Press, New York and Kodansha, Tokyo 1978.

55 Schaller, K. H., Gonzales, J., Thurauf, J., and Schiele, R., Früherkennung von Nierenschäden bei beruflich gegenüber Blei, Quecksilber und Cadmium exponierten Personen. Zbl. Bakt. Hyg. *171* (1980) 320–335.

56 Shaikh, Z. A., and Smith, J. C., Metabolism of orally ingested cadmium in humans, in: Mechanisms of Toxicity and Hazard Evaluation, pp. 569–574. Eds B. Holmstedt, R. Lauwerys, M. Mercier and M. Roberfroid. Elsevier/North Holland Biomedical Press, Amsterdam 1980.

57 Shaikh, Z. A., and Tohyama, C., Urinary metallothionein: a specific test for monitoring occupational cadmium exposure. Vet. hum. Toxic. *24* (1982) 276.

58 Shigematsu, I., and Kawaguchi, T., Environmental pollution and health effects: Akita Prefecture, in: Cadmium Studies in Japan, a Review, pp. 192–199. Ed. K. Tsuchiya. Elsevier/North Holland Biomedical Press, New York and Kodansha, Tokyo 1978.

59 Shiroishi, K., Kjellström, T., Kubota, K., Evrin, P.-E., Anayama, M., Vesterberg, O., Shinada, T., Piscator, M., Iwate, T., and Nishino, H., Urine analysis for detection of cadmium-induced renal changes with special reference to β_2-microglobulin. Envir. Res. *13* (1977) 407–424.

60 Taniguchi, N., Tanaka, M., Kishihara, C., Ohno, H., Kondo, T., Matsuda, I., Fujino, T., and Harada, M., Determination of carbonic anhydrase C and β_2-microglobulin by radioimmunoassay in urine of heavy-metal-exposed subjects and patients with renal tubular acidosis. Envir. Res. *20* (1979) 154–161.

61 Tohyama, C., and Shaikh, Z. A., Metallothionein in plasma and urine of cadmium-exposed rats determined by a single antibody radioimmunoassay. Fundam. appl. Toxic. *1* (1981) 1–7.

62 Tohyama, C., Shaikh, Z. A., Ellis, K. J., and Cohn, S. H., Metallothionein excretion in urine upon cadmium exposure: its relationship with liver and kidney cadmium. Toxicology *22* (1981) 181–191.

63 Tohyama, C., Shaikh, Z. A., Nogawa, K., Kobayashi, E., and Honda, R., Elevated urinary excretion of metallothionein due to environmental cadmium exposure. Toxicology *20* (1981) 289–297.

64 Tohyama, C., Shaikh, Z. A., Nogawa, K., Kobayashi, E., and Honda, R., Urinary metallothionein as a new index of renal dysfunction in «Itai-Itai» disease patients and other Japanese women environmentally exposed to cadmium. Arch. Toxic. *50* (1982) 159–166.

65 Tsuchiya, K. Cadmium in human urine, feces, blood, hair, organs and tissues, in: Cadmium Studies in Japan, a Review, pp. 37–43. Ed. K. Tsuchiya. Elsevier/North Holland Biomedical Press, New York and Kodansha, Tokyo 1978.

66 Tsuchiya, K., Iwao, S., Sugita, M., and Sakurai, H., Increased urinary B_2-microglobulin in cadmium exposure: dose-effect relationship and biological significance of B_2-microglobulin. Envir. Hlth Perspect. *28* (1979) 147–153.

Human health effects of exposure to cadmium

by William H. Hallenbeck

School of Public Health, University of Illinois at Chicago, P.O. Box 6998, Chicago (Illinois 60680, USA)

Summary. The health effects of human exposure to cadmium are discussed with emphases on intake, absorption, body burden, and excretion; osteomalacia in Japan; hypertension; and proteinuria, emphysema, osteomalacia, and cancer in workers. Elevated blood pressure has not been observed as a result of excessive exposures to cadmium in Japan or the workplace. Renal tubular dysfunction and consequent proteinuria is generally accepted as the main effect following long-term, low-level exposure to cadmium. Studies of workers show that proteinuria may develop after the first year of exposure or many years after the last exposure. Proteinuria and deterioration of renal function may continue even after cessation of exposure. The immediate health significance of low-level proteinuria is still under debate. However, there is evidence that long-term renal tubular dysfunction may lead to abnormalities of calcium metabolism and osteomalacia. The few autopsy and cross-sectional studies of workers do not permit conclusions to be drawn regarding the relationship between cadmium exposure and emphysema. Retrospective and historical-prospective studies are needed to settle this important question. No conclusive evidence has been published regarding cadmium-induced cancer in humans. However, there is sufficient evidence to regard cadmium as a suspect renal and prostate carcinogen. Because of equivocal results and the absence of dose-response relationships, the studies reviewed should be used with caution in making regulatory decisions and low-dose risk assessments.

Introduction

The main goal of this critical review was to summarize and evaluate studies of humans which could be useful in assessing the possible health effects of cadmium exposures via air, water, or food. In particular, data were sought which would permit the extrapolation of high dose effects to low levels of exposure. As indicated in the summary, the extrapolation objective was not achieved using the human literature. Whenever there were inconsistencies between the animal and limited human health effects literatures, human data were given precedence in this review.

Intake, absorption, body burden, excretion

The main routes of cadmium intake in man are the lungs and the gastrointestinal tract. The chemical form of ambient airborne cadmium is not known. Although measurements of airborne cadmium concentrations have been made in many countries, the concentrations are not strictly comparable because of different sampling times and different analytical methods. Size distributions of particles containing cadmium are rarely determined. Hence, only rough estimates of lung deposition rates can be made[14].
On the basis of limited cadmium-containing particle size distribution data and the application of a standard lung model, about 25% of cadmium inhaled in ambient air would be deposited in the lower respiratory tract[14]. Using this deposition fraction and an assumed average daily inhalation of 20 m^3, the amount of cadmium deposited in the lower respiratory tract has been estimated: rural areas, 0.0005–0.215 µg/day; urban areas, 0.01–3.5 µg/day; industrialized areas with cadmium emissions, 0.05–25 µg/day. The highest level of 25 µg/day is probably found only in the vicinity of an operation such as a smelter[14]. The rate of absorption through the lungs is a function of the chemical form and size distribution of the inhaled particles. Various rates have been reported.

One model based on human smokers predicts 50% of inhaled cadmium from tobacco smoke is absorbed[17]. One pack of 20 cigarettes can contain 30 µg of cadmium of which 2–4 µg can be inhaled[17,18,21]. In the general environment 13–19% of the cadmium inhaled is absorbed[14].
Ingestion of cadmium occurs via water and food. Tap water which is not particularly contaminated contains < 2 µg/l cadmium. This corresponds to an intake of 2–4 µg/day.
Analyses of the diets characteristic of several countries show that adult cadmium intake from food ranges from 4 to 84 µg/day[14]. Dietary data from Japan are excluded as they constitute a special case where the daily intake of cadmium via food in the endemic area was calculated at 600 µg by assuming an average cadmium concentration in rice of 1 µg/g and a cadmium concentration in other foodstuffs of about 10 times the value for Japan as a whole[21]. Total daily intake from all sources can range from 6 µg for a non-smoker living in a rural area and eating less contaminated food (0.0005 µg from air, 4 µg from food, 2 µg from water) to 115 µg for a 20-cigarette/day smoker living close to a cadmium emitting source and eating more contaminated food (25 µg from air, 84 µg from food, 2 µg from water, 4 µg from cigarettes). The human gastrointestinal absorption rate ranges between 4.7 and 7% and was estimated through experiments on 5 human volunteers (19–50 years old) who were given labeled cadmium orally[14]. Animal studies[14] indicate that diets low in calcium, iron, and protein can stimulate cadmium absorption by a factor of about 2. Also, one study has shown that the rate of gastrointestinal absorption of cadmium is higher in young mice than in adults[14].
Body burden of cadmium ranges from < 1 µg in the human newborn (indicating that the placenta is an effective barrier to cadmium) to 15–30 mg in the normal adult[21]. The placenta is less permeable to cadmium than to lead and mercury. Cadmium is

about 50% lower in newborn vs maternal blood[31]. For the normal adult, about 50% of the body burden is in liver and kidneys and about one-third in the kidneys alone. Kidney cortex concentrations are generally higher than kidney medulla concentrations by a factor of 1.1–9.6[21]. In normal persons, the highest concentration of cadmium is found in the kidney, followed by the liver and other organs. For a recent high-level industrial exposure, the liver will contain a higher proportion of the total body burden than the kidney. In some cases the concentration of cadmium in the liver has exceeded the kidney level[7,21,56]. The pancreas may also contain high concentrations of cadmium[21]. Accumulation in the kidneys peaks at about age 50 when mean renal cortex concentrations range between 11 and 50 µg/g wet weight[17,21]. After about age 50, cortex levels decrease[17]. The report by Elinder et al.[17] is particularly useful in that kidney cortex, liver, and pancreas cadmium concentrations are presented as a function of age. Renal cortex levels of 300 µg/g wet weight have been found in exposed workers. However, normal values have been reported despite signs of cadmium toxicity. It is thought that unexpectedly low values are due to cadmium losses following renal dysfunction. Renal damage may occur at cadmium concentrations over 200 µg/g wet weight of kidney cortex[21].

Several studies have demonstrated significantly higher kidney, liver, and lung cadmium levels in smokers vs non-smokers[17,37,38,42,55,59]. The accumulation in smokers is related to the number of pack-years smoked[38]. These studies indicate that cigarettes are a major source of cadmium and can double a smoker's vs non-smoker's kidney burden of cadmium[17,37,42].

For a non-smoker, the biological half-life of cadmium in the kidney cortex is estimated at 30 years, with an average concentration at age 50 of 11 µg/g wet weight[17]. Smokers have an average cadmium concentration in the kidney cortex of 22 µg/g wet weight at age 50[17].

Given that there are analytical problems in the analysis of blood for cadmium, there appears to be no relation between blood levels of cadmium and body burden or kidney burden[14,21]. In recently exposed workers, cadmium in blood may increase without a corresponding change in urinary cadmium output. Cadmium in blood is probably a reflection of current exposure and not body burden[14]. After cessation of exposure, blood levels decrease slowly. However, for a short, high level exposure, the initial decrease of blood cadmium may be rapid after exposure ceases[21]. Normal concentrations of cadmium in blood are < 1 µg/100 ml whole blood[14,21]. Exposed workers' blood cadmium may range between 1 and 10 µg/100 ml whole blood[21].

Normal urinary levels of cadmium increase with age and are < 2 µg/day. This increase is probably a function of the increase in kidney burden with age. Urinary cadmium is a poor index of body burden or kidney burden. The urinary level of cadmium may remain within normal limits for some time during occupational exposure. If renal tubular dysfunction occurs (signaled by an increase in excretion of low molecular weight proteins) there will be an increase in cadmium excretion which can be dramatic. Urinary excretion in exposed workers can be several hundred µg/day[14,21].

Except for the period right after exposure, fecal excretion is low. Insignificant amounts may be excreted via hair, sweat, breast milk, and saliva[21].

Metallothionein, a low molecular weight (10,000–12,000) metal binding protein rich in cysteine residues, binds with cadmium, zinc, copper, mercury, silver, and tin in vivo. It has been detected in human kidney, liver, heart, brain, testis, and skin epithelial cells. Most of the cadmium in tissues is probably bound to metallothionein[14]. Cadmium and zinc appear to be the only metals which can induce the synthesis of this protein[15,62]. Induction of metallothionein synthesis has been shown in kidney, liver, and intestine[14]. The role of metallothionein in cadmium absorption, transport, storage, and excretion is not well defined in humans. It is especially unclear whether metallothionein plays an overall protective or toxic role.

Association with osteomalacia in Japan (reference 21 unless otherwise indicated)

In 1946 'Itai-Itai byo' or ouch-ouch disease was recognized in Toyama Prefecture, Japan. In 1948 osteomalacia was suspected as the cause. Osteomalcia results from: vitamin D deficiency; malabsorption of vitamin D and bone minerals; or renal tubular dysfunction which results in loss of bone minerals through the kidneys. The last type is called vitamin D-resistant or renal osteomalacia, and Itai-Itai disease is classified as a vitamin D-resistant form of osteomalacia. However, since most of the Itai-Itai patients were postmenopausal women who had an average of 6 deliveries, it is believed that a low vitamin and calcium intake, a high demand for calcium and vitamin D during pregnancy and lactation, and deprivation of UV irradiation were contributing etiological factors. The disease is characterized by: skeletal deformities with a marked decrease in height; lumbar pains; leg muscle pain;pain induced by pressure on bones, especially the femurs, backbone, and ribs; ducklike gait; bones susceptible to multiple fractures after very slight trauma such as coughing; impaired pancreatic function; changes in the gastrointestinal tract; hypochromic anemia; renal tubular dysfunction resulting in proteinuria[54] (low molecular weight proteins such as immune globulins and β_2-microglobulin), glucosuria, and aminoaciduria; low levels of serum iron, calcium, and inorganic phosphorous, and high levels of alkaline phosphatase.

It has been well established that excessive exposure to cadmium can result in renal tubular dysfunction with characteristic proteinuria, aminoaciduria, and glucosuria. Kidney damage seen in Itai-Itai disease has been very similar to that seen in industrial chronic cadmium poisoning[1,5,7,19–21,25,32,47,49,56,58]. Hypercalciuria also occurs as a result of renal tubular damage[25]. However, urinary calcium levels were normal in Itai-Itai patients. This may not seem so surprising if it is assumed that by the time symptoms developed, mobilization and excretion of bone calcium had already occurred.

The most probable source of excessive cadmium intake in the endemic area (Toyama Prefecture) was rice which had been grown in irrigation water contaminated with the effluents of a mining operation. The cadmium content of rice in some areas was more than 10 times the average in Japan.

Association with hypertension

Rat studies clearly indicate that cadmium administered in drinking water at levels in the range of 0.1–20 ppm can produce elevated systolic and diastolic blood pressures and increase mortality[11,29,41,44-46,51,53]. Water concentrations above this range are toxic or decrease blood pressure[29,44,45], while those below seem to have no effect on blood pressure[29,45]. Also rat studies have shown that the blood pressure elevating effect of cadmium can be inhibited by adding selenium (3.6 ppm), zinc (200 ppm), or copper to drinking water or by dissolving the cadmium in hard water rather than deinozed water[45,46]. Rat studies have also shown that there can be a genetic predisposition to the pressor effects of cadmium[41].

In spite of the rather convincing animal data, there is no direct proof of a causal relationship between cadmium and the development of human essential hypertension. Cadmium was first suspected in the early 1950's when effective antihypertensive drugs first became available. The ability to bind transition and related trace metals was a common characteristic of several of these drugs[43]. Experiments with ethylenediamine tetraacetate (EDTA) indicated that cadmium, copper, or zinc could be the metal on which these drugs acted[43]. Of these three, cadmium was most suspect due to its affinity for the kidney, an organ recognized for its critical role in controlling blood pressure. Indeed several studies have shown that humans who have died from hypertensive complications had increased renal cadmium concentrations[14,35,43,52].

However, the results of other human autopsy studies have not shown a significant correlation between renal cadmium accumulation and hypertension[14,42,48,59]. This discrepancy may be a result of uncontrolled differences between test and control groups, e.g. smoking habits, age, nutritional status, and stage of disease. Also there are possible errors due to wrong diagnosis and small sample sizes. Smoking is correlated with elevated renal cadmium. It is likely that the possible association between hypertension and elevated renal cadmium is secondary to a primary association between hypertension and smoking. Also it is extremely important to match autopsy samples not only on the basis of smoking history but also by age. This is necessary due to the 'natural' accumulation of cadmium in the kidneys of non-occupationally exposed people. However, the affect of age may be small since the majority of cadmium accumulates before age 30[52]. Stage of hypertension is important. Patients dying of malignant hypertension with renal

failure have low values of renal cadmium[21,43,52]. Low renal cadmium concentrations due to severe renal damage have also been observed in cadmium-exposed workers and Itai-Itai disease[21,43]. Regarding nutritional status, a recent study[40] was designed to compare the calcium intake in humans with established hypertension to that of a normotensive group matched for age, sex, and race. Compared to 44 normotensive controls, 46 hypertensives reported significantly less daily calcium ingestion (688 ± 55 mg compared to 886 ± 89 mg). The intake of other nutrients, including sodium and potassium, was very similar in the two groups.

An important question arises at this point: if hypertension can be induced in rats at low ingested doses of cadmium, what has been the experience of people living in the cadmium-polluted areas of Japan and those occupationally exposed to cadmium? From animal experiments it is clear that high doses of cadmium do not produce hypertension. Therefore, it is not totally unexpected that the incidence of hypertension is not increased among Japanese suffering from Itai-Itai and presumably exposed to very high levels of cadmium[14,21]. What is surprising is that other Japanese living in the cadmium-polluted areas and not suffering from Itai-Itai did not develop hypertension even though some moderate level of excessive cadmium exposure was almost certain[21,43]. Also, an abnormally high incidence of hypertension has not been observed in workers exposed to cadmium dusts and fumes[1,2,5,7,9,10,14,19-25,32,33,47,49,50,52,56-58,61].

Occupational exposure

One of the earliest reports of industrial cadmium poisoning was published in 1938[13]. This report concentrated on a presentation of the acute responses of 15 workers exposed to high but unspecified levels of cadmium fumes from an annealing furnace. The first symptom was usually throat irritation occurring at the time of exposure. This irritation was not sufficient to compel the workers to leave the exposure even when fatal concentrations were being breathed. Delayed (hours or days) symptoms included chest soreness aggravated by deep breathing, dyspnea, violent coughing, nausea, vomiting, cyanosis, pulmonary edema, elevated temperature (up to 38.9 °C) and elevated pulse. Two deaths occurred after 4 and 8 days. The most distressing symptom was severe attacks of dyspnea which commenced hours or even days after exposure. Due to the delayed appearance of serious symptoms, it is possible to mistake cadmium fume poisoning for some other illness such as influenza. Also, the clinical picture of cadmium poisoning is similar to that caused by nitrous or zinc fumes[6,13]. Both cause severe lung damage which usually manifests itself hours after exposure. No permanent ill effects (such as fibrosis) were observed in the non-fatal cases during the 8-month follow-up period. An attempt was made to quantify the exposure from the annealing furnace[3]. It was concluded that a lethal exposure of thermally generated cadmium oxide, for man doing light work, is less than

2900 min-mg/m^3. Exposures less than this caused incapacitation of all men exposed. A later study in 1966 reported a similar time-concentration of 2589 min-mg/m^3 (8.6 mg/m^3 for 5 h) which had caused the death of a worker exposed to cadmium fume[6].

It was reported in 1940[7] that workers plating metals with cadmium by an electrolytic process had chronic rhinitis and pharyngitis, dryness and irritation of the pharynx, a burning sensation in the nose with nasal hemorrhage, and ulcers in the cartilaginous parts of the nose and the nasopharynx.

Friberg reported on a study of 58 workers employed in the manufacture of storage batteries[19,20]. Workers were exposed to both cadmium-iron dust and nickel-graphite dust. Air analyses showed 3–15 mg cadmium/m^3 and 10–150 mg nickel/m^3 of air. 95% of the cadmium-iron dust and 85% of the nickel-graphite dust particles were less than 5 µm. In the group of 43 workers with 9–34 years of exposure (average age, 44), 50% had pulmonary emphysema. No emphysema was observed in the 15 workers in the short exposure (1–4 years) group (average age, 35). However, a large number of the high-exposure workers had tuberculous lung changes. This finding clouds the interpretation of the emphysema since the tuberculosis may have preceded and caused the emphysema or exposure to the dust may have caused an increased susceptibility to tuberculosis[20]. There was no discussion of smoking habits. Emphysema will be discussed in more detail later. Two-thirds of the workers in the long-exposure group had proteinuria (20,000–30,000 molecular weight). No proteinuria was found in the low-exposure group. One-third of the long-exposure group had anosmia (absence of the sense of smell). A distinct yellow coloring of the front teeth occurred in workers in both the high- and low-exposure groups. Overall the workers complained of tiredness, shortness of breath, cough, and impaired olfactory sense. Friberg suspected that the emphysema, proteinuria, and anosmia resulted from the cadmium component of the dust rather than the nickel. However, it was emphasized that the nickel may have contributed to the emphysema and anosmia[20]. In a follow-up report in 1952, Friberg and Nystrom[7] re-examined the 43 men who had more than 9 years employment in the battery industry. There had been no further exposure to cadmium in the intervening years. Five had died: 2 due to emphysema, 2 due to coronary thrombosis (severe renal damage attributed to cadmium was found at autopsy), and 1 died from acute pancreatitis (lungs were found to be emphysematous). In 9 of the remaining 38, disease had progressed: increased dyspnea in 5, development of proteinuria in 4, deterioration of renal function in 3. The symptoms of 25 were unchanged, and in 4 there was a distinct improvement in the performance of respiratory function tests.

Proteinuria has been reported in cadmium workers many times since Friberg's report[1,5,7,14,21,25,32 47,49,56,58,61]. Exposures of 50–1000 µg cadmium dust/m^3[1], 3–67 µg total cadmium/m^3[5], 75–240 µg time-weighted cadmium fume/m^3[61], and 134 µg cadmium dust/m^3[32], have resulted in proteinuria. The minimal latent period before onset of proteinuria is about 1 year from the beginning of exposure[61]. However, the first sign of disease may develop many years after the last exposure[8]. Urinary protein increases gradually to < 110 mg/100 ml in most cases[1]. Normal adult urinary protein excretion averages 50 mg/day. Cadmium exposed individuals excrete 70–2600 mg/day[47]. Recent work indicates that the kidney lesion is first glomerular and later becomes predominantly tubular[5,32]. The sedimentation and electrophoretic properties of urine proteins from patients with known tubular dysfunction are similar to those found in cases of chronic cadmium poisoning[8,25]. Although Bonnell states that most cases of proteinuria are well compensated and symptoms of renal failure are rare[8], autopsy studies of cadmium-employed workers have shown evidence of severe renal damage[2,7].

Once established, proteinuria persists even after cessation of exposure[1]. There was no evidence in the study by Adams et al.[1] that renal function continues to deteriorate after cessation of exposure. However, others have reported that deterioration of renal function does continue after exposure ceases[7,9,10]. There has been at least 1 fatal case of chronic renal failure in a cadmium worker. The exposure was estimated at several hundred µg cadmium fume per m^3[7].

Some researchers feel that the significance of cadmium-induced proteinuria has not yet been established[8,25,49]. Others take the position that cadmium-induced proteinuria is clinically significant and should be regarded as an early manifestation of renal tubular damage[1,25].

Additional abnormalities suggestive of renal tubular malfunction have been found in cadmium workers: glycosuria, impaired acid excretion, hyperchloremic acidosis, abnormal aminoaciduria, impaired concentrating ability, hypocalcemia, hypophosphatemia, hyperphosphaturia, nephrocalcinosis (renal stones), and hypercalciuria[24,25]. While these biochemical abnormalities may not be of immediate importance to the health of the individual, long-term abnormal calcium metabolism (as indicated by hypercalciuria, nephrocalcinosis, and hyperphosphaturia) may result in osteomalacia[24]. Another possible factor contributing to osteomalacia concerns vitamin D. The results of an animal study showed that cadmium can interfere with the final activation of vitamin D_3 to 1,25-dihydroxycholecalciferol in the renal tubules[24]. Thus, depending on the degree of kidney damage, administration of vitamin D_3 and calcium may or may not lead to improvement in cases of osteomalacia. For example, Itai-Itai appeared to be a vitamin-D resistant form of osteomalcia[21]. However, in the few reported cases of osteomalacia in cadmium workers, administration of vitamin D_3 and other supplements resulted in improvement[1,7,24].

Anemia has been reported in workers exposed to cadmium fumes[20,21,25,61]. Its significance cannot be evaluated at this time.

Experimental animals which survive the acute pneumonitis that follows inhalation of cadmium fumes develop a perivascular and peribronchial fibrosis[6,8,60]. Cadmium-related fibrosis was not described in man[6,7,56,60] until a report by Smith et al. in 1976[57].

Chest X-rays showed mild to moderate fibrosis in cadmium-exposed workers. However, fibrosis in man is also related to tuberculosis, influenza, pneumonia, chronic bronchitis, pneumoconioses (e.g. silica, hematite, silicates, asbestos, coal, aluminium, beryllium, and tungsten). Occupational histories were not discussed.

Since Friberg's early work[19,20], there have been several reports of the occurrence of emphysema in cadmium workers[2,7,9,10,14,21,23,25,30,56]. These have been either autopsy studies[2,30,56] or cross-sectional (prevalance) studies of factory workers[7,9,10,23,25]. None of these studies took smoking habits into account. Furthermore, the diagnosis of emphysema in several studies[7,9,10,23,25] has been disputed[58]. Hence, it is still a matter of controversy whether chronic occupational exposure to cadmium produces emphysema. The concept of cadmium-induced emphysema is based on conclusions drawn in the older literature, and, in many instances, these conclusions have been accepted without any criticism.

Stanescu et al.[58] have critically reviewed the emphysema literature including the work of Friberg[20], Baader[2], Bonnell[7,9,10], Kazantzis et al.[23,25], Princi[50], Potts[49], Hardy and Skinner[22], Suzuki (see Stanescu[58]), Tsuchiya[61], Adams et al.[1], Lauwerys et al.[32], Smith et al.[56,57], and Lane and Campbell[30]. They concluded that either there is no causal relationship between chronic exposure to cadmium and emphysema, or that a mild form of obstructive lung disease affects some workers. This conclusion cannot be extrapolated to acute or subacute inhalation exposure[33,58].

Stanescu et al.[58] studied 18 workers who were exposed to a minimum of 50–356 μg of cadmium oxide dust/m^3 for 22–40 years (average of 32 years). The level of exposure was only a crude estimate. A control group was composed of 20 non-exposed workers, comparable to the exposed group on the bases of age, height, weight, and number of smokers and non-smokers. 33 of the 38 workers exposed and non-exposed were smokers or ex-smokers. However, the number of pack-years of exposure for the nonexposed workers was statistically significantly greater than that for the exposed workers ($p < 0.05$). Proteinuria (88–1740 mg/l) and cadmium concentrations in urine (27.5 μg/g creatinine) and blood (2.47 μg/100 ml) were significantly greater in the exposed group. Grade 1 dyspnea was more frequent in the exposed group, but no difference in the prevalence of other respiratory symptoms was found. The authors suggested that the increased reporting of dyspnea may have been motivated by a desire for compensation for occupational disease. There were only minor differences in lung function between the two groups. No emphysema was reported. This finding of no emphysema was supported by the results of earlier studies[1,22,31,50,57,58,61]. However, the designs of these negative studies were such that the authors may have viewed survivor populations, and those more susceptible to the effects of cadmium may have already disappeared from the work force and possible observation. The autopsy and prevalence studies published to date do not permit conclusions to be drawn regarding the relationship between cadmium exposure and emphysema. Retrospective and historical-prospective epidemiological studies of cadmium workers are needed.

Association with cancer

No conclusive data have been published regarding cadmium-induced cancer in humans[27]. In 1965 Potts[49] reported on 8 deaths in a group of 70 battery workers exposed for more than 10 years to cadmium oxide dust. 3 deaths were due to prostatic cancer, 1 due to bronchial carcinoma, and 1 due to carcinomatosis. Neither autopsy confirmation nor smoking habits were discussed. Definite conclusions cannot be drawn from a study with so few cases and no control group.

Kipling and Waterhouse[26] studied the same battery plants as Potts (see Malcolm[39]). All 248 employees and ex-employees with more than 1 year of exposure were included. There were 12 deaths due to carcinoma. Only the 4 deaths due to carcinoma of the prostate were significantly greater ($p = 0.003$) than the number expected (0.58, calculated from a regional cancer register). The number of observed cases of prostate cancer were too small to permit firm conclusions. Neither autopsy confirmation nor smoking habits were discussed. Malcolm[39] reviewed the Kipling-Waterhouse study[26] and reported that causes of death were not confirmed by autopsy and that the 4 men thought to have died of prostatic cancer had been exposed to cadmium oxide dust and nickel hydroxide, powdered nickel, and ferric hydroxide. However, Malcolm expressed doubt that nickel could be related to prostatic cancer, because the association had never been observed in the nickel industry.

Lemen et al.[34] returned to the same smelter Princi[50] had studied 30 years earlier. However, instead of using a cross-sectional design, Lemen et al. used a much more sensitive historical-prospective design. Employment histories were obtained for 292 white male cadmium workers who had at least 2 years of employment in the plant between January 1, 1940 and December 31, 1969. Vital status follow-up was continued through January 1, 1974. Comparison was made between the observed number of deaths among the study cohort and that expected by use of age-, calendar-time, and cause-specific mortality rates for the total U.S. white male population. Lemen et al. found a significantly increased mortality due to total malignancies (27 observed vs 17.5 expected, $p < 0.05$), lung cancer (12 vs 5.1, $p < 0.05$), and prostatic cancer (4 vs 0.88, $p < 0.05$, for a latency period $\geq$ 20 years). Cause of death was primarily determined by interpretation of death certificates. There was no discussion of autopsy findings. Most of the excess risk for total malignancies was due to the lung cancer deaths. Smoking habits for these 12 men had not been obtained. However, histologic cell type was available for 8 bronchogenic carcinomas: 1 was undifferentiated small cell, 3 were anaplastic, 3 were squamous cell, and 1 was an oat cell carcinoma. No interpretation of this cell type information was made. The number of prostatic cancers were too small to allow definite conclusions to be drawn. As a footnote to this study, it is interesting to note that a long-term cadmium injection study of rats did not show any tumors in the prostate[36].

Kolonel[28] carried out a case-control study wherein 64 patients with renal malignancies were compared to 2 control groups, 72 patients with colon cancer and 197 with non-malignant gastrointestinal diseases. The 3 groups were well-matched for age (50–79), race (white), sex (male), computed dietary intakes of cadmium, smoking habits, interviewer bias (cases and controls were admitted and interviewed under the same tentative diagnosis of malignant tumor), and socioeconomic status. A person was considered to have had potential occupational exposure to cadmium if he had worked 1 or more years at a high-risk job within a high-risk industry (electroplating, alloy-making, welding, manufacture of storage batteries). Herein lies a major weakness in this study. Since there was no exposure data, there was no way of knowing how many patients were misclassified as to possible cadmium exposure. A statistically significant association was found between renal cancer and probable occupational exposure to cadmium ($p < 0.05$). It is notable that similar significant associations were found when the renal cancer cases were compared to either control group. Hence, the association with potential cadmium exposure appears to be specific for renal cancer but not all types of cancer. The association was even stronger between renal cancer and the combined affects of probable occupational exposure and smoking ($p < 0.01$) Hence, synergism between smoking and occupational exposure to cadmium was suggested. This possiblity is reasonable given the cadmium content of cigarettes referred to earlier.

Kjellstrom et al.[27] reported on new cases of cancer (1959–1975) in 228 cadmium-nickel battery workers. This group comprised all workers with 5 or more years of exposure to cadmium. Workers had been exposed to dusts of cadmium oxide and nickel hydroxide. Cadmium levels ranged as follows: before 1947, 1 mg Cd/m^3; 1950's, 200 µg Cd/m^3; 1962–1974, 50 µg Cd/m^3; since 1974, 5 µg Cd/m^3. Expected numbers of cancers (prostate, lung, kidney, bladder, colon-rectum, pancreas, nasopharynx, other, and all sites) were calculated using the life-table method and national average incidence rates. Out of the 9 cancer categories, only the observed number of new cases of nasopharyngeal cancers was statistically significantly greater than expected. However, this finding was based on the observation of only 2 cases, and caution is advised in interpreting such a small number (the negative findings also were based on very small numbers of observed cases). An unusually high occurrence of cancer of the nasal cavity has been reported among nickel smelter workers[27]. Kjellstrom et al. point out that the battery workers had been exposed to higher levels of nickel hydroxide dust than cadmium oxide dust. Finally, Kjellstrom et al. did not discuss smoking habits.

Chromosome aberrations may be related to the development of cancer and/or the inheritance of potentially undesirable traits. Several chromosome analyses have been conducted on the peripheral leucocytes of cadmium workers and Itai-Itai patients[4, 12, 16]. Contradictory results have been obtained, and it is not possible to draw firm conclusions at this time, especially since there were simultaneous exposures to lead and cadmium[4, 16].

1 Adams, R.G., Harrison, J.F., and Scott, P., The development of cadmium-induced proteinuria, impaired renal function, and osteomalacia in alkaline battery workers. Q. Jl Med. *38* (1969) 425–443.

2 Baader, E.W., Chronic cadmium poisoning. Ind. Med. Surg. *21* (1952) 427–430.

3 Barrett, H.M., and Card, B.Y., Studies on the toxicity of inhaled cadmium, the acute lethal dose of cadmium oxide for man. J. ind. Hyg. Toxic. *29* (1947) 286–293.

4 Bauchinger, M., Schmid, E., Einbrodt, H.J., and   J., Chromosome aberrations in lymphocytes after occupational exposure to lead and cadmium. Mutation Res. *40* (1976) 57–62.

5 Bernard, A., Roels, H., Hubermont, G., Buchet, J.P., Masson, P.L., and Lauwerys, R.R., Characterization of the proteinuria in cadmium-exposed workers. Int. Archs Occup. envir. Hlth *38* (1976) 19–30.

6 Beton, D.C., Andrews, G.S., Davies, H.J., Howells, L., and Smith, G.F., Acute cadmium fume poisoning, five cases with one death from renal necrosis. Br. J. ind. Med. *23* (1966) 292–301.

7 Bonnell, J.A., Emphysema and proteinuria in men casting copper-cadmium alloys. Br. J. ind. Med. *12* (1955) 181–197.

8 Bonnell, J.A., Cadmium poisoning. Ann. occup. Hyg. *8* (1965) 45–50.

9 Bonnell, J.A., Kazantzis, G., and King, E., A follow-up study of men exposed to cadmium oxide fume. Br. J. ind. Med. *16* (1959) 135–147.

10 Bonnell, J.A., Ross, J.H., and King, E., Renal lesions in experimental cadmium poisoning. Br. J. ind. Med. *17* (1960) 69–80.

11 Boscolo, P., Porcelli, G., Carmignani, M., and Finelli, V.N., Urinary kallikrein and hypertension in cadmium-exposed rats. Toxic. Lett. *7* (1981) 189–194.

12 Bui, T.H., Lindsten, J., and Nordberg, G.F., Chromosome analysis of lymphocytes from cadmium workers and itai-itai-patients. Envir. Res. *9* (1975) 187–195.

13 Bulmer, F.M.R., and Rothwell, H.E., Industrial cadmium poisoning, a report of fifteen cases, including two ceaths. Can. public Hlth J. *29* (1938) 19–26.

14 Commission of the European Communities, Criteria (Dose/Effect Relationships) for Cadmium, pp. 1–202. Pergamon Press, New York 1978.

15 Cousins, R.J., Metallothionein synthesis and degradation: Relationship to cadmium metabolism. Envir. Hlth Perspect. *28* (1979) 131–136.

16 Deknudt, G., and Leonard, A., Cytogenetic investigations on leucocytes of workers from a cadmium plant. Envir. Physiol. Biochem. *5* (1975) 319–327.

17 Elinder, C.G., Kjellstrom, T., and Friberg, L., Cadmium in kidney cortex, liver, and pancreas from Swedish autopsies. Archs envir. Hlth *31* (1976) 292–302.

18 Fishbein, L., Environmental metallic carcinogens; An overview of exposure levels. J. Toxic. envir. Hlth *2* (1976) 77–109.

19 Friberg, L., Proteinuria and emphysema among workers exposed to cadmium and nickel dust in a storage battery plant. Proc. int. Congr. ind. Med. *9* (1948) 641–644.

20 Friberg, L., Health hazards in the manufacture of alkaline accumulators with special reference to chronic poisoning, a clinical and experimental study. Acta med. scand., suppl. 240, *138* (1950) 94–97.

21 Friberg, L., Piscator, M., Nordberg, G.E., and Kjellstrom, T., Cadmium in the Environment, pp. 1–248. CRC Press, Cleveland 1974.

22 Hardy, H.L., and Skinner, J.B., The possibility of chronic cadmium poisoning. J. ind. Hyg. Toxic. *29* (1947) 321–324.

23 Kazantzis, G., Respiratory function in men casting cadmium alloys, part 1, assessment of ventilatory function. Br. J. ind. Med. *13* (1956) 30–40.

24 Kazantzis, G., Renal tubular dysfunction and abnormalities of

calcium metabolism in cadmium workers. Envir. Hlth Perspect. *28* (1979) 155–159.
25 Kazantzis, G., Flynn, F.V., Spowage, J.S., and Trott, D.G., Renal tubular malfunction and pulmonary emphysema in cadmium pigment workers, Q. Jl Med. *32* (1963) 165–192.
26 Kipling, M.D., and Waterhouse, J.A.H., Cadmium and prostatic carcinoma. Lancet *1* (1967) 730–731.
27 Kjellstrom, T., Friberg, L., and Rahnster, B., Mortality and cancer morbidity among cadmium-exposed workers. Envir. Hlth Perspect. *28* (1979) 199–204.
28 Kolonel, L.N., Association of cadmium with renal cancer. Cancer *37* (1976) 1782–1787.
29 Kopp, S.J., Glonek, T., Perry, H.M., Erlanger, M., and Perry, E.F., Cardiovascular actions of cadmium at environmental exposure levels. Science *217* (1982) 837–839.
30 Lane, R.E., and Campbell, A.C.P., Fatal emphysema after exposure to cadmium. Br. J. ind. Med. *11* (1954) 118–122.
31 Lauwerys, R., Buchet, J.P., Roels, H., and Hubermont, G., Plancental transfer of lead, mercury, cadmium, and carbon monoxide in women, comparison of the frequency distributions of the biological indices in maternal and umbilical cord blood. Envir. Res. *15* (1978) 278–289.
32 Lauwerys, R.R., Buchet, J.P., Roels, H.A., Brouwers, J., and Stanescu, D., Epidemiological survey of workers exposed to cadmium. Archs envir. Hlth *28* (1974) 145–148.
33 Lauwerys, R.R., Roels, H.A., Buchet, J.P., Bernard, A., and Stanescu, D., Investigations on the lung and kidney function in workers exposed to cadmium. Envir. Hlth Perspect. *28* (1979) 137–145.
34 Lemen, R.A., Lee, J.S., Wagoner, J.K., and Blejer, H., Cancer mortality among cadmium production workers. Ann. N.Y. Acad. Sci. *271* (1976) 273–279.
35 Lener, J., and Bibr, B., Cadmium and hypertension. Lancet *1* (1971) 970.
36 Levy, L.S., Roe, F.J.C., Malcolm, D., Kazantzis, G., Clack, J., and Platt, H.S., Absence of prostatic changes in rats exposed to cadmium. Ann. occup. Hyg. *16* (1973) 111.
37 Lewis, G.P., Jusko, W.J., and Coughlin, L.L., Cadmium accumulation in man: Influence of smoking, occupation, alcoholic habit and disease. J. chronic Dis. *25* (1972) 717–726.
38 Lewis, G.P., Coughlin, L.L., Jusko, W.J., and Hartz, S., Contribution of cigarette smoking to cadmium accumulation in man. Lancet *1* (1972) 291–292.
39 Malcolm, D., Potential carcinogenic effect of cadmium in animals and man. Ann. occup. Hyg. *15* (1972) 33–36.
40 McCarron, D.A., Morris, C.D., and Cole, C., Dietary calcium in human hypertension. Science *217* (1982) 267–269.
41 Ohanian, E.V., and Iwai, J., Effects of cadmium ingestion in rats with opposite genetic predisposition to hypertension. Envir. Hlth Perspect. *28* (1979) 261–266.
42 Ostergaard, K., Renal cadmium concentration in relation to smoking habits and blood pressure. Acta med. scand. *203* (1978) 379–383.
43 Perry, H.M., Hypertension and the geochemical environment. Ann. N.Y. Acad. Sci. *28* (1972) 202–216.
44 Perry, H.M., and Erlanger, M.W., Metal-induced hypertension following chronic feeding of low doses of cadmium and mercury. J. Lab. clin. Med. *83* (1974) 541–547.
45 Perry, H.M., Erlanger, M., and Perry, E.F., Increase in the systolic pressure of rats chronically fed cadmium. Envir. Hlth Perspect. *28* (1979) 251–260.
46 Perry, H.M., Erlanger, M.W., and Perry, E.F., Inhibition of cadmium-induced hypertension in rats. Sci. total Envir. *14* (1980) 153–166.
47 Piscator, M., Proteinuria in chronic cadmium poisoning. Archs envir. Hlth *4* (1962) 607–621.
48 Piscator, M., Cadmium and hypertension. Lancet *2* (1976) 370–371.
49 Potts, C.L., Cadmium proteinuria, the health of battery workers exposed to cadmium oxide dust. Ann. occup. Hyg. *8* (1965) 55–61.
50 Princi, F., A study of industrial exposures to cadmium. J. ind. Hyg. Toxic. *29* (1947) 315–320.
51 Schroeder, H.A., Cadmium hypertension in rats. Am. J. Physiol. *207* (1964) 62–66.
52 Schroeder, H.A., Cadmium as a factor in hypertension. J. chronic Dis. *18* (1965) 647–656.
53 Schroeder H.A., and Vinton, W.H., Hypertension induced in rats by small doses of cadmium. Am. J. Physiol. *202* (1962) 515–518.
54 Shiroishi, K., Kjellstrom, T., Kubota, K., Evrin, P.E., Anayama, M., Vesterberg, O., Shimada, T., Piscator, M., Iwata, T., and Nishino, H., Urine analyses for detection of cadmium-induced renal changes, with special reference to beta-2-microglobulin. Envir. Res. *13* (1977) 407–424.
55 Shuman, M.S., Voors, A.W., and Gallagher, P.N., Contribution of cigarette smoking to cadmium accumulation in man. Bull. envir. Contam. Toxic. *12* (1974) 570–576.
56 Smith, J.P., Smith, J.C., and McCall, A.J., Chronic poisoning from cadmium fume. J. Path. Bact. *80* (1960) 287–296.
57 Smith, T.J., Petty, T.L., Reading, J.C., and Lakshiminarayan, S., Pulmonary effects of chronic exposure to airborne cadmium. A. Rev. Resp. Dis. *114* (1976) 161–169.
58 Stanescu, D., Veriter, C., Frans, A., Goncette, L., Roels, H., Lauwerys, R., and Brasseur, L., Effects on lung of chronic occupational exposure to cadmium. Scand. J. Resp. Dis. *58* (1977) 289–303.
59 Syversen, T.L.M., Stray, T.K., Syversen, G.B., and Ofstad, J., Cadmium and zinc in human liver and kidney. Scand. J. clin. Lab. Invest. *36* (1976) 251–256.
60 Townshend, R.H., A case of acute cadmium pneumonitis: Lung function tests during a four-year follow-up. Br. J. ind. Med. *25* (1968) 68–71.
61 Tschuchiya, K., Proteinuria of workers exposed to cadmium fumes. Archs envir. Hlth *14* (1967) 875–880.
62 Webb, M., Protection by zinc against cadmium toxicity. Biochem. Pharmac. *21* (1972) 2767–2771.

Cadmium, the environment and human health: an overview

by K.J. Yost

Institute of Environmental Health, Purdue University, West Lafayette (Indiana 47907, USA)

Cadmium is a relatively difficult environmental substance to assess in terms of its ecological effects and its impact on human health. It occurs naturally in the earths' crust, and as a result is found in varying concentrations in virtually all components of freshwater, marine and terrestrial ecosystems. It is also present in the atmosphere as the result of volcanic activity, and the entrainment of vegetative material.

In addition, anthropogenic sources are many and varied resulting in substantial numbers of environmental pathways connecting them to humans and ecological species.

Concentrations of cadmium in the atmosphere do not appear to pose a human health problem, unless it is shown that the respired metal is preferentially toxic. In particular, the maximum airborne cadmium concentrations in urban areas and/or in the vicinity of intense point sources (zinc and lead smelters, etc.) are rarely higher than 100 ng/m^3. Assuming that an adult breathes 20 m^3 air per day, the resulting cadmium intake is only 2 µg/day. This compares with estimates of 20–50 µg/day intake via the diet, and 2–4 µg/day for smokers consuming 1 pack of cigarettes per day. Given retention rates of 30% and 8% for inhaled and dietary cadmium, respectively, cadmium retained from inhalation is 0.6 µg as compared to 1.6–4 µg for the ingested metal. Given a much more likely maximum urban-industrial cadmium air concentration of 15 ng/m^3, less than 0.1 µg is retained in the body.

The impact of cadmium on aquatic and terrestrial ecosystems is generally sufficiently subtle that its definition is more a matter of philosophy than of observable, substantial changes in indigenous populations. In general, there is a lack of data with which to define relationships between cadmium concentrations in the environment and ecological damage. This is in part due to the fact that cadmium does not manifest itself ecologically in readily discernible ways. It is also a result of the prodigious amount of data required for this task.

In freshwater ecosystems, little is known about the effects of cadmium on species of the various trophic levels. When introduced to a river or lake in an industrial outfall, the great bulk of the dissolved metal typically precipitates out of solution and resides in bottom sediment. From there it may be taken up by bottom feeding animal species and sediment-rooted flora. It then proceeds up the food chain to fish, with the precise pathways dependent upon the species present. Additionally, the small fraction of cadmium present in water in the disolved form may be ingested directly by fish through the gills and/or skin. Once in the fish, the metal typically is retained in the liver and kidney, with far lower concentrations in the muscle which may find its way into the human food chain. Laboratory experiments on fish species tend generally to suggest higher sensitivities to cadmium, especially with respect to reproductive effects, than are observed in field studies.

The foregoing observations apply equally well to the distribution of cadmium in marine systems. It has been observed that most marine organisms, especially bottom feeders such as molluscs and crustaceans, accumulate the metal to much higher concentrations than are present in their environment. As with the freshwater case, however, the metal tends to be retained in organs and exoskeletal compartments as opposed to muscle tissue which may be consumed by humans. This bioaccumulation pattern is evident both in systems exhibiting natural levels of cadmium, and in those receiving the metal from anthropogenic sources. Impact studies in the vicinity of wastewater outfalls containing high levels of cadmium have, in some cases, shown substantial perturbations of natural marine populations. However, such waste streams usually contain many other organic and inorganic pollutants, thus making the correlation of effects with cadmium extremely difficult.

The behavior of cadmium in terrestrial exosystems is a function of its chemical species. The typical translocation pattern involves the contamination of soil via aerial deposition or agricultural practices. Among the latter are the spreading of municipal sewage sludge or phosphate fertilizers to enhance soil nutrients. Solid waste disposal and subsequent leaching from disposal sites is another mechanism for the spread of cadmium in terrestrial systems. Following soil contamination, cadmium may be taken up by plants which, when foraged by animals or consumed by humans, results in its movement into the human food chain. Ecological effects elucidated from field (as opposed to laboratory) studies are difficult to substantiate due to the fact that cadmium sources intense enough to result in significant ecological alterations also emit other pollutants capable of causing the observed changes. For example, vegetation damage in the vicinity of non-ferrous smelters (the most intense sources of cadmium to the atmosphere) appears to be in large part attributable to sulfur dioxide emissions. For this reason, the quantification of the precise role of cadmium, if any, in vegetative blight around smelters is virtually impossible. Laboratory studies involving the growing of plant species in 'pots' or hydroponic solutions suggest that plant physiological effects manifest themselves at levels of cadmium uptake seldom, if ever, encountered in the environment.

Field studies in the United Kingdom at smelter sites which have been inactive since the Middle Ages have uncovered soil cadmium concentrations elevated by factors of hundreds above natural levels. Vegetative biomass in these areas tends to be normal, although there is no way to prove conclusively that plant species diversity has not been affected. Fortunately, the primary mechanisms for cadmium contamination of soils, i.e. aerial deposition, agricultural practices and solid waste disposal, can be readily controlled by modification of these practices and/or controlling point source emissions of cadmium-bearing particulate to the environment.

Cadmium intake by humans is principally effected through inhalation (respiratory exposure) and by the ingestion of food and drink (dietary exposure). Other incidental exposure mechanisms involve such phenomena as 'hand-to-mouth' ingestion in the work place, etc. Episodes of acute cadmium poisoning are extremely rare given the present level of awareness regarding the toxicity of the metal. The primary human health concern is thus related to long term, chronic exposures.

The highest chronic exposures are found in worker populations in facilities where cadmium is either utilized in a production process as a feed material, or is present in by-product streams. The primary mechanism of exposure in the work place is by inhalation with hand-to-mouth also factor where industrial hygiene practices are not stressed. Much of the human toxicity data come from studies on worker populations. Other human toxicity data have been derived from a non-occupational population in Japan subjected to dietary exposure via the contamination of rice paddies by mine run-off. The latter gave rise to the much-heralded Itai-Itai disease which manifested

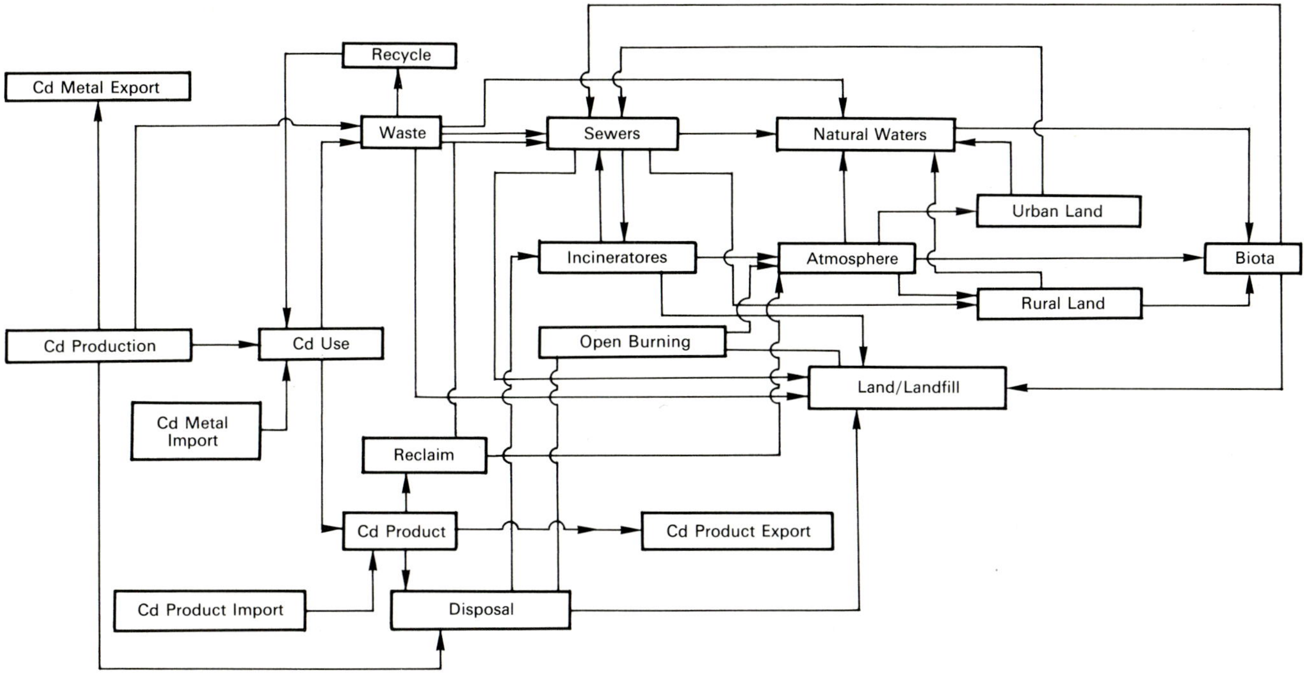

Figure 1. Primary elements of the U.S. cadmium environmental flow system.

itself in the decalcification of skeletal structure, especially in elderly women. Other studies in Europe suggest lowered kidney function in elderly women residing in highly industrialized urban areas.

The principal target organ vis-a-vis long term human exposure to cadmium appears at present to be the kidney. At sufficient exposure levels, renal tubular dysfunction leading to an excessive discharge of proteins in the urine (proteinuria) has been noted. The clinical significance of the condition is not clear, though it has been implicated in the development of osteomalacia detected in the Japanese Itai-Itai population. Other dieseases in which cadmium has been conjectured to play a role include hypertension, emphysema, and renal and prostate cancer. The bases of these conjectural associations consist generally of animal data, much of which is inconclusive. One is of course faced with the problem of extrapolations to humans even where animal data appear to be consistent.

It was pointed out earlier that the primary mechanism for non-occupational cadmium intake is food consumption. Cigarette smoking may also be an important mechanism for individuals consuming twenty or more per day. The least important source is respiratory exposure via airborne cadmium. Upon entering the body the metal is either excreted in the urine and feces, or retained, primarily in the liver and kidneys. For inhaled cadmium, 10-40% may be retained depending upon the size of the airborne particulate of which it is a constituent. Ingested cadmium is retained at a lower rate ranging from 4 to 10%.

Given the inconclusive state of knowledge regarding the health effects of chronic cadmium exposure, it seems prudent to minimize dietary exposure, and to refrain from heavy smoking. Minimizing dietary intake requires a thorough analysis of the environmental flow paths connecting sources to food. It is only on the basis of such an analysis that regulatory policy designed to reduce intake can be realistically undertaken. Such an analysis for the population of the United States has been developed by the author for the U.S. Environmental Protection Agency and a variety of industry trade associations. Details of the study with extensive references are given in a series of papers[1-4].

The objective of the study is to generate estimates of cadmium flow in the environment and (related) human exposures for major cadmium uses and current waste management practices. It is intended to provide a systematic basis for developing regulatory policy decisions.

Cadmium release to various environmental compartments is quantified on a national basis for major uses and inadvertent sources. Source-specific biotic and human exposure projections are made with respect to specific sites in cases where the number of U.S. source locations is low. For the more ubiquitous cadmium

Table 1. Estimated cadmium consumption (tons)[5]

	1976	1977	1978	1979	1980	Average	% Total
Metal finishing	2998	2282	2536	2767	1989	2514	51
Pigments	739	573	650	705	507	635	13
Plastic stabilizers	661	507	551	595	429	549	11
Batteries	1300	981	1091	1190	858	1084	22
Miscellaneous*	233	137	143	172	116	160	3
Totals	5931	4480	4971	5429	3899	4942	100

* Alloys, solders, reactor control rods, etc.

Table 2. Fate of cadmium discharged from use-related sources over 10-year simulation period (MT): Current practice*

Source	Receptors Natural waters	Sewage treatment	Municipal incinerators	Atmosphere	Landfill
Electroplating	95.9	839	–	–	20,135
Pigments**	10.0	116	1387	–	6,320
Plastic stabilizers	1.56	12.5	693	–	6,380
Batteries[†]	–	16.7	2060	145	8,560
Miscellaneous products	3.3	30.4	128	4.5	1,280
Sewage treatment	373	–	–	–	217
Municipal incinerators	72.2	72.2	–	1750	2,200
Sludge incinerators	3.7	3.7	–	21.9	275
Steelmaking	–	14.2	–	342	2,890
Coal-fired power plant	0.07	–	–	203	537
Totals	553	1105	4268	2466	48,794

* Does not include ocean dumping of sludges. [†] Includes chemical processing and battery production. ** Reclaim treated as 'sink'.

sources, exposure projections are developed for representative/prototype regions. Cadmium enrichments in environmental media are compared to naturally occurring levels. Two waste management/environmental control strategies are considered. Cadmium consumptions for various uses incorporated in this analysis are given in table 1[5].

1. Current/achievable waste management scenarios

Two waste management/environmental control scenarios are considered in the present analysis. The 'current practice' scenario incorporates waste treatment and management options thought to be representative of current operations among the various processes analyzed. The 'achievable practice' scenario is based on relatively greater utilization, and more efficient operation, of readily available control technology. Specific aspects of the two scenarios are given below.

Electroplating

Current practice: 25% of electroplating waste is discharged untreated, 90% of which goes to sewage

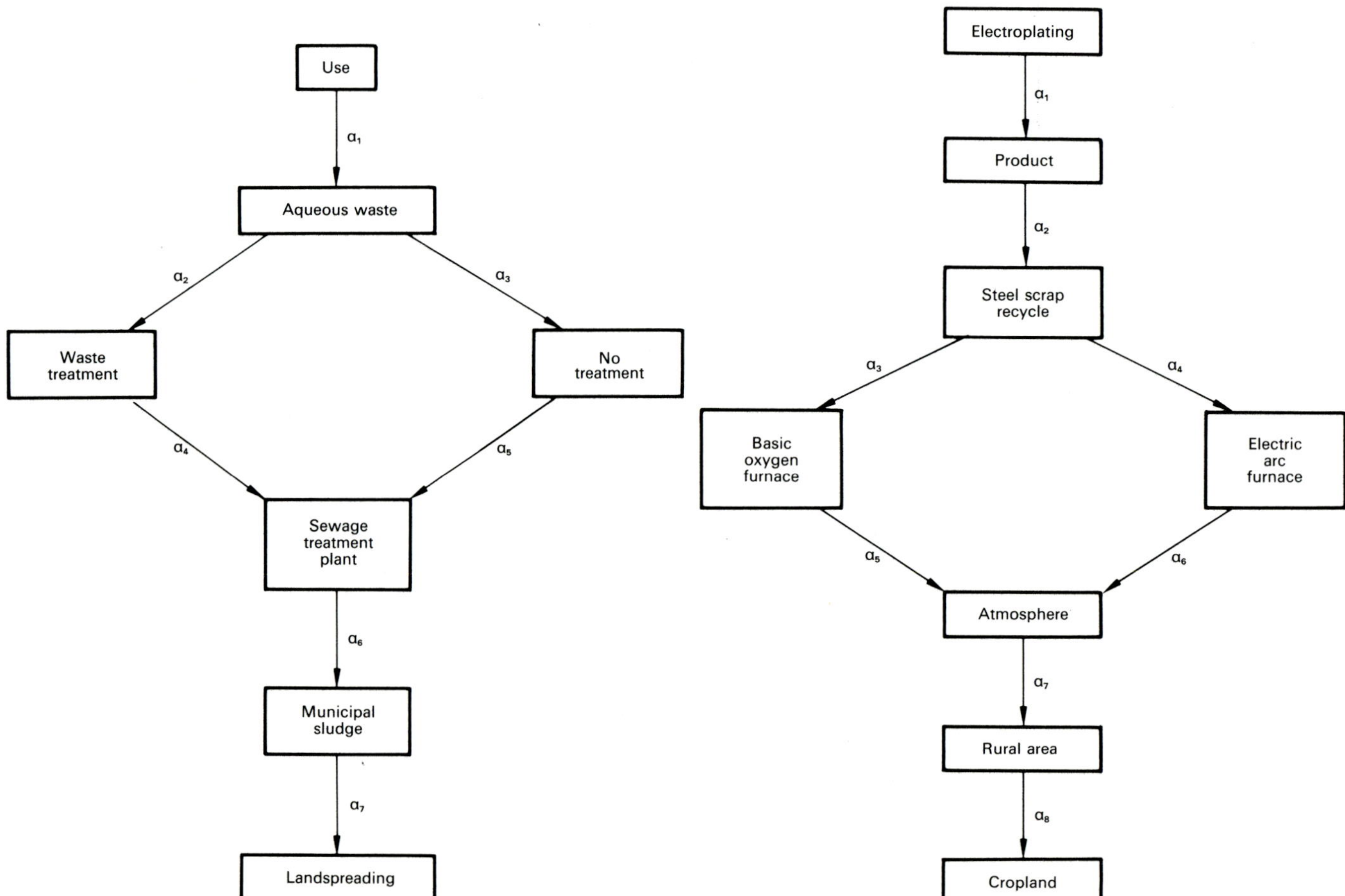

Figure 2. Municipal sludge landspreading-cadmium cropsoil enrichment flow paths for use-related production processes. Coefficient values are given for various uses in table 4.

Figure 3. Coupling between electroplating and atmospheric input of cadmium to cropsoils via recycle of plated steel scrap. Coefficient values are given in table 5.

Table 3. Fate of cadmium discharged from use-related sources over 10-year simulation period (MT): Achievable practice*

Source	Receptors Natural waters	Sewage treatment	Municipal incinerators	Atmosphere	Landfill
Electroplating	36.5	262	–	–	20,770
Pigments**	1.88	21.5	1387	–	6,320
Plastic stabilizers	1.56	12.5	693	–	6,380
Batteries[†]	–	16.7	2060	145	8,560
Miscellaneous products	1.2	9.7	128	4.5	1,300
Sewage treatment	147	–	–	–	85.6
Municipal incinerators	68.9	68.9	–	672	3,330
Sludge incinerators	1.46	1.46	–	8.63	108
Steelmaking	–	14.2	–	342	2,890
Coal-fired power plant	0.07	–	–	203	537
Totals	259	407	4268	1375	50,285

* Does not include ocean dumping of sludges. [†] Includes chemical processing and battery production. ** Reclaim treated as 'sink'.

Table 4. Use-related pathway coefficients for cadmium aqueous discharge input to cropsoil via municipal sludge landspreading (fig. 2): Current practice

Use	Pathway coefficients for cadmium input to cropsoils							% Consumption to cropsoils
	α_1	α_2	α_3	α_4	α_5	α_6	α_7	
Electroplating	0.1	0.55	0.25	0.08	0.9	0.7	0.25	0.47
Pigments	0.059	1.0	0.0	0.22	0.0	0.7	0.25	0.23
Plastic stabilizers	0.002	0.0	1.0	0.0	0.89	0.7	0.25	0.031
Batteries	0.0018	1.0	0.0	1.0	0.0	0.7	0.25	0.032

Table 5. Use-related pathway coefficients for atmospheric cadmium input to cropsoil via steel scrap recycle and refuse incineration (figs 3 and 4): Current practice

Use	Pathway coefficients for cadmium input to cropsoils								% Consumption to cropsoils
	α_1	α_2	α_3	α_4	α_5	α_6	α_7	α_8	
Electroplating (steel scrap recycle)	0.9	0.2	0.43	0.57	0.01	0.08	0.55	0.10	0.05
Pigments (refuse incineration)	0.94	0.18	0.44	0.47	0.1	–	–	–	0.35
Plastic stabilizers (refuse incineration)	0.978	0.1	0.44	0.47	0.1	–	–	–	0.2

treatment plants; chemical destruct waste treatment units are operated with an average Cd removal efficiency of 91%.
Achievable practice: All shops discharging to sewers employ chemical destruct waste treatment units operating at an average 93% removal efficiency.

Pigments

Current practice: Some plants employ settling tanks to recover Cd in wastewater, but have no capability for pH adjustment to precipitate dissolved metal. The resulting industry-wide Cd removal efficiency estimate is only 76%.
Achievable practice: All plants employ pH adjustment, clarification and filtration steps in their waste treatment systems with average Cd removal efficiencies of 97.2%.

Table 6. Expressions for estimating use-related fractions of cadmium destinated for cropland

Cadmium input mechanism	$P(\alpha)$: pathway coefficient configuration
Use-sludge landspreading	$\alpha_1(\alpha_2\alpha_4 + \alpha_3\alpha_5)\alpha_6\alpha_7$
Electroplating-steel scrap recycle	$\alpha_1\alpha_2(\alpha_3\alpha_5 + \alpha_4\alpha_6)\alpha_7\alpha_8$
Cd pigmented/stabilized plastics-incineration	$\alpha_1\alpha_2\alpha_3\alpha_4\alpha_5$

Nickel cadmium batteries

Current practice: 5% of the Cd utilized by chemical processors producing electrode material enters the waste stream, 48% of which is reclaimed. Pocket plate battery Cd production losses are 3%, half of which is reclaimed.
Achievable Practice: Chemical process Cd waste generation is cut to 2%, with 88% being reclaimed. Pocket plate production Cd waste reclaim is increased from 50% to 80%.

Refuse incineration

Current practice: 17% of existing incinerators have no emission control, 23% employ electrostatic precipitators, and 60% have wet scrubbers with an average Cd removal efficiency of 58%.
Achievable practice: The 17% having no control are fitted with electrostatic precipitators; existing scrubbers are upgraded to achieve 75% Cd removal efficiencies on the average.

2. Principal conclusions

Cadmium utilization in the U.S. can continue on an unrestricted basis without harm to human health or

Table 7. Sensitivity of use-related cadmium flow to cropland to changes in pathway parameters for municipal sludge landspreading: Current practice

Use	Sensitivity ($\Delta Cd/\Delta\alpha_i$)						
	α_1	α_2	α_3	α_4	α_5	α_6	α_7
Electroplating	0.047	0.0014	0.016	0.0096	0.0044	0.0067	0.019
Pigments	0.039	0.0023	**	0.01	**	0.0032	0.0091
Plastic stabilizers	0.16	**	0.00031	**	0.00035	0.00045	0.0012
Ni-Cd batteries	0.175	0.00032	**	0.00032	**	0.00045	0.0013

the environment given prudent waste management practices. More specific points include:

Respiratory (inhalation) intake of Cd by urban populations is negligible.

A 10-year, 25% municipal sludge landspreading program with 'Achievable practice' waste management results in a negligible (< 1% of the recommended daily intake) per capita dietary Cd increment. Long term landspreading of sludge from treatment plants receiving uncontrolled Cd discharges may represent a health problem depending upon crop marketing patterns and production levels.

Test results suggest that leaching of Cd from landfilled plastic products does not pose an environmental hazard. Disposal of Cd-rich industrial sludges in nonsecure landfills may result in significant groundwater contamination on a highly localized basis.

A worst case scenario for use-related Cd releases to waterways indicates an average daily Cd intake increment equal to 0.2% of the recommended daily intake by way of shellfish consumption.

Use-related Cd fluxes in U.S. waterways are negligible compared to 'natural' flows associated with background Cd concentrations in river sediments.

Long term (> 50 years) use of high Cd western rock phosphate fertilizers can result in significant increments in the average dietary intake.

The public health impacts of electrolytic refining plants is negligible. Environmental impacts are likewise nimimal given responsible handling (or recycle) of leach residue solid waste.

Tables 2 and 3 summarize the use-related flow and fate of Cd for the 'current practice' and 'achievable practice' scenarios, respectivvely. Figure 1 exhibits the major Cd flow paths encompassed by the analysis.

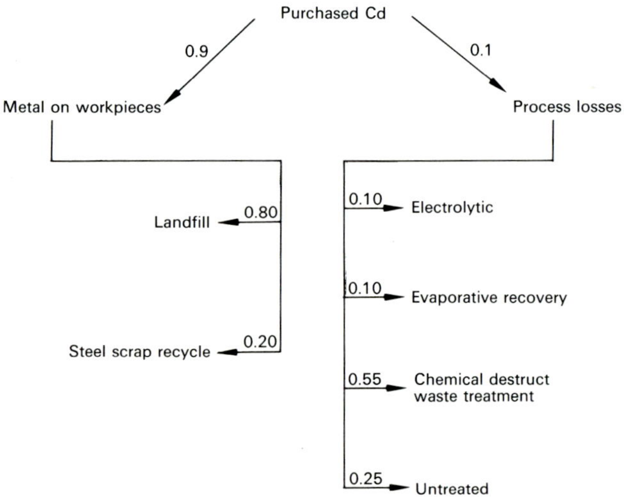

Figure 5. Cd flow in the electroplating process.

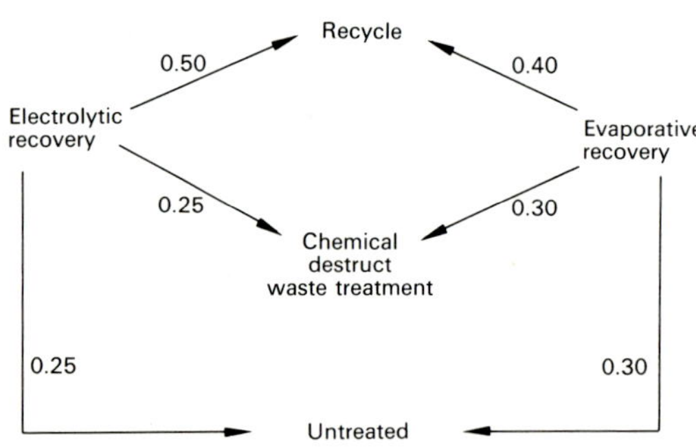

Figure 6. Cd waste management with electrolytic and evaporative recovery options.

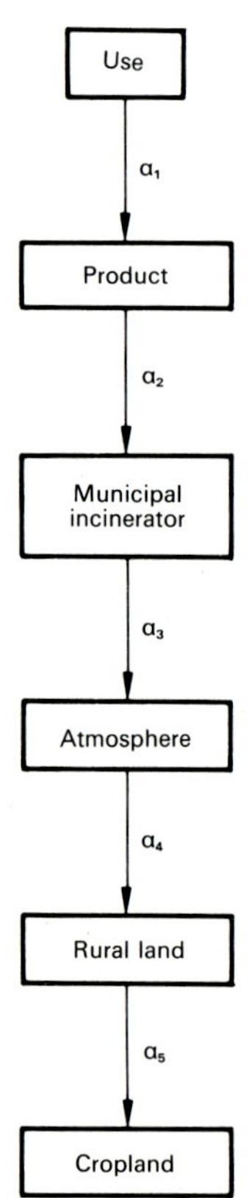

Figure 4. Coupling between a) production of plastics containing cadmium pigments and stabilizers, and b) atmospheric input of cadmium to cropsoils via incineration of municipal refuse. Coefficient values are given in table 5.

Dietary exposure sensitivity analysis

It is well known that the primary cadmium exposure route for humans is through the diet. Respiratory intake is likely to be less than 0.5 µg/day, whereas dietary has been estimated at 35–50 µg/day for U.S. adults[6]. For this reason, the elucidation of use-related exposure mechanism is focused on dietary intake. The two principal pathways for cadmium enrichment of the diet are municipal sludge landsprading, and deposition of airborne cadmium on cropland.

Figure 2 portrays pathways coupling cadmium uses to cropland enrichment via municipal sludge landspreading. The pathway coefficients, a_i, denote fractions of the cadmium in the donor nodes (at tail of arrows) moving to receptor nodes (at head of arrows) per unit time. Their dimensions are thus t^{-1}. Coefficient values for the 4 major cadmium uses, with corresponding fractions of metal consumed destined for cropland via sludge landspreading, are given in table 4.

Cadmium flow diagrams relating to cropland enrichment via deposition of airborne particulate are given in figures 3 and 4. Table 5 exhibits pathway coefficient values for the electroplating-steel scrap recycle and pigment-plastic stabilizer municipal incinerator flow systems. Corresponding fractions of cadmium destined for cropland via atmospheric deposition are also given.

A systems analysis typically evolves a variety of interesting by-product information. For example, it is easy to determine the variability of cadmium exposure with respect to pathway coefficients. This process is often called a 'sensitivity analysis', with the resulting variability measures referred to as 'sensitivities'. It requires the construction of algebraic expressions coupling flow system input and output/exposures. For simple, highly aggregated systems of the type discussed here, the expressions are quite simple. Table 6 gives them for the use-sludge landspreading and (two) atmospheric deposition systems. Denoting these expressions as $P(a)$, the sensitivity associated with the ith pathway coefficient is given by:

$$S_i = \frac{\delta P(a)}{\delta a_i}$$

Table 8. Weighted sensitivities in order of magnitude for cadmium enrichment of cropland via sludge landspreading: Current practice

Use	Coefficient index	Weighted sensitivity
Electroplating	1	5.5
Electroplating	7	2.2
Electroplating	3	1.9
Electroplating	4	1.1
Electroplating	6	0.79
Ni-Cd batteries	1	0.61
Pigments	1	0.59
Electroplating	5	0.52
Plastic stabilizers	1	0.27
Electroplating	2	0.17
Pigments	4	0.15
Pigments	7	0.14
Pigments	6	0.048
Ni-Cd batteries	7	0.046
Pigments	2	0.035
Plastic stabilizers	7	0.002
Ni-Cd batteries	6	0.0016
Ni-Cd batteries	2	0.0011
Ni-Cd batteries	4	0.0011
Plastic stabilizers	6	7.7×10^{-5}
Plastic stabilizers	5	6.0×10^{-5}
Plastic stabilizers	3	5.3×10^{-5}

Table 7 gives sensitivities of use-related cadmium destined for cropland via sludge landspreading. The sensitivities are useful for evaluating the relative effectiveness of policies which modify coefficients common to a single use. It is more difficult to compare the impact of policy decisions reflected in coefficient changes where comparisons must be made between measures affecting different uses. To accomplish this, it is useful to weight sensitivities by the product of a) annual cadmium consumption for the pertinent use, and b) the fraction of annual consumption destined for the environmental compartment of interest given baseline/nominal coefficient values (e.g. '% consumption to cropland' column in table 4). Expressing annual average consumption for major uses in hundreds of metric tons (table 1), the weighting factors for electroplating, pigments, plastic stabilizers and Ni-Cd batteries are 118, 15, 1.7, and 3.5, respectively. Table 8 gives a ranking of weighted sensitivities for cadmium input to cropland via municipal sludge landspreading.

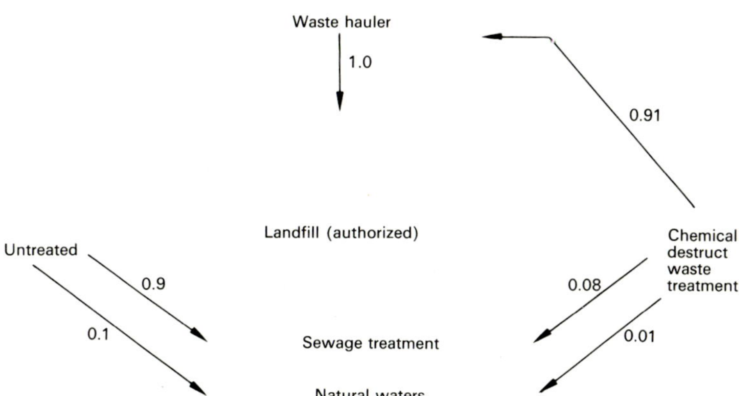

Figure 7. Fate of Cd in electroplating wastewater.

Table 9. Sensitivity of use/waste management practice-related cadmium flow to cropland vs changes in pathway coefficients for atmospheric deposition: Current practice

Use-waste management practice	Sensitivity ($\Delta Cd/\Delta \alpha_i$)							
	α_1	α_2	α_3	α_4	α_5	α_6	α_7	α_8
Electroplating-steel scrap recycle	0.00055	0.0025	$(-)$0.0007	0.0007	0.0043	0.0056	0.0009	0.005
Pigments-refuse incineration	0.0037	0.019	0.008	0.0074	0.035	–	–	–
Plastic stabilizers-refuse incineration	0.0021	0.02	0.0046	0.0043	0.02	–	–	–

Note that electroplating pathway coefficients dominate the top of the list. From the policy standpoint, this suggests that regulations designed to reduce cadmium discharges by this industry would be most effective for reducing cadmium input to cropland. To get an idea of how electroplating-related regulations should be formulated, refer to figures 5–7 which portray the flow of cadmium through the plating process. It is seen that the most sensitive coefficient in table 8 (5.5) denotes cadmium lost in the plating process to tank spills and barrel/rack 'dragout'. A possible regulations might involve a requirement for spill containment and counter-current rinses following plating, pickling and acid cleaner baths to reduce dragout. The next most sensitive process-ralated coefficient is α_3 which denotes plating waste discharged without treatment. Figure 5 indicates an estimated 25% of process losses are presently discharged without treatment. An obvious measure would be to require some minimum level of treatment for all plating waste. This can be accomplished indirectly by imposing effluent control guidelines, or directly by writing a technology requirement.

A similar sensitivity analysis for pathway coefficients defining cadmium flow to cropland via atmospheric deposition may be generated by applying the preceding weighting factors to the coefficients in table 5. In this case, of course, the cropsoil-destined fractions of table 5 are factored in with annual major cadmium use consumptions in units of 10^2 MT. Unweighted sensitivities for the cadmium atmospheric deposition path to cropland are given in table 9. Note that,

unlike the pathway analysis for sludge landspreading, pigments and plastic stabilizers exhibit larger sensitivities than does electroplating by way of steel scrap recycle. This observation is borne out strongly in the weighted sensitivities ranked in table 10, where the largest plating-scrap recycle sensitivity is 6–10 times smaller than the pigments/plastic stabilizer-refuse incineration maxima. Thus, to reduce the deposition of airborne cadmium on cropland the plastic-incineration flow system should be studied. Major process-related coefficients in the simplified schematic of figure 4 are a_2 and a_3, the fraction of municipal waste incinerated and fraction of cadmium released to the atmosphere from combusted plastics containing pigments/stabilizers, respectively. The latter can be elucidated by considering the cadmium flow pattern for refuse incinerators given in figure 6. It indicates the 17% of U.S. refuse incinerators are uncontrolled, and that electrostatic precipitators are an estimated 7 times more efficient for capturing cadmium-bearing particulate from top gas than are wet scrubbers. Regulations to reduce cadmium input to the atmosphere might include the installation of electrostatic precipitators on uncontrolled incinerators, or perhaps the retro-fitting of precipitators on uncontrolled incinerators, or perhaps the installation of electrostatic precipitators on uncontrolled incinerators, or perhaps the retro-fitting of precipitators on units equipped with low pressure/efficiency scrubbers.

Table 10. Weighted sensitivities in order of magnitude for cadmium enrichment of cropland via atmospheric deposition: Current practice

Use-waste management practice	Coefficient index	Weighted sensitivity
Pigments-refuse incineration	5	0.078
Pigments-refuse incineration	2	0.042
Plastic stabilizers-refuse incineration	5	0.022
Plastic stabilizers-refuse incineration	2	0.022
Pigments-refuse incineration	3	0.018
Pigments-refuse incineration	4	0.016
Pigments-refuse incineration	1	0.0082
Electroplating-steel scrap recycle	6	0.0071
Electroplating-steel scrap recycle	8	0.0063
Electroplating-steel scrap recycle	5	0.0054
Plastic stabilizers-refuse incineration	3	0.0051
Plastic stabilizers-refuse incineration	4	0.0047
Electroplating-steel scrap recycle	2	0.0032
Plastic stabilizers-refuse incineration	1	0.0023
Electroplating-steel scrap recycle	7	0.0011
Electroplating-steel scrap recycle	3	8.8×10^{-4}
Electroplating-steel scrap recycle	5	8.8×10^{-4}
Electroplating-steel scrap recycle	1	6.9×10^{-4}

1 Yost, K.J., and Miles, L.J., Environmental Health Assessment for Cadmium: a Systems Approach. J. envir. Sci. Hlth *A14* (1979) 285–311.
2 Yost, K.J., Miles, L.S., and Parsens, T.W., A Method for Estimating Dietary Intake of Environmental Trace Contaminants: Cadmium, a Case Study. Envir. int. *3* (1980) 473–484.
3 Yost, K.J., Miles, L.J., and Greenkorn, R.A., Cadmium: Simulation of Environmental Control Strategies to Reduce Exposure. Envir. Management *5* (1981) 341–352.
4 Yost, K.J., Source-Specific Exposure Mechanisms for Environmental Cadmium. Environmental keynote talk; Fourth International Cadmium Conference. Munich, FRG, Feb. 28–March 4, 1983.
5 FDA Compliance Program, U.S. Food and Drug Administration, FY78 Total Diet Studies – Adult (7305.003). 1981.
6 Zinc Institute, U.S. Zinc and Cadmium Industries: Annual Review 1980. Zinc Institute, Inc., 292 Madison Ave., New York, N.Y., June 1981.